BIOTECHNOLOGY IN AGRICULTURE, INDUSTRY AND MEDICINE

CELLULASE: TYPES AND ACTION, MECHANISM AND USES

BIOTECHNOLOGY IN AGRICULTURE, INDUSTRY AND MEDICINE

Additional books in this series can be found on Nova's website under the Series tab.

Additional E-books in this series can be found on Nova's website under the E-book tab.

BIOTECHNOLOGY IN AGRICULTURE, INDUSTRY AND MEDICINE

CELLULASE: TYPES AND ACTION, MECHANISM AND USES

ADAM E. GOLAN
EDITOR

Nova Science Publishers, Inc.
New York

For permission to use material from this book please contact us:
Telephone 631-231-7269; Fax 631-231-8175
Web Site: http://www.novapublishers.com

Additional color graphics may be available in the e-book version of this book.

Library of Congress Cataloging-in-Publication Data

Cellulase : types and action, mechanism, and uses / editor, Adam E. Golan.
p. ; cm.
Includes bibliographical references and index.
ISBN 978-1-61761-983-0 (hardcover)
1. Cellulase. I. Golan, Adam E.
[DNLM: 1. Cellulases--physiology. 2. Biotechnology--methods. 3. Cellulases--chemical synthesis. QU 136]
QP609.C37C45 2010
572'.756--dc22

2010036034

Published by Nova Science Publishers, Inc. † New York

CONTENTS

Preface

Cellulase refers to a class of enzymes produced chiefly by fungi, bacteria, and protozoans that catalyze the cellulolysis (or hydrolysis) of cellulose. The enormous potential that cellulases have in biotechnology is the driving force for continuous basic and applied research on these biocatalysts from fungi and bacteria. Cellulases are found in many fields, such as animal feeding, brewing and wine, food, textile and laundry, pulp and paper products. The growing interest toward the conversion of lignocellulosic biomass into fermentable sugars has generated an additional request for cellulases and their related enzymes. This book presents research in the study of cellulase, including biotechnological applications of microbial cellulases; using agro-industrial by-products as raw material for cellulase production; and the enzyme saccharification of cereal crop residues using dilute alkali pretreatment.

Chapter 1 - Cellulases (EC 3.2.1.4) catalyze the hydrolysis of 1,4-β-D-glucosidic linkages in cellulose, and play a significant role in nature by recycling this polysaccharide which is the main component of plant cell wall. Cellulases work in synergy with other hydrolytic enzymes in order to obtain the full degradation of the polysaccharide to soluble sugars, namely cellobiose and glucose, which are then assimilated by the cell.

The enormous potential that cellulases have in biotechnology is the driving force for continuous basic and applied research on these biocatalysts from Fungi and Bacteria. Nowadays, cellulases found application in many fields, such as animal feeding, brewing and wine, food, textile and laundry, pulp and paper industries. Moreover, the growing interest toward the conversion of lignocellulosic biomass into fermentable sugars has generated an additional request for cellulases and their related enzymes. In fact, bioconversion of biomass has significant advantages over other alternative energy production strategies because lignocellulose is the most abundant and renewable biomaterial on our planet. Bioconversion of lignocellulose is initiated primarily by microorganisms which are capable of degrading lignocellulosic materials. Several Fungi produce large amounts of extracellular cellulolytic enzymes, whereas bacterial and few anaerobic fungal strains mostly produce cellulolytic enzymes in a complex associated with the cell wall which is called "cellulosome". However, the heterogeneous and recalcitrant nature of lignocellulosic waste represents an obstacle for an efficient saccharification process, and pretreatment techniques are required to make the polysaccharide more accessible to the enzymatic action.

Thermostable enzymes can offer potential benefits in the hydrolysis of pretreated lignocellulosic substrates because the harsh conditions often required by several pretreatments can be harmful for conventional biocatalysts. The enhanced stability of thermostable enzymes to high temperature and extreme operative parameters allows improved hydrolysis performance and increased flexibility compared to process configurations, all leading to enhancement of the overall economy of the process.

The present review gives an outline of several mesophilic and thermophilic cellulases from Fungi and Bacteria that have been characterized in the last years. Moreover, applications of these enzymes in some biotechnological fields, with particular regard to lignocellulosic biomass bioconversion, will be illustrated.

Chapter 2 - Cellulases, responsible for the hydrolytic cleavage of cellulose, are composed of a complex mixture of enzymes with different specificities to hydrolyse glycosidic bonds. Cellulases can be grouped into three major enzyme classes viz. endoglucanase, exoglucanase and β-glucosidase. Endoglucanases, often called carboxy methyl cellulases (CMCase), are proposed to initiate random attack at multiple internal sites in the amorphous regions of the cellulose fiber to open up sites for subsequent attack of cellobiohydrolases. Exoglucanase, better known as cellobiohydrolase, is the major component of the microbial cellulase system accounting for 40-70% of the total cellulase proteins and can hydrolyse highly crystalline cellulose. It removes mono-and dimers from the end of the glucose chain. β-glucosidase hydrolyse glucose dimers and in some cases cello-oligosaccharides to release glucose units. Generally, the endo- and exoglucanase work synergistically in cellulose hydrolysis but the underlined mechanism is still unclear. Microorganisms generally appear to have multiple distinct variants of endo- and exoglucanases. A diverse spectrum of cellulolytic microorganisms have been isolated and identified over the years and this list still continues to grow. Cellulases play a paramount role in natural carbon cycle by hydrolysing the lignocellulosic structures. Besides their applications in pharmaceutical industry, cellulases are also widely used in textile industry, in laundry detergents and in pulp and paper industry for various purposes. The cost of enzyme preparation is a major impediment in its commercial application. Recombinant DNA technology (RDT) and protein engineering have great potential for making significant improvements in increased production and higher specific activity of cellulases. The role of cellulases holds the key for transformation of organic wastes especially agricultural residue into biofuels through fermentation. although the process is at its infant stage, this is an important aspect for sustainable development.

Chapter 3 - Cellulases are well established in different industrial areas, and are currently the third largest industrial enzyme worldwide, by dollar volume, mainly because of their use in cotton processing and paper recycling, as detergent industry enzymes, and in juice extraction and animal feeding additives as well. Nowadays the cellulases are the most important enzyme group for studies aiming at the so called second generation ethanol production and others chemicals products. The cellulase group involves three different enzymes: β-1,4-endoglucanase (EC 3.2.1.4), β-1,4-exoglucanase (EC 3.2.1.91) and cellobiase (EC 3.2.1.21), that are produced by an array of microorganisms, including bacteria and fungi. For cellulase production economically viable the raw material needs to be cheap. There are many types of low cost carbon sources that could be used for cellulases production, such as sugar cane bagasse, sugar cane straw, wheat straw, wheat bran, corn cobs, etc, reducing the costs effects and being friendly environmentally. In this chapter, the importance of using agro-industrial by-products as raw material for cellulase production will be addressed, as well as its biotechnological application in industry.

Chapter 4 - Cellulose present in renewable lignocellulosic material is considered to be the most abundant organic substrate on earth for the production of hexoses and pentoses, for fuel and other chemical feed stock. Research on cellulase has progressed very rapidly in the past few decades, emphasis being on enzymatic hydrolysis of cellulose to hexose sugars. The enzymatic hydrolysis of cellulose requires the use of cellulase [1,4-(1,3:1,4)-β-D-glucan glucanohydrolase, EC 3.2.1.4], a multiple enzyme system consisting of endo-1,4,-β-D-glucanases [1,4-β-D-glucanases (CMCase, EC 3.2.1.4)], exo-1,4,-β-D-glucanases [1,4-β-D glucan cellobiohydrolase, FPA, EC 3.2.1.91] and β – glucosidase (cellobiase) (β-D-glucoside glucanohydrolase, EC 3.2.1.21). Major impediments to exploiting the commercial potential of cellulases are the yield, stability, specificity, and the cost of production. In the past few decades focus has been on submerged fermentation (SmF) and very little attention has been given to solid-state fermentation (SSF). SSF refers to the process whereby microbial growth and product fermentation occurs on the surface of the solid materials. This process occurs in the absence of "free" water, where the moisture is absorbed to the solid matrix. The direct applicability of the product, the high product concentration, lower production cost, easiest product recovery and reducing energy requirement make SSF a promising technology for cellulase production. This review highlights the research carried out on the production of cellulase in SSF using various lignocellulosic substrates, microorganisms, cultural conditions, process parameters (i.e., moisture content and water activity, mass transfer processes: aeration and nutrient diffusion, substrate particle size, temperature, pH, surfactant, etc), bioreactor design, and the strategies to improve enzyme yield. Also, the biotechnological potentials of microbial cellulases produced in SSF for bioconversion of agricultural wastes –providing a means to a "greener" technology, have been discussed.

Chapter 5 - Mild alkali pretreatment of lignocellulosic biomass is an effective pretreatment method which improves enzymatic saccharification. Alkaline pretreatment successfully delignifies biomass by disrupting the ester bonds cross-linking lignin and xylan, resulting in cellulose and hemicellulose enriched fractions. Here the authors report the use of dilute alkaline (NaOH) pretreatment followed by enzyme saccharification of cereal crop residues for their potential to serve as feedstock in the production of next-gen biofuels in Australia. Specifically, the authors discuss the impacts of varying pretreatment parameters on enzymatic digestion of residual solid materials. Following pretreatment, both solids and lignin content were found to be inversely proportional to the severity of the pretreatment process. Higher temperatures and alkali strength were also shown to be quintessential for maximising sugar recoveries from enzyme saccharifications. Essentially, pretreatment at elevated temperatures led to highly digestible material enriched in both cellulose and hemicellulose fractions. Increasing cellulase loadings and tailoring enzyme activities with additional β-glucosidases and xylanases delivered greater rates of monosaccharide sugar release and yields during saccharification. Sugar conversion efficiency of alkali treated sorghum and wheat straw residues following enzyme saccharification, approached 80 and 85%, respectively. Considering their abundance and apparent ease of conversion with high sugar yield, cereal crop residues are ideally suited for the production of second generation biofuels and/or use as feedstock for future biorefineries.

Chapter 6 - The plant cell wall consists of cellulose, hemicelluloses and pectin as well as the phenolic polymer lignin. Cellulose is the most abundant polysaccharide in nature and the major constituent of a plant cell wall providing its rigidity. Cellulose consists of β-1,4 linked D-glucose units that form linear polymeric chains of about from 8000 to 12000 glucose units.

In crystalline cellulose, these polymeric chains are packed together by hydrogen bonds to form highly insoluble structures. Hemicelluloses, the second most abundant polysaccharides in nature, have a heterogeneous composition of various sugar units. Hemicelluloses are usually classified according to the main sugar residues in the backbone of the polymer such as xylan, (galacto)glucomannan, arabinan, galactan found in cereals and hardwood, softwood and hardwood, The main chain sugars of hemicelluloses are modified by various side groups such as 4-O-methylglucuronic acid, arabinose, galactose, and acetyl, making hemicelluloses branched and variable in structure. Pectins are a family of complex polysaccharides containing a backbone of α-1,4 linked D-galacturonic acid. Pectins contain two different types of regions. In the region of pectin classified as a smooth region, D-galacturonic acid residues can be methylated or acetylated, whereas the region classified as a hairy one consists of two different structures, D-xylose substituted galacturonan and rhamnogalacturonan to which long arabinan and galactan chains are linked via rhamnose. The cellulose wall is strengthened by lignin, a highly insoluble complex branched polymer of substituted phenylpropane units joined together by carbon–carbon and ether linkages forming an extensive cross-linked network within the cell wall.

Chapter 7 - Cellulases are key industrial enzymes used to breakdown agriculture biomass to fermentable sugars. Cellulase has been used on the market as an industrial enzyme preparation and used as a main component of various products, such as detergents, fiber treating agents, paper pulp, additives for feed, and digestants. Cellulase is also used for commercial food processing in coffee. It performs hydrolysis of cellulose during drying of beans. Due to increasing environmental concerns and constraints being imposed on textile industry, cellulase treatment of cotton fabrics is an environmentally friendly way of improving the property of the fabrics. Furthermore, Cellulases are being used also in textiles for removing excess dye from denim fabric in pre-faded blue jeans (biostoning), also in removing the microfibirle which stick out from cotton fabrics after several washing. Restoring the softness and color brightness of cotton fabric could be achieved by using the cellulases. Cellulases can be used as a supplement in animal feed to decrease the production of fecal waste by increasing the digestibility of the feed. Cellulases can also be used to increase the efficiency of alcoholic fermentations (e.g., in beer brewing) by converting undigestible biomass into fermentable sugars. Ethanol is an alcohol made by fermenting and distilling simple sugars. As a result, ethanol can be produced from any biological feedstock that contains appreciable amounts of sugar or materials that can be converted into sugar such as starch or cellulose. Biofuels are liquid fuels produced from agriculture biomass using cellulases and other different enzymes. Agriculture biomass is available on a renewable or recurring basis, including agricultural crops and trees, wood and wood wastes and residues, plants (including aquatic plants), grasses, residues, fibers, and animal wastes, municipal wastes, and other waste materials. Biofuels (Types of biofuels include ethanol, biodiesel, methanol, and reformulated gasoline components) are primarily used as transportation fuels for cars, trucks, buses, airplanes, and trains. As a result, their principal competitors are gasoline and diesel fuel. Cellulase produced by the organisms isolated from Rumen Fluid of Cattle was used for biopolishing.

Cellulases are used also, in removing microbial slime in slime covered surfaces and maintaining a slime-free surface as in exposed cooling tower surfaces and in waste water treatment and paper making. This method comprises utilizing an enzyme blend in 2 to 100 parts per million (ppm) of cellulase, α-amylase and protease. Such enzyme blends have been found specifically to digest microbial slime and reduce microbial attachment and biofilm.

Chapter 8 - Cellulose is the most abundant component of plant biomass found in nature that is almost exclusively in plant cell walls, whereas it cannot be effectively converted into the usable sugars due to the lower cellulase activity. Although various classes of cellulase have been isolated and the synergism between them has been studied in detail, the thorough degradation of natural cellulose cannot be observed in the depolymerization by cellulase system, presumably it is due to the cellulase assay methods and substrates used for determining cellulases. Assay using cellulose as substrate is useful for assessing the potential enzyme system but it cannot be used for searching novel individual cellulase because the total activity is determined with such substrates. Considering that the natural cellulose can be degraded by living microbes whereas cannot by the secreted cellulases, the authors can conclude that there are the true cellulases degrading natural cellulases not to be isolated yet. It is obvious that the substrate is the vital determinant for cellulase assay and the key problem for seeking the true cellulase is how to obtain single cellulose chains. It is believed that the thoughtful substrate for cellulase assay should be amorphous cellulose molecule in the form of single chains.

Chapter 9 - Cellulases are celluloytic enzymes (EC 3.2.1.4) produced mainly by microbes including fungi, bacteria, and also by protozoans. However, plants and animals also produce cellulases. Several different kinds of cellulases differing in structure and mechanism of action are known. Cellulases catalyze the hydrolysis of 1, 4-beta-D-glycosidic linkages in cellulose, lichenin and cereal beta-D-glucans. Other names of cellulase are endoglucanase, endo-1,4-beta-glucanase, carboxymethyl cellulase (CMCase), endo-1,4-beta-D-glucanase, beta-1,4-glucanase, beta-1,4-endoglucan hydrolase and celludextrinase. Excocellulases and beta-glucosidases are other types of cellulases. Avicelase refers to the total cellulase activity of a given sample of the enzyme(s). The cellulase activity may be the consequence of the action of more than one type of enzymes.

Chapter 10 - To evaluate the synergism of cellulases from animal and microorganism, mixture of cellulases from snail (CES) and Trichoderma reesei (CET) was used to enzymatic hydrolysis and ethanol fermentation of lignocellulose. When the mixed cellulase was used to enzymatically hydrolyze Pennisetum hydridum, the optimal ratio of CES and CET was 3:1, and the glucose yield using the mixed enzyme was 100.3% and 50.2% higher than that produced individually by CES and CET, respectively. For ethanol fermentation of lignocellulose, the optimal ratio of CES and CET was 1:3, the ethanol yield using the mixed enzyme was 42.5% and 20.1% higher than that produced individually by CES and CET, respectively. Our results showed that mixed cellulase from animal and microorganism is a potential approach for improving enzymatic hydrolysis and ethanol fermentation of lignocellulose.

Chapter 11 - The importance of cellulases in the production of fuels from biomass makes understanding their catalytic mechanisms on crystalline cellulose important in order to design more active enzymes. Seven modular cellulases from Thermobifida fusca have been purified and characterized; of which, three inverting cellulases: endocellulase Cel6A, exocellulase Cel6B and processive endocellulase Cel9A have been studied extensively. Each one has an atypical catalytic mechanism: two Asp residues hold the nucleophilic water in Cel9A while no single catalytic base was found in the family-6 enzymes, suggesting that several residues might be involved in catalysis and form a network that functions as the catalytic base in these enzymes. Site-directed mutagenesis and removal of domains demonstrate the important role of cellulose-binding modules in crystalline substrate hydrolysis and processivity. To investigate if independent enzymes could function effectively in a cellulosome, the catalytic domains of the two family-6 T. fusca cellulases were attached to dockerin domains and then the chimeric enzymes were used to form designer cellulosomes. Additionally, Cel6B enzymes have been fluorescence-labeled, providing another way to measure binding and processivity. These studies have created several enzymes with higher activity on crystalline cellulose; however, better strategies are necessary to produce more active engineered cellulases that will be able to lower the cost of cellulases for biomass hydrolysis.

In: Cellulase: Types and Action, Mechanism, and Uses
Editor: Adam E. Golan
ISBN: 978-1-61761-983-0

Chapter 1

Cellulases from Fungi and Bacteria and Their Biotechnological Applications

A. Morana[1], L. Maurelli[1], E. Ionata[1], F. La Cara[1] and M. Rossi[2]
Institute of Protein Biochemistry, Italian National Research Council,
Via P. Castellino 111, 80131, Naples, Italy[1]
Dipartimento di Biologia Strutturale e Funzionale, Complesso Universitario
Monte S. Angelo, University of Naples "Federico II", Naples, Italy[2]

Abstract

Cellulases (EC 3.2.1.4) catalyze the hydrolysis of 1,4-β-D-glucosidic linkages in cellulose, and play a significant role in nature by recycling this polysaccharide, which is the main component of the plant cell wall. Cellulases work in synergy with other hydrolytic enzymes in order to obtain the full degradation of the polysaccharide to soluble sugars, namely cellobiose and glucose, which are then assimilated by the cell.

The enormous potential that cellulases have in biotechnology is the driving force for continuous basic and applied research on these biocatalysts from Fungi and Bacteria. Nowadays, cellulases have found applications in many fields, such as animal feeding, brewing, wine, food, textile and laundry, pulp and paper industries. Moreover, the growing interest toward the conversion of lignocellulosic biomass into fermentable sugars has generated an additional request for cellulases and their related enzymes. In fact, bioconversion of biomass has significant advantages over other alternative energy production strategies because lignocellulose is the most abundant and renewable biomaterial on our planet. Bioconversion of lignocellulose is initiated primarily by microorganisms which are capable of degrading lignocellulosic materials. Several Fungi produce large amounts of extracellular cellulolytic enzymes, whereas bacterial and a few anaerobic fungal strains mostly produce cellulolytic enzymes in a complex associated with the cell wall which is called "cellulosome". However, the heterogeneous and recalcitrant nature of lignocellulosic waste represents an obstacle for an efficient saccharification process, and pretreatment techniques are required to make the polysaccharide more accessible to the enzymatic action.

Thermostable enzymes can offer potential benefits in the hydrolysis of pretreated lignocellulosic substrates because the harsh conditions often required by several pretreatments can be harmful for conventional biocatalysts. The enhanced stability of thermostable enzymes to high temperature and extreme operative parameters allows improved hydrolysis performance and increased flexibility compared to process configurations, all leading to enhancement of the overall economy of the process.

The present review gives an outline of several mesophilic and thermophilic cellulases from Fungi and Bacteria that have been characterized in the last years. Moreover, applications of these enzymes in some biotechnological fields, with particular regard to lignocellulosic biomass bioconversion, will be illustrated.

1. INTRODUCTION

Plant biomass is the only predictable sustainable source of organic fuels, chemicals, and materials. As the primary component of the biosphere, biomass is also an industrial raw material uniquely compatible with human and other forms of life. The complex structure of the plant cell wall consists of lignocellulosic material mainly constituted by cellulose fibers strictly linked with hemicellulose and lignin, thus complicating their hydrolysis and which composition differs considerably according to the source (Table 1) [Deobald et al., 1997].

Cellulose is a linear polymer of β-D-glucose units linked through 1,4-β-linkages with a degree of polymerization ranging from 2,000 to 25,000 [Kuhad et al., 1997]. More in detail, cellulose chains form numerous intra- and intermolecular hydrogen bonds, which account for the formation of rigid, insoluble, crystalline microfibrils. Natural cellulose compounds are structurally heterogeneous and have both amorphous and highly ordered crystalline regions. The degree of crystallinity depends on the source of the cellulose and the highly crystalline regions are more resistant to enzymatic hydrolysis.

Cellulosic materials are particularly attractive because of their relatively low cost and abundant supply. As the most abundant polysaccharide in nature, cellulose decomposition plays not only a key role in the carbon cycle of nature, but also provides a great potential for a number of applications, most notably biofuel and chemical production [Lynd et al., 2002]. The central technological impediment to more widespread utilization of this important resource is the general absence of low-cost technology for overcoming the recalcitrance of cellulosic biomass.

A promising strategy to overcome this impediment involves the production of cellulolytic enzymes, hydrolysis of cellulose, and fermentation of resulting sugars in a single process step via free cellulolytic enzymes or consortium. In general, two systems occur in regard to plant cellulose degradation by microorganisms. In the first, the organism produces a set of free enzymes that work synergistically to degrade the plant cell wall. In the second, the degradative enzymes are organized into enzymatic complexes. Aerobic Bacteria and Fungi join weakly or not to cellulose and produce free cellulases, while anaerobic Bacteria and Fungi show high tendency to adhere to the polysaccharide, thus producing cellulases included in enzymatic complexes called "cellulosome".

However, full degradation of cellulose involves a complex interaction between different cellulolytic enzymes. It has been widely accepted that three types of enzymes including endoglucanases (EC 3.2.1.4), exoglucanases (EC 3.2.1.91) and β-glucosidases (EC 3.2.1.21) act synergistically to convert cellulose into β-glucose [Lynd et al., 2002]. Extensive evidence obtained from aerobic cellulolytic microorganisms supports a hydrolysis mode mediated by the synergistic action of endoglucanases and exoglucanases with cellobiose as the main product [Zhang et al., 2004]. Then, cellobiose is further hydrolyzed by β-glucosidases to glucose.

Table 1. Lignocellulose composition of several agricultural waste

Lignocellulosic materials	Cellulose (%)	Hemicellulose (%)	Lignin (%)
Hardwood	40-55	24-40	18-25
Softwood	45-50	25-35	25-35
Nut shell	25-30	25-30	30-40
Chestnut shell	27.4	10	44.6
Grape stalk	38	15	33
Corn stover	36.7	13.33	33
Wheat straw	30	50	15
Rice straw	32.1	24	18
Brewer's spent grain	16.8	28.4	27.8
Paper	85-99	0	0-15
Leaves	15-20	80-85	0
Cotton seeds hairs	80-95	5-20	0
Newspaper	40-55	25-40	18-30
Waste paper from chemical pulps	60-70	10-20	5-10
Switch grass	45	31.4	12.0

Modified from [Jorgensen et al., 2007]

This review will consider only the enzymes involved in the first step of cellulose degradation, namely the endoglucanases. Interest in these enzymes has grown markedly because of the potential of the substrates for yielding marketable products. Cellulose-hydrolyzing enzymes are widespread in Fungi and Bacteria [Tomme et al., 1995], and they have found application in various biotechnological fields. The most effective enzymes of commercial interest are the cellulases from aerobic cellulolytic Fungi, such as *Trichoderma reesei* (*Hypocrea jecorina*), *Aspergillus niger* and *Humicola insolens* [Nakari-Seta La and Penttila, 1995; Okada, 1988; Davies et al., 2000]. This is due to the ability of engineered strains of these microorganisms to produce large amounts of crude cellulases which possess high specific activity on crystalline cellulose. In general, cellulases can be used to improve color extraction and the yield of juices. Their presence in detergents causes color brightening and softens and improves particulate soil removal. Cellulases are also used for the “biostoning” of jeans instead of the classical stones in stonewashed jeans. Other applications of cellulases include the pretreatment of forage crops to improve nutritional quality and digestibility, and the production of fine chemicals. In addition, the growing interest in the last years toward the conversion of lignocellulosic biomass to fermentable sugars for bioethanol production has generated an increasing demand for cellulases and related enzymes. In fact, bioconversion of lignocellulosic biomass has significant advantages over other alternative energy production strategies because lignocellulose is the most abundant and renewable biomaterial on our planet, and moreover, it is not in competition with food sources.

The aim of the present review is to give an overview of several cellulases from Bacteria and Fungi that have been characterized in the last years. Thermophilic enzymes will also be considered, as their elevated stability to high temperatures and extreme operative parameters allows improved hydrolysis performance, leading to enhancement of the overall economy of the biotechnological process. Molecular and biotechnological aspects of these enzymes, with particular regard to their application in lignocellulosic biomass saccharification, will be illustrated.

2. Structure of Cellulose

2.1 Chemical Structure

Payen first used the term cellulose for this plant constituent which is the most widespread organic compound on Earth [Payen 1938; Guo et al., 2008]. The total amount of this polysaccharide on our planet has been estimated at 7×10^{11} tons [Coughlan, 1985] and constitutes the most abundant and renewable polymer resource available today. Cellulose is an insoluble crystalline substrate, flavorless, odorless, hydrophilic, insoluble in water and in most organic solvents, chiral, and with a wide chemical variability. It is a structural component of the cell wall of green plants accounting for almost 33% of the total biomass. It is also biosynthesized in other living systems such as Bacteria and Algae.

Cellulose produced by plants usually exists within a matrix of other polymers primarily hemicellulose, lignin, pectin and other substances, forming the so-called lignocellulosic biomass, while microbial cellulose is quite pure, has a higher water content, and consists of long chains. It is a carbohydrate polymer with formula $(C_6H_{10}O_5)_n$, consisting of a linear chain of several hundred to over ten thousand 1,4-β-D-glucose units linked through acetal functions between the equatorial -OH group of C4 and the C1 carbon atom [Jagtap and Rao, 2005]. The high stability of this conformation leads to a reduced flexibility of the polymer, so this is usually described as a real tape.

There are two different types of intra- and one interchain hydrogen bonds in the structure, and it has been considered that the intrachain hydrogen bonds determine the single-chain conformation and the stiffness of cellulose, while the interchain hydrogen bond is responsible for the sheetlike nature of cellulose [Watanabe, H; Tokuda, 2001; Klemm et al., 2002; Klemm et al., 2005]. The chains are arranged parallel to each other and form elementary fibrils that have a diameter between 1.5 and 3.5 nm (microfibrils), the length of the microfibrils is about of several hundred nm.

The chain length of cellulose, expressed as degree of polymerization (DP) in relation to the number of monomers, varies with the origin and treatment of the raw material. In case of wood pulp, the values are typically 300 and 1,700; cotton and other plant fibers have DP values in the 800-10,000 range; bacterial cellulose has a similar DP value.

In relation to the amount of hydrogen bonds into and between cellulose molecules, a model of cellulose structure with two states in plant cell wall has been proposed: amorphous and crystalline.

2.2 Crystalline Structure

The high degree of hydrogen bonds within and between cellulose chains can form a 3-D lattice-like structure, while amorphous cellulose lacks this high degree of hydrogen bonds and the structure is less ordered.

The physical and chemical properties of cellulose are defined by intermolecular interactions, cross-linking reactions, polymer lengths, and distribution of functional groups on the repeating units and along the polymer chains.

Initially, crystalline structure of native cellulose (cellulose I) has been studied by X-ray diffraction and has been defined as monoclinic unit cells with two cellulose chains with a twofold screw axis in a parallel orientation forming slight crystalline microfibrils [Gardner and Blackwell, 2004; Klemm et al., 2005].

Afterwards, it was discovered by using 3C-CP/MAS NMR spectroscopy, that the native cellulose was present in two crystalline forms obtained by modifications of cellulose I: Iα with triclinic unit cells and Iß with monoclinic unit cells. The ratio between Iα and Iß changes in relation to the source of the cellulose [Atalla and Vanderhart, 1984].

Moreover, there are other types of crystal structures: cellulose II, III, and IV [Gardner and Blackwell, 2004]; the cellulose I, result the less stable thermodynamically, while the cellulose II is the most stable structure [Klemm et al., 2005; Fengel and Wegener, 1984]. The cellulose I can turn into other forms using different treatments; for example, by mercerization, using aqueous sodium hydroxide or dissolution_followed by precipitation and regeneration (formation of fiber and film) [O'Sullivan, 1997; Nishiyama et al., 2002].

However, additional information on the structure of noncrystalline random cellulose chain segments are needed because it is very important for the accessibility and reactivity of the polymer and the characteristics of cellulose fibers [Paakkari et al., 1989].

3. Cellulose Hydrolysis

Cellulases belong to a class of enzymes that catalyze the hydrolysis of cellulose and are produced chiefly by Fungi, Bacteria, and Protozoa, as well as other organisms like plants and animals. The cellulolytic enzymes are inducible since they can be synthesized by microorganisms during their growth on cellulosic materials [Lee and Koo, 2001].

A variety of different kinds of cellulose-degrading enzymes are known, with different structure and mechanism of action. In general, the "cellulolytic enzyme complex" breaks down cellulose to β-glucose, and involves the following types of enzymes: endoglucanases

(EC 3.2.1.4), exoglucanases (EC 3.2.1.74), cellobiohydrolases (EC 3.2.1.91), and β-glucosidases (EC 3.2.1.21) [Li et al., 2010].

Endoglucanases (EC 3.2.1.4) hydrolyze randomly internal glycosidic linkages in soluble and amorphous regions of the cellulose, and produce new ends by cutting into long cellulose strands. This action results in a rapid decrease of the polymer length and in a gradual increase of reducing sugars concentration.

Exoglucanases (EC 3.2.1.74) hydrolyze cellulose chain and oligosaccharides with high DP removing successive β-glucose units.

Cellobiohydrolases (EC 3.2.1.91) hydrolyze cellulose chains by removing processively 2 units (cellobiose) either from the non-reducing and reducing ends. This action results in rapid release of reducing sugars but little changes in polymer length occurr. Cellobiohydrolases with specificity for the reducing and the non-reducing end have to work together.

β-Glucosidases (EC 3.2.1.21) convert the resulting oligosaccharide products to glucose [Bhat and Bhat, 1997].

These enzymatic components act sequentially in a synergistic system to facilitate the breakdown of cellulose and the subsequent biological conversion to β-glucose (Figure 1) [Beguin and Aubert, 1994].

All these enzymes hydrolyze the 1,4-β-glycosidic bonds in cellulose, but they are different in their specificities based on the macroscopic features of the substrate. There are progressive (also known as processive) enzymes when they interact with a single polysaccharide strand continuously, and non-progressive types when they interact once and then, the polypeptidic chain disengages to attack another polysaccharide strand.

The enzymatic hydrolysis of cellulose requires a carbohydrate binding module (CBM) that binds and arranges the catalytic components on the surface of the substrate. Cellulases from Fungi have a two-domain structure with one catalytic domain, and one cellulose binding domain, that are connected by a flexible linker. However, there are also cellulases that lack cellulose binding domain.

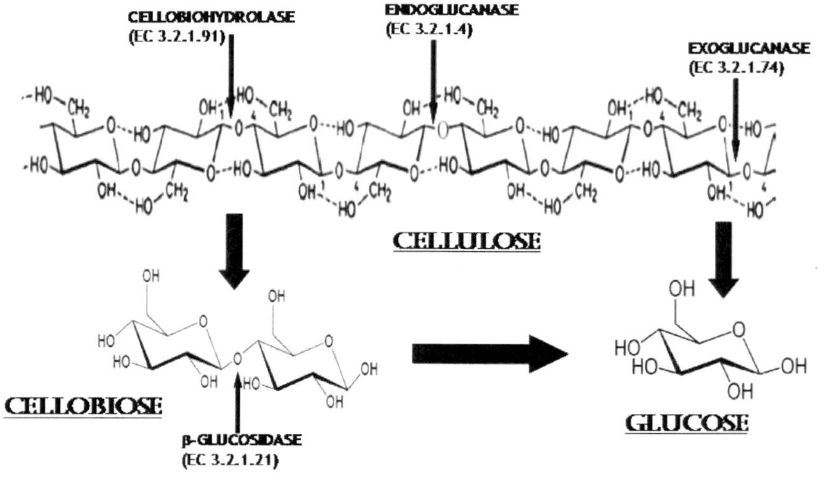

Figure 1. Enzymatic hydrolysis of cellulose.

4. CELLULASES CLASSIFICATION

Enzymes are designated according to their substrate specificity, based on the guidelines of the International Union of Biochemistry and Molecular Biology (IUBMB). The cellulases are grouped with many of the hemicellulases and other polysaccharidases as *O*-glycoside hydrolases (EC 3.2.1.x). Since the substrate specificity classification is sometime little informative, because the complete range of substrates is only rarely determined for individual enzymes, an alternative classification of glycoside hydrolases (GH) into families based on amino acid sequence similarity has been suggested [Henrissat, 1991; Henrissat and Bairoch, 1993; Henrissat and Bairoch, 1996]. In addition, Henrissat *et al.* [1998] have proposed a new type of nomenclature for glycoside hydrolases in which the first three letters designate the preferred substrate, the number indicates the glycoside hydrolase family, and the following capital letter indicates the order in which the enzymes were first reported. For example, the enzymes CBHI, CBHII, and EGI of *Trichoderma reesei* are designated Cel7A (CBHI), Cel6A (CBHII), and Cel6B (EGI).

Due to the great increase of identified glycoside hydrolases, Coutinho and Henrissat have created an integrated database which is continuously updated (http://www.cazy.org/) [Coutinho and Henrissat, 1999]. At the latest update (13 July 2010), glycoside hydrolases were grouped into 118 families. In addition, 876 glycoside hydrolases have not yet assigned to a family (Glycoside Hydrolase Family "Non-Classified") because some of them display weak similarity to established GH families, but they are too distant to allow a reliable assignment. Cellulases are found in several different GH families (5, 6, 7, 8, 9, 12, 44, 45, 48, 51, 61, and 74), suggesting convergent evolution of different folds resulting in the same substrate specificity. GH family 9 contains cellulases from bacteria (aerobic and anaerobic), fungi, plants and animals (protozoa and termites). Other families only group hydrolases from a specific origin, as GH family 7 which contains only fungal hydrolases and GH family 8 which contains only bacterial hydrolases. At last, cellulases from the same microorganism can be found in different families (e.g. the *Clostridium thermocellum* cellulosome contains endoglucanases and exoglucanases from families 5, 8, 9, 44, and 48) [Shoham et al., 1999].

Where necessary, GH families have been subclassified. It is the case of GH family 9 that has been divided into two subfamilies: E1 and E2. Members of the subfamily E1 show a tight association of an Ig-like domain with a catalytic domain, while members of subfamily E2 are associated with a CBM classified in family 3c [Beguin, 1990].

The study of cellulolytic enzymes at the molecular level has revealed some of the features that contribute to their activity. Within each GH family, available data suggest that the various cellulases share a common folding pattern, the same catalytic residues, and the same reaction mechanism, i.e. either single substitution with inversion of configuration or double substitution resulting in retention of the β-configuration at the anomeric carbon [Beguin and Aubert, 1994]. As observed, cellulases are a composite group of enzymes, and the diversity within the cellulase families could reflect the heterogeneity of cellulose within plant materials and the variety of niches where hydrolysis takes place. The insoluble, recalcitrant nature of cellulose represents a challenge for cellulase systems. In addition to catalytic domains, many cellulolytic enzymes contain domains not involved in catalysis, but participating in substrate binding (cellulose-binding domains, CBDs), or to the attachment to the cell surface. Most probably, these domains facilitate cellulose hydrolysis by bringing the catalytic domain in close proximity to the insoluble cellulose, and assisting in the degradation of crystalline cellulose. CBDs are generally located at the -COOH or $-NH_2$ terminus of the polypeptidic chain, and are often separated from the catalytic domains by glycosylated Pro/Thr/Ser-rich linker segments. CBDs are now better classified as carbohydrate-binding modules (CBMs). The previous definition was based on the early discovery of a number of modules that bound cellulose [Tomme et al., 1995].

The binding efficiency of the cellulase is much enhanced by the presence of the CBM and the enhanced binding correlates with better hydrolytic activity toward insoluble cellulose. For example, the presence of CBM in *T. reesei* is reported to enhance the enzymatic hydrolysis of insoluble cellulose and chemical pulp [Suurnakki et al., 2000].

However, additional modules are continuously found able to bind other carbohydrates besides cellulose. Hence, the need to reclassify these polypeptides using more inclusive terminology. Until now, carbohydrate-binding modules have been divided into 59 families. Additionally, 35 carbohydrate-binding modules have not yet assigned to a family (Carbohydrate-Binding Module Family "Non Classified").

5. Reaction Mechanism of Cellulase Enzymes

The cellulase catalytic mechanism, which has been optimized by many years of evolution, has by far attracted the interest of many researchers. On the basis of kinetic and chemical modification studies some authors have suggested that several cellulases operate by a lysozyme-type mechanism. These conclusions are strengthened by the finding that specific amino acid sequences in these cellulases were homologous with the active centers of known lysozymes [Coughlan, 1991].

Lysozymes were the first glycoside hydrolases to have their three-dimensional structures solved [Blake et al., 1965]. The two catalytic amino acids of the active site were identified, like in most glycoside hydrolases, as aspartate and glutamate residues.

In particular, the mechanism of the lysozyme catalyzed reaction may be described as a double displacement acid hydrolysis in which a non-ionized glutamic acid and an ionized aspartic acid residues participate as proton donor and acceptor, respectively [Lehninger, 1982].

Bacterial enzymes utilizing a lysozyme-type mechanism were identified in *Cellulomonas fimi* endo- and exoglucanases [Gilkes et al., 1988] and *Clostridium thermocellum* endoglucanases A and D [Schwarz et al., 1988] by sequences comparison with the active center of lysozyme. In fact, the homology studies confirmed, for these enzymes, putative active centers containing the glutamic and aspartic acid residues. However, other studies demonstrated that may be an error to assign, on the basis of homology data alone, a catalytic role at any glutamic acid/aspartic acid pair residues [Yablonsky et al., 1988].

Site-directed mutagenesis experiments have been performed to elucidate the question, and an example of chemical modification study on the bacterial cellulases is that of Claeyssens and Tomme who provided evidence for the involvement of an histidyl residue in the reaction catalyzed by the endoglucanase D from *C. thermocellum* [Claeyssens and Tomme, 1989].

An explanation of the mechanism by which the glycoside hydrolases catalyze the cleavage of the glycosidic linkage was provided already in 1953 by Koshland that proposed two different stereospecific reaction mechanisms namely the inverting and retaining [Koshland, 1953].

The Figure 2 shows a representation of the two types of mechanisms hypothesized for cellulase enzymes. The retaining mechanism (a), in which the first residue acts as an acid catalyst (AH) that protonates the glycosidic oxygen and the nucleophilic assistance to leaving group departure is provided by the second residue, the base B^{-}. The resulting glycosyl-enzyme is hydrolyzed by a water molecule and this second nucleophilic substitution at the anomeric carbon generates a product with the same stereochemistry as the substrate, similarly to the reaction of lysozyme [Kelly et al., 1979]. The inverting mechanism (b), in which there is also a protonation of the glycosidic oxygen by the acid residue and the leaving group departure is accompanied by a concomitant attack of a water molecule activated by the base residue B^{-}. This single nucleophilic substitution yields a product with opposite stereochemistry to the substrate as observed in the case of ß-amylase. A detailed description of the catalytic mechanism for glycoside hydrolase enzymes can also be found in several excellent reviews [Vasella, et al., 2002; Zechel and Withers, 2000; Zechel and Withers, 2001].

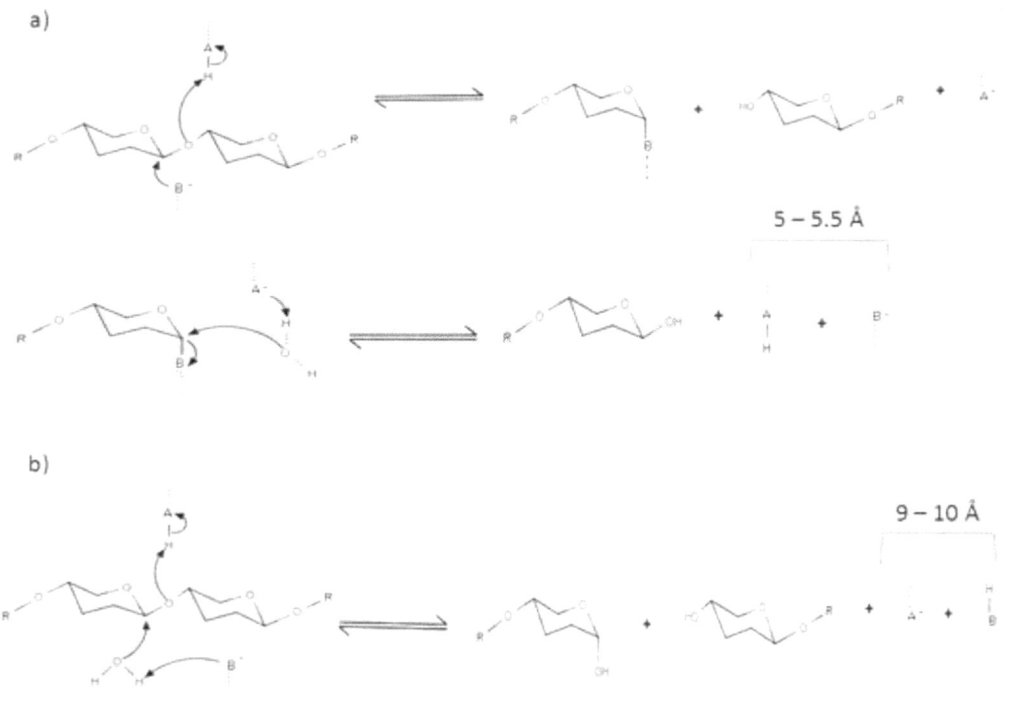

Figure 2. Schematic representation of the retaining (a) and inverting (b) reaction mechanisms.

The main structural difference that occurs in glycosidase enzymes which perform the hydrolysis of the glycosidic linkage in the retaining or inverting manner consists in the distance separating their respective carboxyl groups. In the enzymes that operate with the inverting mechanism, the base and acid carboxyl residues are separated, on average, by 9-10 Å, whereas in retainers the nucleophile and general acid-base catalyst are only ~ 5-5.5 Å apart. The explanation of the greater span found in inverters is justified by the necessity to accommodate the nucleophilic water molecule.

The knowledge of these structural differences in glycosidases has made possible to perform experiments to convert inverters enzymes into retainers and vice-versa, with the appropriate substitution (i.e. mutation) at the nucleophile position [Vocadlo and Davies, 2008].

Therefore, on the basis of these differences, glycosidases can be classified as enzymes that catalyze the glycosidic linkages hydrolysis with retention or inversion of the anomeric configuration at the hemiacetal center of the newly formed product.

The stereochemical course of hydrolyses catalyzed by cellulolytic enzymes from various sources have been investigated. Characteristic examples of several retention mechanisms are the double-inversion of glycosyl enzyme intermediate utilized by lysozyme [Lehninger, 1982] and the carbohydrate hydrolysis catalyzed by Cex from *C. fimi*, whereas the reaction catalyzed by *C. thermocellum* CenA enzyme proceeds by inversion of configuration [Withers et al., 1986].

In 1991, Coughlan [1991] underlined some aspects of cellulases which bear the catalytic domain at the N-terminus and utilize lactosides and cellobiosides as substrates of the hydrolytic reaction that proceedes with the retention of the β-configuration. By contrast, those enzymes in which the catalytic domain is located at the C-terminus cannot utilize lactosides and cellobiosides as substrates, obtaining products with an inversion of the β-configuration.

6. The Cellulosome Concept

All the microorganisms capable of plant cell wall degradation produce complex cellulase enzymes systems; however, two different types of strategy occur between aerobic and anaerobic groups [Tomme et al., 1995]. Aerobic cellulose degraders, both bacterial and fungal, apart from few exceptions [Wachinger et al., 1989], produce a set of free enzymes which are released in the extracellular environment and work synergistically to degrade the plant cell walls [Schwarz, 2001]. Instead, anaerobic microorganisms degrade cellulosic substrates primarily through the cell-bound multienzyme systems known as the "cellulosomes". These structures, which are quite stable cellulolytic complexes, show considerable dimensions that can vary from 2.0 to 16.0 MDa and even up to 100.0 MDa in the case of polycellulosomes [Béguin and Lemaire, 1996]. The occurrence of the cellulosome was firstly observed in the thermophilic bacterium *Clostridium thermocellum* [Bayer et al., 1983]; successively, a range of anaerobic bacteria such as *C. cellulovorans* [Shoseyov et al., 1992], *C. cellulolyticum* [Pages et al., 1999], *C. acetobutylicum* [Sabathe et al., 2002], *C. josui* [Kakiuchi et al., 1998], *C. papyrosolvens* [Pohlschröder et al., 1994], *Bacteroides cellulosolvens*, *Acetivibrio cellulolyticus* [Pages et al., 1997], *Ruminococcus flavefaciens* [Rincon et al., 2003] and the anaerobic fungi of the genera *Neocallimastix*, *Piromyces*, and *Orpinomyces* [Bayer et al., 2004] were shown to produce cellulosomal systems. The cellulosomes are characterized by the presence of two general components: a) a large non catalytic scaffoldin protein with enzyme binding sites called cohesins, and b) the catalytic components that contain, at the C-terminus, highly conserved noncatalytic modules, called dockerins, which bind to the cohesin modules.

The cellulosome enzymatic components contain not only cellulases but also a large array of hydrolytic activities such as hemicellulases [Kosugi et al., 2002], pectinases [Tamaru and Doi, 2001], chitinases, lichenases, mannanases and esterases. This extraordinary enzymes diversity reflect the chemical and structural complexity of the cellulosome substrate, the plant cell wall, that can be efficiently attacked and degraded only by the concerted action of different enzymatic activities [Fontes and Gilbert, 2010]. Moreover, in the cellulosome assembly, the cohesin domains are unable to discriminate among the dockerins present in the various catalytic modules due to the high level of conservation in the same species of both cohesins and dockerins domains [Yaron et al., 1995]. This leads, through the induction of specific genes by plant cell wall polymers, to different and temporally evolving cellulosome enzyme combinations, which allow a successful cell wall structures degradation. In addition to the cohesins, the scaffoldin also bears a cellulose-binding module (CBM) that targets the cellulosomal enzymes as well as the entire cell to the cellulosic substrates. In fact this domain, interacting with crystalline cellulose, brings the cellulosome into close proximity with the plant cell wall and concentrates the hydrolytic enzymes to a particular site of the substrate [Gilbert, 2007]. In the simplest cellulosome system, there is a single scaffoldin protein with a CBM and 6 to 9 catalytic components in dependence of the cohesin number that varies with the different species [Lynd et al., 2002]. Moreover, several cellulosome-producing microbes express more than one type of scaffoldin: this is the case of the bacteria with cell-surface anchored cellulosomes, such as *C. thermocellum*, *Acetivibrio cellulolyticus*, *Bacterioides cellulosolvens* and *Ruminococcus flavefaciens* [Bayer et al., 2008]. After its discovery in the mid-1980s, the first polyscaffoldin cellulosome structure, resolved through a combination of biochemical, immunochemical, ultrastructural, and genetic techniques was that of *C. thermocellum* [Mayer et al., 1987]. This cellulosome revealed an highly ordered structure with sets of polypeptides arranged in parallel chain-like arrays. The cellulosomal system consists of a large scaffoldin protein (CipA) of 147.0 kDa, whose encoding sequence is part of an operon, called "scaffoldin gene cluster", containing several other genes coding for the secondary scaffoldins (see below) [Fujino et al., 1993]. CipA, that contain a CBM module and 9 cohesin domain, termed of type I, is defined as primary scaffoldin [Bayer et al., 1998]. Cohesin domains are folded in 9-stranded β-barrel like families II and III CBDs, in spite of the total absence of homology. The catalytic components, bearing at their C-terminus a dockerin domain named of type I, are bound in presence of Ca^{++} to the type I cohesins onto CipA [Salamitou et al., 1994]. A total of 22 catalytic modules, at least 9 of which exhibiting endoglucanase activity (CelA, CelB, CelD, CelE, CelF, CelG, CelH, CelN, and CelP), 4 exoglucanase activity (CbhA, CelK, CelO and CelS), 5 hemicellulase activity (XynA, XynB, XynV, XynY and XynZ), 1 chitinase activity (ManA), and 1 lichenase activity (LicB), are grafted into the cohesin sites of CipA to form the cellulosome complex.

Three different types of cell surface proteins, named SdbA, Orf2p and Olpb, defined as secondary or anchoring scaffoldins, contain different numbers (1, 2, or 7 respectively) of type II cohesins that, interacting with a type II dockerin located at the C-terminus of the primary scaffoldins, allow their attachment to the cell envelope [Leibovitz and Beguin, 1996]. The type II dockerin-cohesin affinity is further enhanced by the stabilizing effect of an hydrophilic domain, named X module, located immediately upstream the type II dockerin. The anchoring scaffoldins also contain a C-terminal threefold reiterated SLH domains, normally found in the

S-layer proteins [Rincon et al., 2003] which mediate the anchoring of these structural proteins to the bacterial cell wall. In the case that seven primary scaffoldins are assembled onto the 7 cohesin II of an OlpB anchoring scaffoldin, a polycellulosome bearing 63 catalytic units may be produced.

The *C. thermocellum* cellulosome structure has been considered the paradigm for such enzymatic nanomachines and several subsequent studies were aimed to verify if the cellulosomes from other bacteria would follow the *C. thermocellum* paradigm (Figure 3).

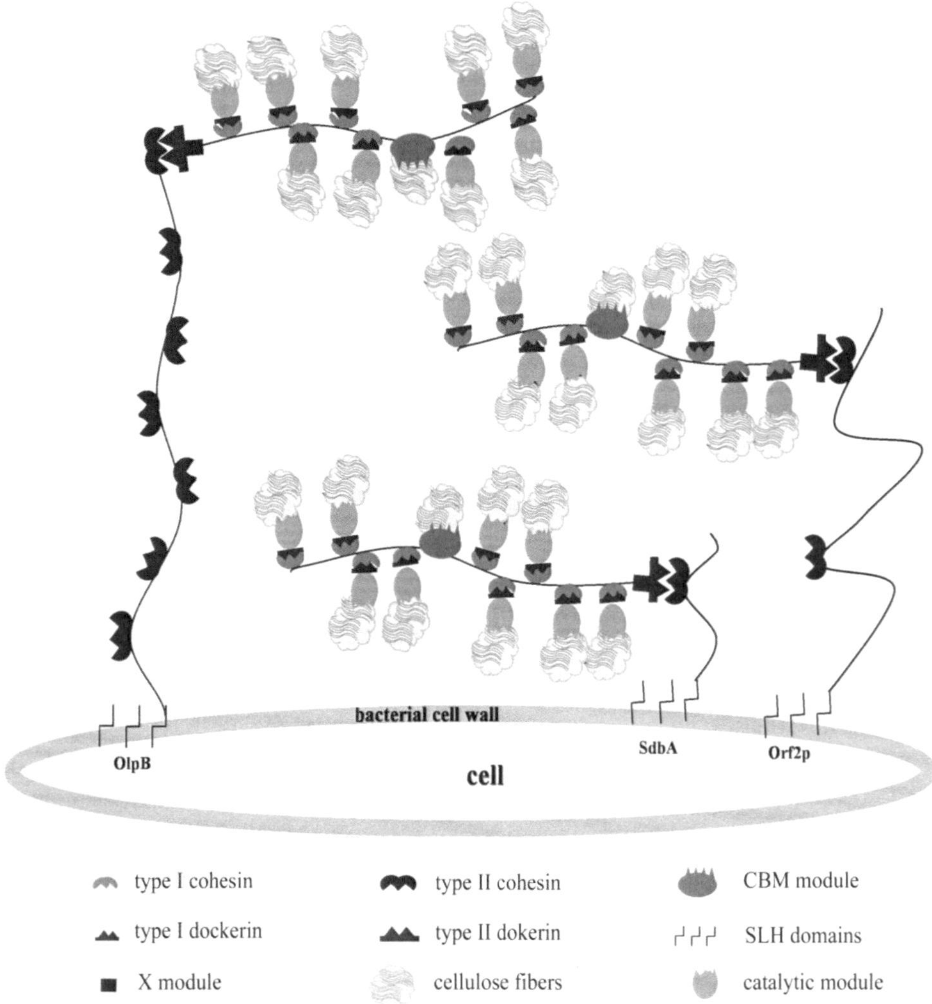

Figure 3. Schematic representation of C. themocellum cellulosome structure. The cell surface proteins, Sdba, Orf2 and OlpB, act as anchoring scaffoldins. They bind, with their type II cohesins, the type II dockerins domains located at the C-terminus of single CipA primary scaffoldin. Each primary scaffoldin can bind, with its type I cohesions, up to nine type I dockerin appended at the C-terminus of the catalytic modules. In the primary scaffoldins, a CBM mediates the attachment of the cellulosome and the entire cell to the cellulose fibers. The cellulosome is anchored the to the cell surface by the SLH C-terminal modules of the secodary scaffoldins.

Surprisingly, the results obtained showed very divergent cellulosome structures. This is the case of the aforementioned bacteria *R. flavefaciens*, *A. cellulolyticus*, and *B. cellulosolvens,* which revealed very different and versatile cellulosome architecture and modular arrangement [Bayer et al., 2008]. The versatile nature, composition and arrangement of cellulosome is best elucidated by the complex structure of *R. flavescens* cellulosome that is characterized by at least four types of scaffoldin with different functions. The whole cellulosomic system is attached to the cell surface by the ScaE scaffoldin through a sortase-mediated transpeptidation reaction [Rincon et al., 2005]. The primary scaffoldin ScaB is bound to ScaE and through its 9 cohesin with different specificity can accommodate up to 4 catalytic modules and up to 5 ScaA adaptor scaffoldins, each endowed with 3 cohesins [Fierobe et al., 2005]. Another adaptor scaffoldin, ScaC, that bear a single cohesin with specificity for an unknown group of dockerins is also accommodated onto the ScaB [Jindou et al., 2008]. In addition to this structural characteristic, the *R. flavefaciens* system shows also several unique features respect to the other cellulosomes that significantly increase its degradative capabilities in the rumen environment. In fact, the equipment of the cellulosomal catalytic modules also include putative pectate lyase, rhamnogalacturan lyase, mannanase, arabinase, transglutaminase, proteinase and peptidase activities. Moreover, the different *R. flavefaciens* strains show distinct scaffoldin sequences [Jindou et al., 2008] that implies a strain-specific cellulosome organization. This cellulosome strain variety provides a very useful functional diversity in the degradative capacities that are required by the complexity and the heterogeneity of the lignocellulosic substrates found in the rumen [Bayer et al., 2008].

Several studies have also been conducted on the cellulosomes of mesophilic clostridia such as *C. cellulovorans* [Doi et al., 1994], *C. cellulolyticum* [Bagnara-Tardif et al., 1992], *C. acetobutylicum* [Cornillot et al., 1997], *C. josui* [Kakiuchi et al., 1998], and *C. papyrosolvens* [Pohlschröder et al., 1994], which also revealed a structure different from that of *C. thermocellum* [Doi, 2008]. These microorganisms in fact express cellulosomes characterized by a less complex architecture, where the enzymatic modules are grafted on a single primary scaffoldin and no anchoring scaffoldins have been identified [Bayer et al., 2008]. As opposed to the "scaffoldin gene cluster" of *C. thermocellum* and the other multiple scaffoldin cellulosome producers, in these species an "enzyme linked gene cluster" has been individuated that comprises a primary scaffoldin gene followed by the genes encoding the dockerin-bearing enzymes [Bayer et al., 2004; Doi, 2008]. It is just this collection of enzymes and their coordinated expression regulation, strictly dependant from the carbon sources [Han et al., 2005], that makes the cellulosome system so versatile and efficient in attacking and degrading the plant cell walls. It is important to underline that even the mesophilic clostridium cellulosomes, which exhibit the lowest complexity levels, result in a more efficient system in deconstructing plant structural polysaccharides respect to the "free" enzymes produced by the aerobic microorganisms. This is particularly apparent in the case of *C. thermocellum* where the cellulosome is reported to display a specific activity against crystalline cellulose 50-fold higher than the corresponding *Trichoderma* system [Demain et al., 2005]. The evolutionary drivers that brought to the undoubted cellulosome success are not clear but most probably were just the energetic constraints imposed by the anaerobic environment that led to a necessary improvement of the microorgamisms degradative capabilities. In fact, the grafting of plant cell wall-degrading enzymes onto a macromolecular complex brings to a spatial enzyme proximity that potentiates the synergistic interactions among the cellulosomal catalytic units. The catalytic cellulosome efficiency is further increased by enzyme-substrate targeting that allows a close proximity of the cell to the substrate. In fact, due to the restricted extracellular diffusion rate of the degradation products, these are readily removed by an enhanced cell uptake, leading to an increased cellulose hydrolytic rate [Bayer et al., 1998]. Moreveor, the cellulosomes action in concert with noncellulosomal glycosidic hydrolases, further amplify the whole degradative process yield. This has been clearly demonstrated by the concerted actions of cellulosomal hemicellulase XynA and noncellulosomal hemicellulases ArfA and BgaA [Kosugi et al, 2002] in *C. cellulovorans*.

Table 2. Some mesophilic cellulolytic Bacteria

Microorganism	Gram reaction	Growth T (°C)	Growth conditions	References
Acetivibrio cellulolyticus	-	37	Anaerobic	Sanchez et al., 1999 Patel et al., 1980
Bacillus megaterium	+	30	Aerobic	Beukes et al., 2006
Bacillus pumilus	+	30	Aerobic	Kotchoni et al., 2003
Bacteroides cellulosolvens	-	35	Anaerobic	Murray et al., 1984
Butyrivibrio fibrisolvens	+	37	Anaerobic	Bryant, 1959
Cellulomonas fimi	+	30	Aerobic	Langsford et al., 1984
Cellulomonas fermentans	+	30	Aerobic	Bagnara et al.,1985
Cellulomonas flavigena	+	30	Aerobic	Van Leeuwenhoek, 1984
Cellulomonas gelida	+	30	Aerobic	Stackebrandt and Kandler, 1979
Cellulomonas iranensis	+	28	Aerobic	Elberson et al., 2000
Cellulomonas persica	+	28	Aerobic	Elberson et al., 2000
Cellulomonas uda	+	30	Aerobic	Stoppok et al., 1982
Cellvibrio mixtus	-	20	Aerobic	Blackall et al., 1985
Clostridium acetobutylicum	+	37	Anaerobic	Sabathe et al., 2002
Clostridium cellulolyticum	+	35-37	Anaerobic	Petitdemange et al., 1984
Clostridium cellulofermentans	-	40	Anaerobic	He et al., 1991
Clostridium cellulovorans	-	37	Anaerobic	Sleat et al., 1984
Clostridium herbivorans	+	37	Anaerobic	Varel et al., 1995
Clostridium hungatei	-	30	Anaerobic	Monserrate et al., 2001
Clostridium josui	-	45	Anaerobic	Sukhumavasi et al., 1988
Clostridium papyrosolvens	-	25	Anaerobic	Madden et al., 1982
Cytophaga hutchinsonii	-	30	Aerobic	Li and Gao, 1997
Erwinia carotovora	-	26	Aerobic	Barras et al., 1994
Fibrobacter succinogenes	-	37	Anaerobic	Stewart and Flint, 1989 Chen and Wang, 2008
Halocella cellulolytica	-	39	Anaerobic	Simankova et al., 1993
Prevotella ruminicola	-	37	Anaerobic	Chen and Wang, 2008
Pseudomonas fluorescens	-	30	Aerobic	Hazlewood et al., 1992
Ruminococcus albus	+	37	Anaerobic	Bryant, 1959
Ruminococcus flavefaciens	+	37	Anaerobic	Bryant, 1959
Streptomyces antibioticus	+	28	Aerobic	Enger and Sleeper, 1965
Streptomyces cellulolyticus	+	28	Aerobic	Li, 1997
Streptomyces lividans	+	28	Aerobic	Wittmann et al., 1994
Streptomyces reticuli	+	28	Aerobic	Wachinger et al., 1989
Zymomonas mobilis	-	30	Anaerobic	Rajnish et al., 2008

7. CELLULASES OF MESOPHILIC ORIGIN

Microorganisms growing best at moderate temperatures (between 10 and 45°C) are named mesophiles. They represent the majority of microbial species on Earth, and their habitats include the soil, the human body, the animals, etc. There are many mesophilic Bacteria and Fungi that play a significant role in the carbon cycle on Earth, and there is increasing interest in the enzymes from these microorganisms, since they have a key function in the conversion of plant biomass into useful products.

The ability to digest cellulose is widely distributed among Bacteria and Fungi and some of them are listed in Tables 2 and 3.

As already described, the different strategy of degradation between the anaerobic and aerobic groups resides in the production of complex cellulase systems, exemplified by the well-characterized cellulosome from the *Clostridium* genus [Beguin and Lemaire, 1996; Schwarz, 2001], or the extracellular cellulases freely released in the culture supernatant, respectively [Wachinger et al., 1989].

7.1 Bacterial Cellulases

Active research on cellulases and related polysaccharidases began in the early 1950s, owing to their enormous potential to convert lignocellulose to glucose [Mandels, 1985]. In this review we have limited the description of cellulases of mesophilic origin to the last ten years, because of the considerable amount of literature that has been produced up to now, including reviews. Some representatives of bacterial cellulases described before 2000 are reported in Table 4.

Identification, purification and characterization of cellulases are continuously increasing and always in progress, with incessant research and isolation of new microorganisms able to produce novel cellulolytic activities. As an example, a bacterial strain, TR7-06(T), showing high sequence similarity (98.5 %) to *Cellulomonas uda* DSM 20107(T), was isolated from compost at a cattle farm near Daejeon, Republic of Korea. The isolated type strain of a novel *Cellulomonas* species, named *Cellulomonas composti* sp. nov., possesses endoglucanase and β-glucosidase activities [Kang et al., 2007]. A microorganism capable of hydrolyzing rice hull, one of the major cellulosic waste materials in Korea, was isolated from soil and identified as *Bacillus amyloliquefaciens* DL-3. Basing on the characteristics of this novel strain of *Bacillus*, Lee *et al.* [2008] aimed to develop an economical process for production of cellulases by using cellulosic waste as inexpensive and widely distributed carbon source. The new isolate produced an extracellular cellulase with an estimated molecular mass of about 54.0 kDa. The deduced amino acid sequence of the cellulase from *B. amyloliquefaciens* DL-3 showed high identity to cellulases from other *Bacillus* species, a modular structure containing a catalytic domain of the GH family 5, and a cellulose-binding module type 3 (CBM3). The purified enzyme was optimally active at 50°C and pH 8.0, and showed broad thermal and pH stability ranging from 40 to 80°C and from 4.0 to 9.0, respectively.

Table 3. Some mesophilic cellulolytic Fungi

Microorganism	Growth T (°C)	Growth conditions	References
Acremonium cellulolyticus	24	Aerobic	Yamanobe et al., 1987 Ikeda et al., 2007
Anaeromyces mucronatus	37	Anaerobic	Lee et al., 2001
Aspergillus glaucus	30	Aerobic	Tao et al., 2010
Aspergillus niger	30	Aerobic	Hasper et al., 2002
Aspergillus terreus	35	Aerobic	Elshafei et al., 2009
Caecomyces communis	37	Anaerobic	Orpin, 1976
Ceratocystis paradoxa	20	Aerobic	Olutiola, 1976
Chalara (Syn. Thielaviopsis) paradoxa	27	Aerobic	Lucas et al., 2001
Chrysosporium lucknowense	25-43	Aerobic	Bukhtojarov et al., 2004
Cyllamyces aberensis	37	Anaerobic	Oziose et al., 2001
Fusarium solani	25	Aerobic	Wood and McCrae, 1977
Neocallimastix frontalis	37	Anaerobic	Wood et al., 1985 Li and Calza, 1991
Neocallimastix patriciarum	37	Anaerobic	Denman et al., 1996
Orpinomyces sp.	37	Anaerobic	Chen et al., 1998
Penicillium funiculosum	24	Aerobic	Machado de Castro et al, 2010
Penicillium pinophilum	24	Aerobic	Bhat et al., 1989
Phanerochaete chrysosporium (Sporotrichum pulverulentum)	35	Aerobic	Henriksson et al., 1999 Eriksson and Pettersson, 1975
Piptoporus betulinus	25	Aerobic	Valaskova and Baldrian, 2006
Piromyces sp.	39	Anaerobic	Ali et al., 1995
Piromyces equi	39	Anaerobic	Eberhardt et al., 2000
Pycnoporus cinnabarinus	24	Aerobic	Sigoillot et al., 2002
Rhizopus oryzae	30	Aerobic	Moriya et al., 2003
Rhizopus stolonifer	24	Aerobic	Pothiraj et al., 2006
Serpula lacrymans	25	Aerobic	Hastrup et al., 2006
Trichoderma koningii	22	Aerobic	Wang et al., 2007
Trichoderma reesei	24	Aerobic	Kuhls et al., 1996

Although cellulases have been isolated from many microorganisms, no functional cellulase genes were reported for *Vibrio* genus until now. Gao *et al.* [2010] isolated from mangrove soil a new bacterium belonging to the *Vibrio* genus, *Vibrio* sp. G21, and a novel endoglucanase gene, *cel5A*, was cloned. The mature Cel5A enzyme was overexpressed in *Escherichia coli* and purified to homogeneity. It was stable over a wide range of pHs and retained more than 90% of activity after incubation at pHs 7.5-10.5 for 1 h. Moreover, the enzyme was activated after pretreatment with mild alkali, a novel characteristic that has not been previously reported in other cellulases. The deduced protein contained a catalytic domain of the GH family 5, followed by a cellulose-binding module type 2 (CBM2).

Table 4. Properties of some cellulases from mesophilic Bacteria

Microorganism	Enzyme	Mol mass (kDa)	Optimal T (°C)	Optimal pH	References
Bacillus circulans	Avicelase I	75.0	50	4.5	Kim, 1995
Bacillus pumilus	EglA	71.3	60	8.0	Lima et al., 1995
Cellulomonas flavigena	CMCase 1 CMCase 2	20.4 20.4	50 50	6.5 7.0	Sami and Akhtar, 1993
Clostridium acetobutylicum	EG	66.7	55	5.0	Warner et al., 2010
Clostridium cellulovorans	EngK EngL EngH EngM EngY	97.0 57.6 79.3 96.4 80.3	37 42 42 42 37	7.0 7.0 8.0 6.0 6.5	Arai et al., 2006
Clostridium josui	EG	39.0	65-70	7.2-7.5	Fujino et al., 1990
Erwinia chrysanthemi	Cel5Z	42.0	40	6.0	Park et al., 2000
Fibrobacter succinogenes	EG1 EGF	65.0 118.3	39 39	6.4 5.8	McGavin and Forsberg, 1988
Paenibacillus sp.	EGI-659	58.36	55	6.0-8.5	Ogawa et al., 2007
Ruminococcus albus	EGV	42.0 (truncated form)	40	7.0	Ohara et al., 2000
Sinorhizobium fredii	CMCase	94.0	35	7.0	Chen et al., 2004
Synechocystis PCC6803	SsGlc	112.0	42	7.0	Tamooi et al., 2007
Zymomonas mobilis	CelA	37.0	30	6.0	Rajnish et al., 2008

The discovery of alkaline cellulases has generated new industrial applications of cellulases as laundry detergent additives [Ito, 1997]. Many microorganisms belonging to *Bacillus* sp. are producers of alkaline cellulases even if other microorganisms also possess cellulases active at high pH value. Since the discovery of an alkaline cellulase by Horikoshi *et al.* [1984], many other alkaline cellulases from alkaliphilic *Bacillus* strains have been identified. A novel strain of *Bacillus sphaericus* JS1 was isolated from soil. The strain produced an extracellular carboxymethylcellulase (CMCase) with a molecular mass of 183.0 kDa, and a single band of about 42.0 kDa was estimated by SDS-PAGE. The enzyme was active over a broad range of pH (7.0-10.5), with a half-life of 18 h at pH 8.0 and 4.5 h at pH

10.0 at 60°C [Singh et al., 2004]. Endo *et al.* [2001] purified to homogeneity from the culture broth of the alkaliphilic *Bacillus* sp., strain KSM-N252, a highly alkaline endoglucanase (Egl-252) with a molecular mass of approx. 50.0 kDa. The enzyme exhibited reasonable homology to other alkaline endoglucanases belonging to GH family 5. In fact, the deduced amino acid sequence of Egl-252 showed moderate homology to that of NK-1 [Fukumori et al., 1986], and to Cel5A from *B. agaradherens* (accession no. AF067428) with 75.6% and 64.3% identity, respectively. This suggested that also Egl-252 belongs to GH family 5. The optimal temperature for activity was 55°C, and the optimal pH was 10.0 with more than 80% of the maximal activity retained between pH 8.0 and 11.0. The enzyme was very stable between pH 6.0 and 11.5 at 30°C.

An alkaline endoglucanase with a molecular mass of 43.0 kDa (Egl-257) was purified and crystallized from *B. circulans* KSM-N257 [Hakamada et al., 2002]. The enzyme, showing 76.3% amino acid identity with a lichenase from *B. circulans* WL-12 which belongs to GH family 8, hydrolyzed carboxymethylcellulose (CMC) as well as lichenan. Egl-257 showed optimal temperature and pH at 55°C and 8.5, respectively. It was stable over a range of pH between 5.0 and 11.0 after incubation at 30°C for 1 h retaining the nearly total activity.

A novel alkaline cellulase from the alkalophilic *Bacillus* sp. HSH-810 was purified and characterized by Kim *et al.* [2005]. The purified enzyme was optimally active at pH 10.0 and showed about 60% activity at pH 12.0. In contrast, enzyme from *Bacillus* sp. strain KSM-N252 showed similar optimum pH but retained only 35% activity at pH 12.0 [Endo et al., 2001].

As already mentioned, microorganisms different from *Bacillus* are also capable of producing alkaline cellulases. *Marinobacter* sp. (MSI032), isolated from the marine sponge *Dendrilla nigra*, produces an extracellular alkaline cellulase at 27°C and pH 9.0 [Shanmughapriya et al., 2009]. Usually, cellulase production by Bacteria occurs during the late growth phase. Thus, maintenance of the culture conditions for long times causes economic disadvantages for the development of industrial processes. Unexpectedly, the production of cellulase by *Marinobacter* MSI032 occurs at an earlier stage of growth suggesting the usefulness of the strain in industrial processes. The purified enzyme displayed maximum activity at pH 9.0 and at temperature between 27 and 35°C. In addition, it was stable over a broad range of pH, with residual activity higher than 80% between pH 8.0 and 12.0, indicating that this alkaline cellulase has a very high pH stability. *Paenibacillus* sp., strains KSM-N115, KSM-N145, KSMN440, and KSM-N659 produces cellulases (Egls) that hydrolyze Avicel, filter paper and amorphous cellulose, to cellotriose, cellobiose, and glucose by endo-fashion cleavage at alkaline pH [Ogawa et al., 2007]. The optimal temperature and pH of one representative recombinant enzyme (Egl-659) for degrading CMC and Avicel were 45-55°C and 6.0-8.5, respectively. Even at pH 9.0 the enzyme showed more than 75% relative activity. Egl-659 was very stable over a pH range between 5.0 and 11.0 after incubation at 50°C for 20 h.

Among the anaerobic cellulase-producing bacteria, the genera *Clostridium* is without doubt the most studied. It numbers mesophilic and thermophilic representatives and multienzyme complexes having high activity against crystalline cellulose, known as the cellulosome, have been identified and characterized in many of these Bacteria as reported above.

C. phytofermentans was isolated by Warnick *et al.* from forest soil [2002]. The essential component of the *C. phytofermentans* cellulolytic system (Cel9) is a processive endoglucanase that shows activities on both soluble CMC and crystalline cellulose. In order to obtain high-purity cellulase and facilitate its production, the *cel9* gene was recently expressed in *E. coli*, and the recombinant protein was purified and characterized [Zhang et al., 2010a]. The pH and temperature optima for activity were 6.5 and 65°C, respectively. The unusual high optimal temperatures for Cel9 and for the noncellulosomal Cel48 (60°C) [Zhang et al., 2010b] are somewhat surprising, but can be explained by possible acquisition of the *cel9–cel48* gene cluster from a thermophilic microorganism through horizontal gene transfer.

7.2 Fungal Cellulases

Fungal cellulases are well-studied enzymes used in various industrial processes [Bhat, 2000], and the properties of several of them, not considered in this text, are listed in Table 5. A variety of aerobic and anaerobic Fungi are producers of cellulose-degrading enzymes. The aerobic Fungi play a major role in the degradation of plant materials and are found on the decomposing wood and plants, in the soil, and on the agricultural residues. The cellulase systems of the aerobic Fungi *Trichoderma reesei*, *T. koningii*, *Penicillium pinophilum*, *Phanerochaete chrysosporium*, *Fusarium solani*, *Talaromyces emersonii*, and *Rhizopus oryzae* are well characterized [Bhat and Bhat, 1997]. Much of the knowledge on enzymatic depolymerization of cellulosic material has come from *Trichoderma* cellulase system. In particular, the cellulase system of *T. reesei* (initially called *T. viride*) has been the focus of research for 50 years [Reese et al., 1959; Reese and Mandels, 1971]. A lot of work on cellulases has been directed toward this fungus since it produces readily, and in large quantities, a complete set of extracellular cellulases, and consequently, it has a high commercial value [Claeyssens et al., 1998; Miettinen-Oinonen and Suominen 2002].

In fact, *T. reesei* is capable of secreting more than 30 g/L of protein into the extracellular medium [Conesa et al., 2001]. It has been reported that *T. reesei* possesses two CBH (cellobiohydrolase) genes, *cbh1-2*, and eight EG (endoglucanase) genes, *egl1-8*, and that CBH I–II and EG I–VI are secreted proteins [Foreman et al., 2003]. Altough the present review essentially concerns cellulases from the last ten years, the authors like to give short signal about endoglucanases from *T. reesei* as they represent very attractive biocatalysts for industrial applications [Schuster and Schmoll, 2010].

EGI (Cel7B) hydrolyzes both cellulose and xylan and has optimal temperature and pH at 30°C and 5.0, respectively [Biely et al., 1991]. The structure of EGI was resolved to reveal the presence of short loops that create a groove rather than a tunnel.

The catalytic domain resembles an open substrate-binding cleft, thus enabling the enzyme to interact more effectively with the amorphous or disordered crystalline cellulose [Kleywegt et al., 1997]. A similar groove was shown for the structure of EGIII (Cel 12A) that lacks a CBM [Sandgren et al., 2000]. The glycosylation profile of EGI and EGII (Cel5A) was determined by Hui *et al.* combining enzymatic digestion with mass spectrometry techniques, and the analyses indicated that glycosylation accounted for 12-24% of the molecular mass of the enzymes [Hui et al., 2002]. Saloheimo *et al.* [1988] isolated and determined the primary structure of the gene *egl3* coding for the EGIII endoglucanase from *T. reesei*. The protein was purified, and its amino acid composition and N-terminal sequence supported the data obtained from the gene sequence. The enzymatic properties of EGIII and EGV (Cel45A) have been investigated by Karlsson *et al.* [2002]. Adsorption studies on Avicel and phosphoric acid swollen cellulose (PASC) showed that Cel45A and Cel45A catalytic core adsorbed to these substrates. On the contrary, Cel12A adsorbed weakly to both Avicel and PASC. Cel12A showed maximal activity at pH 5.0, while pH 4.0 was the best value for Cel45A maximal activity. The optimal temperature for Cel12A was 50°C. Interestingly, Cel45A showed the highest activity at 70°C. EGIV (Cel61A) was homologously expressed in high amounts with a histidine tag on the C-terminus, purified by metal affinity chromatography and characterized [Karlsson et al., 2001]. The only activity exhibited by Cel61A was the endoglucanase activity toward substrates containing 1,4-β-glycosidic bonds (CMC, hydroxyethylcellulose and β-glucan).

Table 5. Properties of some cellulases from mesophilic Fungi

Microorganism	Enzyme	Mol mass (kDa)	Optimal T (°C)	Optimal pH	References
Chalara paradoxa	EG	35.0	37	5.0	Lucas et al., 2001
Chrysosporium lucknowense	Cel45A	25.0	65	4.5	Bukhtojarovet al., 2004
	Cel12A	28.0	60	5.5	
	EG44	44.0	70	5.5	
	EG47	47.0	65-70	5.0-6.0	
	EG51	51.0	70	5.0	
	EG60	60.0	60	4.5-5.0	
Daldinia eschscholzii	EG	46.4	70	6.0	Karnchanatat et al., 2008
Fomitopsis palustris	EGII	32.0	55	3.5	Shimokawa et al., 2008
Fomitopsis pinicola	EG	32.0	60	5.0	Yoon et al., 2008
Fusarium oxysporum	EG	23.2	50	6.0	Christakopoulos et al., 1995
Gloeophyllum sepiarium	EGS	45.1	59	4.1	Mansfield et al., 1998
Gloeophyllum trabeum	EGT	40.5	62	4.2	Mansfield et al., 1998
Orpinomyces joyonii	CelA	98.3	40	4.0	Li uet al., 1997
Penicillium pinophilum	EGI	25.0	50-60	4.0-5.0	Bhat et al., 1989
	EGII	39.0	50-60	4.0-5.0	
	EGIII	62.5	50-60	4.0-5.0	
	EGIV	54.0	50-60	4.0-5.0	
	EGV	44.5	65-70	4.0-5.0	
Phanerochaete chrysosporium	Cel12A	28.0	37	5.0	Henriksson et al., 1999
Rhizopus stolonifer	PCE1	45.0	50	6.0	Shimonaka et al., 2004

In recent years, research on *Trichoderma* has been facilitated significantly by sequencing of the genomes of three strains representing the most important applications of this genus. The genome of *T. reesei* has been fully sequenced and published on the http://genome.jgipsf.org/ Trire2/Trire2.home.html website [Martinez et al., 2008]. Analyses and annotation of the genomes of *T. atroviride* and *T. virens*, (http://genome.jgipsf. org/Triat1/ Triat1.home.html; http://genome.jgi-psf.org/ Trive1/Trive1.home.html), are still in progress.

The Fungus *Acremonium cellulolyticus*, isolated in 1987, is known to be a potent producer of cellulases as *T. reesei*, even if many cellulases and β-glucosidases have not been as well characterized as those produced by *T. reesei* [Yamanobe et al., 1987; Ikeda et al., 2007]. Since enzymatic saccharification using cellulases has proven to be a powerful method in the production of bioethanol, a comparison between cellulase activity from the two fungi against three lignocellulosic materials (eucalyptus, Douglas fir wood chip and rice straw) has been performed by Fujii *et al.* [2009]. Saccharification efficiency of both culture supernatants and commercial preparations (AC derived from *A. cellulolyticus* and Accellerase 1000 derived from *T. reesei*) was investigated. The culture supernatant from *A. cellulolyticus* produced higher glucose yield from lignocellulosic materials than the *T. reesei* supernatant. In the same way, AC produced a greater amount of glucose from lignocellulosic materials than Accellerase 1000.

In the last years, a great deal of attention has been focused on enzymes capable of degrading biomass for a number of applications, and on their potential to be produced industrially. However, the cost of producing sugars from lignocellulosic waste for fermentation into bioethanol is still high to attract industrial attention, mainly due to low enzyme yields from microorganisms. Cellulases produced by Fungi such as the *Aspergillus* and *Penicillium* species have been widely studied by numerous researchers, in addition to cellulases from *T. reesei* [van Peij et al., 1998; Jun et al., 1992]. Recently, Hassan *et al.* [2008] demonstrated that six filamentous Fungi, including *A. terreus* DSM 826, produce big amounts of different enzymes involved in the degradation of cellulose (namely endoglucanase and cellobiohydrolase) when grown on media containing corn cobs, corn stalks, rice straw or sugar cane bagasse as carbon sources. Sugar cane bagasse is a very low-cost substrate for endoglucanases production from different microorganisms. An endoglucanase from *A. terreus* DSM 826 was purified and characterized after growth on sugar cane bagasse as a carbon source [Elshafei et al., 2009]. The purified enzyme showed a high specific activity toward CMC with its optimal activity at pH 4.8 and 50°C. A similar optimal temperature was reported for enzymes from *Melanocarpus* sp. MTCC 3922 [Kaur et al., 2007] and *Bacillus amyloliquefaciens* DL-3 [Lee et al., 2007]. When heated at 50°C for 1 h, the endoglucanase from *A. terreus* DSM 826 did not show loss of activity, seeming to be more thermostable than endoglucanases from other microorganisms such as that from *Sinorhizobium fredii* which retained 96% of its activity at 40 °C [Chen et al., 2004].

Aspergillus glaucus XC9, grown on 0.3% sugar cane bagasse as a carbon source, produced an extracellular cellulase with a molecular mass of 31.0 kDa [Tao et al., 2010]. The optimum of pH and temperature for enzyme activity were 4.0 and 50°C, respectively. This enzyme was stable over a wide pH range (3.5-7.5) and at temperatures below 55 °C. It retained only 60% activity after incubation at 60°C for 1 h. The newly isolated endoglucanase from *A. glaucus* XC9 shares common characteristics with those from industrial cellulase-producing Fungi, such as *A. niger* and *T. reesei* suggesting its possible use in industry.

In Brazil, sugar cane bagasse is one of the major residues of first-generation bioethanol production, and this residue has been greatly taken into consideration as a carbon source for low-cost cellulase production by several microorganisms [Barros et al., 2010]. Several substrates were generated after different pretreatment of sugar cane bagasse, and they were used as carbon source for *Penicillium funiculosum* growth. The best results, in terms of cellulolytic enzymes production, were observed when sugar cane bagasse was treated with acid and subsequently, alkali in order to obtain partially delignified cellulignin. The culture filtrate of *P. funiculosum* contained several cellulolytic activities. The optimal temperature for cellulase action was comprised between 52 and 58°C and the best pH for maximal activity was 4.9. Cellulases from *P. funiculosum* grown on sugar cane bagasse were highly stable at 37°C, as they retained more than 85% activity at this temperature [de Castro et al., 2010]. Two new fungal strains from subtropical soils, *Penicillium* sp. CR-316 and *Penicillium* sp. CR-313, were identified and selected because they secreted high levels of cellulases [Picart et al., 2007]. The culture filtrate from the two strains, analyzed by SDS-PAGE and zymography, showed several bands, indicating that both strains produced a multisystem of cellulases. Zymograms from *Penicillium* sp. CR-316 showed four activity bands of 35.0, 37.0, 48.0 and 71.0 kDa, respectively, while zymograms from *Penicillium* sp. CR-313 showed three activity bands of 35.0, 37.0 and 50.0 kDa. Multiple enzyme systems are frequently produced by cellulose-degrading microorganisms, as the cooperation of different cellulases acting in a coordinated manner enhance the degradation of the cellulose [Lynd et al., 2002]. The activity produced by *Penicillium* sp. CR-316 was higher than that produced by *Penicillium* sp. CR-313, and for this reason, this activity was better characterized. Crude cellulase of *Penicillium* sp. CR-316 exhibited optimum of temperature and pH at 65°C and 4.5, respectively, and the activity remained stable after incubation at 60°C and pH 4.5 for 3 h. The high yield of cellulases from *Penicillium* sp. CR-316, active and stable at high temperatures, should facilitate their use in biotechnological applications to improve the manufacture of recycled paper, and the transformation of cellulosic materials.

Species of the genus *Rhizopus* are known to have strong starch-degrading activity and this type of enzyme is extensively studied in this fungus [Li et al., 2010]. Conversely, there are few reports describing the production of cellulases by this filamentous fungus. Two extracellular endoglucanases, named RCE1 and RCE2, produced by *Rhizopus oryzae* FERM BP-6889 isolated from soil, were identified and purified by Murashima *et al.* [2002]. The molecular masses of the two enzymes were 41.0 and 61.0 kDa, respectively. The optimal pH for the activity of both enzymes was found to be between 5.0 and 6.0, and the optimum of temperature was 55°C. RCE1 and RCE2 did not hydrolyze hemicelluloses such as xylan, galactan, arabinan, or mannan. The amino acid sequences of some fragments obtained from internal regions of RCE1 and RCE2 were found to be homologous to the catalytic domain of EGV from *H. insolens* which belongs to GH family 45 [Schulein, 1997]. Thus, these findings supported the assumption that the enzymes belong to GH family 45. A novel gene, encoding for an endoglucanase was isolated from *R. stolonifer* var. *reflexus* TP-02, sequenced and expressed in *E. coli*. The recombinant enzyme exhibited an apparent molecular mass of 40.0 kDa, and the phylogenetically analysis on the sequence demonstrated that it grouped with *Aspergillus niger* (AJ224451), but they only shared 49% identity [Tang et al., 2009].

The white rot Fungus *Phanerochaete chrysosporium* has been used as a model organism for lignocellulose degradation since it produces a set of cellulases, hemicellulases, and lignin-degrading enzymes for an efficient degradation of the three major components of plant cell wall [Broda et al., 1996]. Several cellulases have been purified by Eriksson *et al.* [1975a;1975b] more than 20 years ago. Then, two GH family 5 isozymes (Cel5A, previously indicated as EG44) and Cel5B (previously indicated as EG38) and a 28-kDa endoglucanase (Cel12A) have been reported [Uzcategui et al., 1991; Henriksson et al., 1999]. The endoglucanase gene, *cel61A,* has been characterized although the corresponding protein has not yet been identified [Vanden Wymelenberg et al., 2002].

The gene that encodes the GH family 45 endoglucanase from the Fungus has been identified, cloned, and heterologously expressed in the yeast *Pichia pastoris*, and the recombinant protein has been characterized [Igarashi et al., 2008]. The enzyme has not carbohydrate binding module and the analysis of its amino acid sequence has revealed that the protein has low similarity (<22%) to known fungal EGs belonging to the GH family 45, thus suggesting that the protein should be classified into a new subdivision of this family. The recombinant protein shows hydrolytic activity toward amorphous cellulose, lichenan, and glucomannan but not xylan.

Cellulases have also been isolated from several brown rot Fungi as *Gloeophyllum trabeum*, *G. sepiarium*, and *Serpula incrassata* [Mansfield et al., 1998; Kleman-Leyer and Kirk, 1994]. The brown rot Fungus *Piptoporus betulinus* is a parasite for birch (*Betula* specie) and exhibits a high rate of wheat straw degradation accompanied by a high production of hydrolytic enzymes (endoglucanase, endoxylanase, endomannanase, β-glucosidase, and additional glycolytic activities) which makes it interesting for potential biotechnological applications. The major glycoside hydrolase produced and purified was an endoglucanase (EG1) with a molecular mass of 62.0 kDa [Valaskova and Baldrian, 2006]. This molecular mass was higher than that of cellulases from the majority of brown rot Fungi with the exception of Cel25 from *S. incrassata* [Kleman-Leyer and Kirk, 1994] and Cel12a from *G. trabeum* [Cohen et al., 2005]. In fact, the typical molecular mass ranges between 35.0 and 50.0 kDa [Clausen 1995]. EG1 was active in a broad pH range (2.5-6.0) with maximal activity at pH 3.0 and 70°C. The enzyme exhibited the highest substrate specificity for CMC, and it cleaved also xylan and galactomannan at lower rates. Scarce activity was showed toward crystalline cellulose.

Although a number of filamentous Fungi, such as *Trichoderma* and *Aspergillus*, are well known as producers of cellulases and accessory cellulolytic enzymes, the search for new strains and new enzymes has become a priority with the increase in diversity of biomass sources. A series of marine sponge-derived Fungi were isolated and screened for cellulolytic activity by Baker *et al.* [2010] in order to determine the potential of this environmental niche as a source of novel cellulase activities. Several strains of Fungi isolated from the marine sponge *Haliclona simulans* produced high levels of extracellular cellulases with significant activity at low temperatures. Moreover, a potent endoglucanase-producing Fungus was isolated and identified as a strain of *Penicillium pinophilum* (*P. pinophilum* KMJ601). Maximal production of cellulase (Eng5) was observed when the fungus was grown on rice straw or cellulose as a carbon source [Jeya et al., 2010]. The secreted and purified enzyme, with a molecular mass of 37.0 kDa, showed optimal pH and temperature of 5.0 and 70°C, respectively, and a $t_{1/2}$ value of 15 h at 70°C. Eng5 showed broad substrate specificity, exhibiting maximum specific activity toward lichenan and also acting on xylan. The partial gene sequence of *P. pinophilum* Eng5 contained the EG-like domain that is found in the GH family 5. NCBI BLAST analysis confirmed that the endoglucanase belongs to the GH family 5.

Anaerobic Fungi were first isolated by Orpin from the rumen of a sheep [Orpin, 1975]. Since then, they have been recovered from the digestive tracts of many species of herbivores, including both ruminants and non-ruminants, where they are believed to be responsible for the digestion of 50-70% of the ingested plant material [Trinci et al., 1994]. Six genera of anaerobic Fungi are recognized: *Anaeromyces*, *Caecomyces* (formerly *Sphaeromonas*), *Cyllamyces*, *Neocallimastix*, *Orpinomyces*, and *Piromyces*, and excellent information about them can be found in the book edited by Mountford and Orpin [Barr et al., 1989; Mountford and Orpin, 1994]. Unlike aerobic Fungi, anaerobic Fungi produce large multienzyme complexes similar to bacterial cellulosomes, and these complexes can degrade both amorphous and crystalline cellulose [Dijkerman et al., 1997]. Cellulosome-type complexes with endoglucanase, xylanase, mannanase, and β-glucosidase activities containing at least 10 proteins have been found in *Neocallimastix frontalis*, *Piromyces*, and *Orpinomyces* [Wilson and Wood, 1992; Steenbakkers et al., 2002; Ljungdahl, 2008]. The cellulose/hemicellulose degrading system of *Piromyces equi*, so called because this Fungus was isolated from the caecum of a pony [Orpin, 1981], consists of a large multienzyme complex, which accounts for up to 90% of the cellulase, mannanase and xylanase activities produced by the Fungus. Eberhardt *et al.* [2000] described the elucidation of the primary structures and enzyme features of two endoglucanases from *P. equi*, Cel5A and Cel45A. The two endoglucanase cDNAs, *cel5A* and *cel45A*, were isolated from a cDNA library of the Fungus. Sequence analysis revealed that *cel5A* encodes a 1714 amino acid modular enzyme, Cel5A, with a molecular mass of 195.0 kDa. The cDNA *cel45A* encodes a 410 amino acid modular enzyme, Cel45A, with a molecular mass of 44.3 kDa. Cel45A represented the first GH family 45 endoglucanase to be isolated from an anaerobic organism. The catalytic domains of Cel5A and Cel45A were hyperexpressed as thioredoxin fusion proteins, (Trx-Cel5A' and Trx-Cel45A'), and subjected to biochemical analysis. Trx-Cel5A'exhibited optimum pH at 5.0, but it retained about 78% activity at pH 6.4. Trx-Cel45A' exhibited optimum pH at 6.5, and it retained 65% activity between pH 5.2 and 7.9. The influence of temperature on Trx-Cel5A' and Trx-Cel45A' activity showed that the initial rate of hydrolysis of CMC, selected as substrate, by Trx-Cel5A' was highest at 45°C, while the initial rate of Trx-Cel45A' activity increased with increasing temperature between 40 and 70°C. Trx-Cel45A' was more thermostable than Trx-Cel5A'. The enzyme was stable for at least 1 h at 65°C, whereas an almost complete loss of activity was observed with Trx-Cel5A' after 20 min preincubation at 55°C.

While fungal cellulolytic and hemicellulolytic enzymes have been well studied for *Neocallimastix*, *Orpinomyces*, and *Piromyces*, information regarding cellulose-degrading enzymes from *Caecomyces* is still limited [Gerbi et al., 1996]. In 2008, Matsui and HBan-Tokuda isolated from bovine rumen a new anaerobic Fungus which was classified, after phylogenetical analysis, as *Caecomyces* CR4 [Matsui and Ban-Tokuda, 2008]. Elevated levels of cellulase activity were found in the culture supernatant of the CR4 isolate when it was grown in medium added with xylose as carbon source. Zymogram analysis showed that cellulase activity could be associated to three protein bands with molecular masses of 64.0, 89.0, and 95.0 kDa, respectively.

8. Cellulases of Thermophilic Origin

The (hyper) thermophilic microorganisms represent a unique group growing at temperatures that may exceed 100°C. More precisely, thermophilic microorganisms thrive at temperatures from 65 to 85°C, and hyperthermophiles grows at temperatures of above 85°C.

Hyperthermophiles are microorganisms within the Archaea domain although some Bacteria are able to tolerate temperatures around 100°C. An extraordinary heat-tolerant hyperthermophile is *Methanopyrus kandleri*, discovered on the wall of a black smoker from the Gulf of California at a depth of 2000 m, at temperatures of 84-110°C. It can survive and reproduce at 122°C [Takai et al., 2008]. Thermophilic and hyperthermophilic microorganisms have received considerable attention as sources of thermostable cellulolytic enzymes, as the properties of these biocatalysts make them interesting candidates for industrial applications. Running biotechnological processes at elevated temperatures has many advantages. High temperature has a significant influence on the solubility of the substrates (especially if viscous or polymers) and on the reaction rate. Moreover, problems of microbial contamination can be avoided when a reaction is performed at elevated temperature.

Degradation of cellulosic and hemicellulosic substrates among thermophiles is mostly due to Eubacteria species, e.g. *Rhodothermus marinus*, *Thermotoga* sp., *Caldibacillus cellulovorans*, *Alicyclobacillus acidocaldarius* [Halldórsdóttir et al., 1998; Bronnenmeier et al., 1995], while these activities are present in only a few representatives of Archaea (*Pyrococcus* sp. and *Sulfolobus* sp.) with cellulases belonging to GH families 5 and 12 (Table 6). Production of cellulases in a small number of thermophilic Fungi has also been reported (Table 7).

8.1 Bacterial and Archaeal Cellulases

Thermostable cellulases are of great biotechnological interest [Hongpattarakere, 2002]. A number of cellulolytic thermophilic Bacteria have been isolated, and many cellulose-degrading enzymes have been identified, characterized, cloned and expressed [Bergquist et al., 1999]. Conversely, screening of hyperthermophilic Bacteria for cellulose-degrading enzymes has revealed that the presence of such enzymes is rather rare in this group. In addition, among the Archaea, only the genus *Pyrococcus* and *Sulfolobus* have been found to process thermoactive cellulases (Table 8).

Few aerobic thermophilic microorganisms have been described to produce cellulases in comparison with the anaerobic ones. *Acidothermus cellulolyticus*, isolated from 55-60°C acidic water and mud samples collected in Yellowstone National Park, produces at least three thermostable endoglucanases [Mohagheghi, 1986]. One of them, E1 belonging to GH family 5, was crystallized, while properties and application of the other enzymes are protected by patents [Sakon et al., 1996].

The aerobic thermophilic bacterium *Rhodothermus marinus*, isolated from a submarine hot spring at Reykjanes, NW Iceland [Alfredsson et al., 1988], produces one higly thermostable cellulase (Cel12A) which retains 50% activity after 3.5 h at 100°C [Hreggvidsson et al., 1996].

Table 6 - Some (hyper) thermophilic cellulolytic Bacteria and Archaea

Microorganism	Domain	Gram reaction	Growth T (°C)	Growth conditions	References
Acidothermus cellulolyticus	Bacteria	+	55	Aerobic	Mohagheghi et al., 1986
Alicyclobacillus acidocaldarius	Bacteria	+	60	Aerobic	Darland and Brock, 1971 Wisotzkey et al., 1992
Anaerocellum thermophilum	Bacteria	+	75	Anaerobic	Svetlichnyi et al., 1990
Aquifex aeolicus	Bacteria	-	85-95	Aerobic	Deckert et al., 1998
Caldibacillus cellulovorans	Bacteria	+	68	Aerobic	Bergquist et al., 1999
Caldicellulosiruptor saccharolyticus	Bacteria	-	70	Anaerobic	Rainey et al., 1994
Clostridium stercorarium	Bacteria	+	65	Anaerobic	Madden, 1983
Clostridium thermocellum	Bacteria	+	60	Anaerobic	Saraswathy et al., 1993
Dictyoglomus thermophilus	Bacteria	-	73	Anaerobic	Saiki et al., 1985
Dictyoglomus turgidus	Bacteria	-	72	Anaerobic	Svetlichny and Svetlichnya, 1988
Pyrococcus abyssi	Archaea	-	96	Anaerobic	Cohen et al., 2003
Pyrococcus furiosus	Archaea	-	98	Anaerobic	Fiala and Stetter, 1986
Pyrococcus horikoshii	Archaea	-	98	Anaerobic	Gonzalez et al., 1998
Pyrodictium abyssi	Archaea	-	98	Anaerobic	Stetter et al., 1983
Rhodothermus marinus	Bacteria	-	65	Aerobic	Alfredsson et al., 1988
Spirochaeta thermophila	Bacteria	+	70	Anaerobic	Aksenova et al., 1992
Sulfolobus solfataricus	Archaea	-	85	Aerobic	She et al., 2001
Thermoanaerobacter cellulolyticus	Bacteria	+	75	Anaerobic	Taya et al., 1988
Thermobifida fusca (Themomonospora fusca)	Bacteria	+	50-55	Aerobic	Zhang et al., 1998
Thermococcus kodakaraensis	Archaea	-	85	Anaerobic	Fukui et al., 2005
Thermotoga maritima	Bacteria	-	80	Anaerobic	Huber et al., 1986
Thermotoga neapolitana	Bacteria	-	80	Anaerobic	Jannasch et al., 1988
Thermotoga petrophila	Bacteria	-	80	Anaerobic	Takahata et al., 2001
Thermotoga thermarum	Bacteria	-	77	Anaerobic	Windberger et al., 1989

Table 7. Some (hyper)thermophilic cellulolytic Fungi

Microorganism	Growth T (°C)	Growth conditions	References
Chaetomium thermophilum	45-55	Aerobic	Eriksen and Goksoyr, 1977
Humicola grisea var. thermoidea	45	Aerobic	Takashima et al., 1996
Humicola insolens	40-50	Aerobic	Schulein, 1997
Melanocarpus albomyces	45-55	Aerobic	Miettinen-Oinonen et al., 2004
Scytalidium thermophilum (Torula thermophila)	40	Aerobic	Cooney and Emerson, 1964
Sporotrichum thermophilum (Myceliophthora thermophila)	45-50	Aerobic	Loginova et al., 1983
Talaromyces emersonii	40-45	Aerobic	Stolk, 1965
Thermoascus aurantiacus	45-50	Aerobic	Stolk, 1965
Thermomyces lanuginosus (Humicola lanuginosa)	45-50	Aerobic	Olutiola, 1982

The thermophilic filamentous bacterium *Thermobifida fusca* (formerly *Thermomonospora fusca*) is one of the most extensively studied aerobic, thermophilic, cellulose degrading bacterium, and a major cellulose degrader in soil. This Actinomycete secretes three endoglucanases Cel9B, Cel6A, Cel5A (formerly named E1, E2, and E5), two exoglucanases Cel6B and Cel48A (formerly E3 and E6), and an endo/exoglucanase Cel9A (formerly E4) which have been characterized in detail [Ghangas and Wilson, 1988; Lao et al., 1991; Irwin et al., 1993; Irwin et al., 2000].

Deepened studies aimed to understand the catalytic mechanism of the cellulase Cel6A by computational and experimental investigation have been performed by Andrè *et al.* [2003]. In addition, the crystal structure of this enzyme, in complex with substrate and inhibitor has been solved [Larsson et al., 2005]. Zymogram analysis revealed that additional cellulases are produced by *T. fusca*. A new endoglucanase gene, *Tf cel5B,* was identified, and heterologous Cel5B was produced in *Streptomyces lividans* [Posta et al., 2004]. The novel cellulase has a molecular mass of about 67.0 kDa and optimum of pH and temperature of 8.2 and 77°C, respectively. It retained more than 60% of the maximal activity after 24 h incubation at 60°C. The temperature optimum of 77°C is comprised in the temperature optimum range of the other endoglucanase, endoxylanase and endomannanase enzymes from *T. fusca* (70-80°C), suggesting that the new endoglucanase has similar advantages for industrial application as the other thermostable hydrolases from this thermophilic microorganism. Alignments of the Cel5B catalytic domain to similar regions of other cellulases revealed 67% identity with CEND from *Cellulomonas fimi*, 66% identity with Cel5A from *C. flavigena* and 60% identity with CelB from *Caldicellulosiruptor saccharolyticus*.

The genus *Alicyclobacillus* was first established by Wisotzkey *et al.* [1992], and it is characterized by the presence of alicyclic fatty acids as major components of the membrane lipids. All *Alicyclobacillus* species are highly thermoacidophilic (defined as optimal growth conditions at 45-60°C and pH 2.0-5.0) and may be a good source of acidic glucanases. *Alicyclobacillus acidocaldarius* is known to produce four cellulases: CelA, CelB, CelG and

CelA4. All of them are highly active at acidic pH and at temperatures between 65 and 80°C. The gene encoding for CelA, belonging to GH family 9, was cloned and sequenced, and the recombinant protein, expressed in *E. coli*, was characterized [Eckert et al., 2002]. The molecular mass was 59.0 kDa and in agreement with that deduced from the ORF encoding a putative protein of 537 amino acids. CelA exhibited temperature and pH optima of 70°C and 5.5, respectively, and it was active against CMC, lichenan and also oat spelt xylan and cello-oligosaccharides, suggesting a role as a cytoplasmic enzyme for the degradation of short-chain sugars imported from the medium. CelB was a membrane-bound protein of 100.0 kDa with pH and temperature optima of 4.0 and 80°C, respectively [Eckert and Schneider, 2003].

Table 8. Properties of some cellulases from (hyper)thermophilic Bacteria and Archaea

Microorganism	Enzyme	Mol mass (kDa)	Optimal T (°C)	Optimal pH	References
Acidothermus cellulolyticus	E1	72.0	81	5.0	Sakon et al., 1996
Anaerocellum thermophilum	CelA	230.0	85-95	5.0-6.0	Zverlov et al., 1998
Caldibacillus cellulovorans	CMCase	174.0	80	6.5-7.0	Huang and Monk, 2004
Clostridium thermocellum	CelI	98.5	70	5.5	Gilad et al., 2003
Pyrococcus furiosus	EglA	35.9	100	6.0	Bauer et al., 1999
Rhodothermus marinus	Cel12A	28.8	100	6.0-7.0	Halldòrsdòttir et al., 1998
Thermobifida fusca (Thermomonospora fusca)	Cel6A	42.0	55	6.0-8.5	Irwin et al., 1993
Thermomonospora curvata	EG	22.0	70	6.0	Lin and Stutzenberger, 1995
Thermotoga maritima	CelA CelB	29.7 31.7	90 85	4.5 6.0	Liebl et al., 1996
Thermotoga neapolitana	CelA CelB	29.0 30.0	95 106	6.0 6.0-6.6	Bok et al., 1998

The enzyme was extremely stable at acidic pH and temperatures up to 80°C, and was provided with activity toward CMC and oat spelt xylan. High sequence similarity to arabinofuranosidases belonging to GH family 51 was displayed, whereas, in terms of substrate specificity, CelB was comparable only to an endoglucanase from *Fibrobacter succinogenes* (EGF), also belonging to the GH family 51. Another membrane-bound cellulase was identified and characterized from *A. acidocaldarius* when it was grown in liquid medium added with CMC as carbon source [Morana et al., 2008]. The enzyme was demonstrated to be glycosylated (therefore named CelG) with a molecular mass of 56.2 kDa. It was active over a broad range of pH (3.0-7.0) with optimal activity at pH 4.0, and showed activity between 40 and 80°C with optimal temperature at 65°C. Themostability was evaluated at 65 and 75°C, and CelG retained 100% activity after 2 h at 65°C and 1 h at 75°C. This enzyme was more thermostable than CelA, which displayed small loss of activity after 1 h at 60°C and an half-life of 30 minutes at 75°C. Moreover, it was found to be specific for CMC hydrolysis, while

CelA and CelB were also active toward other polysaccharides. The most recently identified cellulase in *A. acidocaldarius* is the secreted 48-kDa protein CelA4. It was isolated, purified and characterized by Bai *et al.* [2010]. The enzyme was the more acidophilic among the cellualses from *A. acidocaldarius*, being its optimal pH 2.6. In addition, it was stable under acid conditions, retaining more than 80% activity in the pH range of 1.8-6.6 after incubation for 1 h at 37°C. The optimal temperature was 65°C, but the total activity was lost after incubation at this temperature for 1 h. Purified CelA4 was protease-resistant and exhibited the highest enzymatic activity toward barley β-glucan. The high activity at low pH and protease resistance make CelA4 a good candidate as commercial cellulase able to improve nutrient use in the animal feed industry. In fact, barley β-glucan cannot be digested by monogastric animals and, furthermore, the pig gastrointestinal tract is highly acidic and contains a lot of proteolytic enzymes. As consequence, an ideal glucanase should have high activity in extremely acidic conditions and be protease-resistant to function under the conditions present in the digestive tract of animals [Li et al., 1996].

Aquifex aeolicus, a hyperthermophilic microorganism belonging to the Bacteria domain, was originally isolated in the Aeolic Islands at north of Sicily near underwater volcanic vents. It represents the deepest branch in the bacterial phylogeny and its growth temperature can reach 95°C [Pitulle et al., 1994]. Only one thermostable cellulase has been reported to be produced by this microorganism to date, and the gene *cel8Y* has been cloned and expressed in *E. coli* by Kim *et al.* [2000]. The gene product (Cel8Y) has been purified and characterized. It has a molecular mass of 36.7 kDa and it was able to hydrolyze β-1,4 linkages in CMC but not in Avicel. Optimal pH and temperature for activity were 7.0 and 80°C, respectively. At 90 and 100°C, the activity of the recombinant Cel8Y decreased by 50% after 4 and 2 h, respectively. On the basis of sequence comparisons, the enzyme is very closely related to other cellulases of GH family 8.

Three genes (*sso1354*, *sso1949*, and *sso2534*) encoding three endoglucanases belonging to GH family 12, have been annotated in the fully sequenced genome of the hyperthermophilic and acidophilic crenarchaeote *Sulfolobus solfataricus* strain P2 [She et al., 2001]. This microorganism, originally isolated from a solfataric field in the area of Naples, Italy, grows at temperatures ranging from 80 to 87°C and at very low pH values (2.0-4.0) [De Rosa et al., 1975]. The gene *sso2534* was identified in *S. solfataricus*, strain MT4, and its product (CelS) showed 100% amino acid sequence identity with CelB from *S. solfataricus* strain P2 [Limauro et al., 2001]. CelS (38.4 kDa) was detected as an extracellular cellulase. It was active at pH 5.8, but no further indication about optimal conditions for activity or about pH and temperature stability, and substrate specifity were reported. In 2005, the gene *sso1949* was cloned in *E. coli* and expressed by Huang *et al.* [2005]. The purified recombinant enzyme hydrolyzed CMC as well as cello-oligomers, whereas Avicel, xylan and lichenan, appeared not to be substrates for the enzyme. The failure to detect activity toward Avicel could be ascribed to the lack of a cellulose-binding module. The same feature was observed for the SSO1354 protein. SSO1949 protein exhibited pH and temperature optima of 1.8 and 80°C, respectively, and showed a half-life of approx. 8 h at 80°C and pH 1.8. When incubated at 95°C and pH 1.8, SSO1949 was rapidly inactivated. Analysis by tandem mass spectrometry led to identify the endoglucanase precursor, encoded by the *sso1354* gene, as the protein provided with xylanase activity isolated from *S. solfataricus* strain Oα, a strain capable of growing in minimal medium supplemented with xylan as sole carbon source [Cannio et al., 2004]. SSO1354 protein was membrane-bound and possessed both cellulolytic and xylanolytic activities. It was also active toward arabinan and debranched arabinan in contrast to SSO1949 protein which was inactive toward xylan and others polysaccharides. The enzyme was glycosylated with a molecular mass of 57.0 kDa, and exhibited optimal temperature and pH of 95°C and 3.5, respectively [Maurelli et al., 2008]. Among the characterized cellulases from thermophilic microorganisms, only EglA from *P. furiosus* [Bauer et al., 1999] and CelB from *T. neapolitana* [Bok et al., 1998] showed higher optimal temperatures. SSO1354 was highly thermophilic and thermostable, retaining 71% of the initial activity at 100°C and 49% at 110°C, and showing a half-life of 53 min at 95°C.

Clostridium thermocellum is a thermophilic, strictly anaerobic bacterium, and in 1983 the cellulosome concept was established in this microorganism as cellulases were found to be organized into a high molecular weight, cellulolytic complex [Lamed et al., 1983]. *C. thermocellum* produces a number of cellulases both free and assembled in cellulosome [Fauth et al., 1991; Lemaire and Béguin, 1993; Ashan et al., 1997]. Kurokawa *et al.* [2002] identified an additional endoglucanase (CelT) in 2002. The mature form of CelT consists of a GH family 9 cellulase domain and a dockerin domain responsible for cellulosome assembly, but has neither an Ig-like domain domain nor a family 3c CBM. However, it is more similar to the family E2 cellulases than the family E1 with respect to sequence similarity, and has a catalytic domain that is homologous with the catalytic domains of *T. fusca* E4, *C. cellulolyticum* CelG and *C. thermocellum* CelQ. The enzyme exhibited optimal pH and temperature at 7.0 and 70°C, respectively, but no data concerning pH and temperature stability have been reported. Immunological analysis indicated that CelT is a catalytic component of the *C. thermocellum* F1 cellulosome. To identify the predominant catalytic components of *C. thermocellum* cellulosome, the cellulolytic complexes were purified and the components were separated and identified by MALDI-TOF/TOF [Zverlov et al., 2005]. Ten of the components were already known, and in addition, three hitherto unknown genes, which were named according to their sequences *celR*, *xghA* and *xynD*, were detected. According to the putative function of their catalytic modules, the corresponding gene products were called Cel9R, Xgh73A and Xyn10D, with catalytic modules of GHF9 (endoglucanase), GHF74 (xyloglucanase) and GHF10 (endoxylanase), respectively [Coutinho and Henrissat, 1999]. The new putative endoglucanase Cel9R gave one of the most prominent protein spots of all cellulosomal components.

Highly cellulolytic microorganisms are found among the anaerobic (hyper) thermophiles, e.g., *Thermotoga* [Liebl, 2001], *Caldicellulosiruptor* [Rainey et al., 1994], and *Pyrococcus* [Bauer et al., 1999]. Several highly thermophilic and thermostable cellulases have been isolated and chacracterized from the hyperthermophilic anaerobic bacterium *Thermotoga* sp. Two thermostable endoglucanases (CelA and CelB) were purified and characterized from *T. neapolitana*, an extremophilic bacterium firstly isolated and described in 1986 as coming from the vicinity of a black smoker in the bay of Naples, Italy [Jannasch et al., 1988; Bok et al., 1998]. Highly active cellulases were also characterized from *T. maritima*, which was originally isolated from a hot marine sediment at Vulcano, Italy [Huber et al., 1986]. Its genome sequence indicates the presence of a number of endoglucanases that can be classified according to the GH families to which they belong as Cel5A, Cel5B, Cel12A, Cel12B and Cel74, and their biochemical properties have been reported [Bronnenmeier et al., 1995; Liebl et al., 1996; Liebl, 2001]. The most recently described was the cellulase Cel74, which was expressed in *E. coli* as a 77.0 kDa protein. The enzyme exhibited optimal pH and temperature of 6.0 and 90°C, and was highly thermostable with a half life of 5 h at 90°C. It was most active toward barley β-glucan and, to a lesser extent, toward CMC, glucomannan and xyloglucan. The endo-mode of action of Cel74 was confirmed by the failure in hydrolyzing oligosaccharides with a degree of polymerization smaller than six units [Chhabra and Kelly, 2002].

Among the three species belonging to the genus *Pyrococcus*, one endoglucanase showing regions of homology with GH family 12 (EglA or Cel12) has been found in *P. furiosus*, a hyperthermophilic anaerobic archaeon isolated from geothermally heated marine sediments with temperatures between 90°C and 100°C at the beach of Porto Levante, Vulcano Island, Italy [Fiala and Stetter, 1986]. It has been reported that the purified recombinant Cel12 hydrolyzes β-1,4 but not β-1,3 glucosidic linkages [Bauer et al., 1999]. There has been no report so far on endoglucanases from *P. abyssi*, while the gene coding for an endoglucanase from *P. horikoshii*, originally isolated from a hydrothermal vent at a depth of 1395 meters in the Okinawa Trough in the Pacific Ocean, has been cloned and expressed in *E. coli* [Gonzalez et al., 1998]. The enzymatic characteristics of the gene product were examined. This enzyme (EGPh) was the first endoglucanase belonging to GH family 5 found from *Pyrococcus* species, and it hydrolyzed CMC, Avicel, and lichenan. The pH optimum was between 5.4 and 6.0, and the temperature optimum was higher than 97°C. The endoglucanase from *P. horikoshii* was highly thermostable since the residual activity was 80% after 3 h at 97°C [Ando et al., 2002]. The features of thermophilicity and thermostability make this enzyme of potential use for industrial hydrolysis of cellulose at high temperatures, particularly in biopolishing of cotton products. In fact, the desizing, that is the step to remove starch from the fabrics, is performed at temperatures of 70°C or above as amylases active at these temperatures are already available. If a hyperthermostable cellulase will be introduced, it will be possible to combine desizing and biopolishing in a single step. Very recently, the X-ray crystal structure of EGPh has been determined [Kim and Ishikawa, 2010a].

Table 9. Properties of some cellulases from thermophilic Fungi

T	Enzyme	Mol mass (kDa)	Optimal T (°C)	Optimal pH	References
Chaetomium thermophile	EG	67.8	60	4.0	Ganiu et al., 1990
Humicola grisea var. thermoidea	EGI	58.0	55-60	5.0	Takashima et al., 1996
Myceliophthora thermophila	EG	100.0	65	4.8	Roy et al., 1990
Talaromyces emersonii	EGI-III	68.0	75-80	5.5-5.8	Moloney et al., 1985
Thermoascus aurantiacus	EGI EGII	78.0 49.0	75 68	5.0 5.0	Tong et al., 1980

The potential for glucose production from cellulosic materials using hyperthermophilic cellulolytic enzymes from *Pyrococcus* sp. (endoglucanase from *P. horikoshii*, EGPh and β-glucosidase from *P. furiosus*, BGLPf) has been investigated. The findings obtained indicated that the EGPh and BGLPf mixture was a successful enzyme cocktail in totally converting phosphoric acid swollen Avicel into glucose at extremely high temperature [Kim and Ishikawa, 2010b].

A cellulolytic and thermophilic anaerobic bacterium closely related to *Moorella thermoacetica* (99% identity) was isolated from soil by Karita *et al.* [2003]. The microorganism, identified as *Moorella* strain F21, secretes cellulolytic enzymes after growth on ball-milled cellulose or Avicel as carbon source, and ferments cellulosic materials to organic acids.

8.2 Fungal Cellulases

Among the thermophilic Fungi, only a few number is described to be cellulase-producer (Table 9).

The thermophilic filamentous fungus *Humicola* sp. has been known to produce several cellulases, and some of the genes have been cloned, sequenced and expressed [Takashima et al., 1997]. The cellulase system of the thermophilic fungus *Humicola insolens* possesses a battery of enzymes that allows the efficient utilization of cellulose. This system, homologous to that of *T. reesei*, contains five endoglucanases: EGI (Cel7B), EGII (Cel5), EGIII (Cel12), EGV (Cel45A), and EGVI (Cel6B) in addition to two cellobiohydrolases: CBHI (Cel7A), and CBHII (Cel6A) [Schulein, 1997]. All the endoglucanases showed optimal activity between pH 5.5 and 9.0. Cel7B was highly active in a broad pH range, retaining more than 60% activity between pH 5.0 and 10.0. Its optimal pH was 5.5. Cel45A showed optimum of pH at 9.0, whereas the remaining three cellulases had pH optima between 6.0 and 8.0. No data about optimal temperature and stability were reported. The reason why *H. insolens*, as well as *T. reesei* and many other cellulolytic microorganisms, produces so many enzymes is not completely clear. The most accredited theory recognizes a different role to each enzyme in relation to the diversity of the substrate (solubility, degree of substitution or polymerization). In order to better understand the various role of several cellulases produced by a same microorganism, the efficiency in hydrolyzing soluble cellulose (CMC) by the endoglucanases Cel5A, Cel7B and Cel45A from *H. insolens* was investigated by Karlsson *et al.* [2002], and compared with the catalytic efficiency showed by the endoglucanases Cel7B, Cel12A, and Cel45Acore from *T. reesei*. The hydrolysates were analyzed for production of substituted and non-substituted oligosaccharides with size exclusion chromatography and with matrix-assisted laser desorption/ionization mass spectrometry. The authors demonstrated that the enzymes clearly differ in their capacity in hydrolyzing CMC and in product formation. Cel5A form *H. insolens* and Cel7B from *T. reesei* were the most efficient enzymes able to hydrolyze the substrate significantly, whereas others were more affected by the presence of substituents on the polymeric chain. So, the use of pure cellulases allows the selective cleavage of cellulose derivatives. Besides to the above mentioned endoglucanases, *H. insolens* produces another cellulase of about 51.2 kDa, Avi2 (avicelase 2), which was purified from the culture medium of *H. insolens* FERM BP-5977, an industrial cellulase producing strain [Jensen, 2002]. The *avi2* gene was cloned and sequenced. The cDNA of *avi2* contained an ORF encoding 476 amino acids residues. The catalytic domain of Avi2 (residues 117-476) was very similar to the catalytic domain of cellulases belonging to GH family 6 [Moriya et al., 2003]. No data about the biochemical characterization of the enzyme were depicted.

The thermophilic fungus *Chaetomium thermophile* var. *dissitum*, was able to produce in the culture medium all the enzymes involved in cellulose breakdown, namely endoglucanase (41.0 kDa), exoglucanase (67.0 kDa) and β-glucosidase [Eriksen and Goksoyr, 1977]. In 2002, Lu *et al.* [2002] reported that *C. thermophile* secreted in the culture medium a glycosylated endocellulase with an apparent molecular weight of 67.8 kDa, as determined by SDS-PAGE. The enzyme was optimally active at pH 4.0-4.5 and 60°C, and it retained 30% activity after 60 min at 70°C.

Melanocarpus albomyces, a rare true thermophilic Ascomycete capable of growing copiously at 50°C, has been documented to produce high levels of endoglucanases under optimized culture conditions [Jatinder et al., 2006]. The endoglucanases from this Fungus have been recognized as potentially important in denim washing. In fact, the supernatant from *M. albomyces* worked well in biostoning, with low backstaining. Three cellulases were identified and purified to homogeneity, and two of them were endoglucanases with apparent molecular masses of 20.0 kDa (Cel45A) and 50.0 kDa (Cel7A) [Miettinen-Oinonen et al., 2004]. Cel45A was relatively heat stable retaining 70% activity after 1 h at 80°C. The temperature and pH optima were 70°C and 6.0-7.0, respectively. The enzyme had a broad range of pH activity exhibiting 80% or more of its maximum activity throughout the pH interval 4.0-9.0. Cel45A has been crystallized [Hirvonen and Papageorgiou, 2002]. Cel7A was glycosylated and showed optimal temperature between 65 and 70°C and optimal pH at 6.0. Recently, a new strain isolated from composting soils and identified as *Melanocarpus* sp. MTCC 3922, was demonstrated to secrete two endoglucanases (EGI and EGII). The enzymes were purified and characterized by Kaur *et al.* [2007]. The molecular mass of EGI was 40.0 kDa, and the enzyme showed optimal temperature and pH at 50°C and 6.0, respectively. It was highly active in the 5.0-7.0 pH range, but loss of activity was observed as the temperature was increased from 50 to 80°C. EGII has a molecular mass of 50.0 kDa, optimum of temperature for activity at 70°C, ad optimum of pH at 5.0. It was active over pHs 4.0-6.0 and 40-80°C. Similar range of pH and temperature optima were reported for endoglucanases from the thermophilic Fungi, *Chaetomium thermophile* var. *coprophile* [Ganju et al., 1990] and *Myceliophthora thermophila* [Roy et al., 1990]. Because of their properties, EGI and EGII could be suitable under different conditions. For example, acidic cellulases found application in the non-ionic surfactant-assisted acidic deinking of old magazines [Xia et al., 1996]. Cellulases active in the range of pH 6.0-10.0 are useful in textile industry in biostoning [Kochavi et al., 1990], and in laundry [Suominen et al., 1993].

The thermophilic Fungus *Thermoascus aurantiacus* produces high levels of cellulase components when grown on lignocellulosic carbon sources such as corncob and cereal straw [Khandke et al., 1989]. As these enzyme components are remarkably stable over a wide range of pH and temperatures, they appear to have great commercial potential. A major extracellular endoglucanase, with a molecular mass of 34.0 kDa, was purified and characterized [Parry et al., 2002]. It was optimally active at 70-80°C and pHs 4.0-4.4, and it was stable at pH 5.2 and up to 60°C for 48 h. At 70°C and pH 5.2 the enzyme retained 40% of the original activity after 48 h. The cellulase exhibited the highest activity toward CMC; barley β-glucan and lichenan were also hydrolyzed, but the enzyme was inactive on laminarin, confirming that it was an endoglucanase and was specific toward β-1,4 linked polysaccharides. Sequence alignment of the first 33 amino acid suggested that the endoglucanase from *T. aurantiacus* is a member of the subfamily A6 of the GH family 5 [Lo

Leggio et al., 1997]. The gene *egl* encoding for the endoglucanase was cloned and expressed in *S. cerevisiae* by Hong *et al.* [2003].

A cellulase complex capable of degrading both soluble and insoluble cellulose has been found in the culture filtrate of the thermophillic fungus *Talaromyces emersonii*. When grown on media containing cellulose, this microorganism produces a complete extracellular cellulase system containing seven endocellulases, four exocellulases and three β-glucosidases [McHale and Coughlan, 1980; McHale and Coughlan, 1981]. Moloney *et al.* [1985] isolated and characterized the endoglucanases EGI-EGIV. Successively, McCarthy *et al.* [2003] identified and characterized three novel endoglucanases, namely EGV (22.9 kDa), EGVI (26.9 kDa) and EGVII (33.8 kDa). These enzymes work in an endoacting mode, exhibiting greatest activity against mixed 1,3;1,4-β-D-glucans. EGVI and EGVII displayed also activity against 1,3-β-glucan (laminarin) and therefore, are likely to belong to EC 3.2.1.6.

Since the production of microbial enzymes has a large impact on the overall microbial process economy, and *T. emersonii* is capable to produce high level of cellulases in shaken cultures, optimization experiments have been carried out in the last few years to further improve the cellulase production by this Fungus through the addition, to the culture medium, of cheap and readily available substrates as sugar cane bagasse. The aim was to obtain large quantities of enzymes to test their effectiveness in the textile field [Gomes et al., 2007]. Speaking more in general, *T. emersonii* is a Fungus able to produce an enzyme cocktail attractive for biotechnological applications since it comprises not only cellulolytic, but also amylolytic and hemicellulolytic enzymes. These fungal enzymes work at temperatures 10-20°C higher than the commercially available *Trichoderma* sp. enzymes, and moreover, like these mesophilic Fungus, *T. emersonii* also has GRAS (generally regarded as safe) status, making it safe for use in food processing [Waters et al., 2010]. Many enzyme systems from this Fungus have been described, and patents have been developed for key applications [Tuohy et al., 2007].

9. Cloning and Expression of Cellulase Genes

Recombinant DNA techniques offer powerful means to solve various problems which arise both in the development of efficient cellulose producers for commercial applications and in the studies of complex cellulolytic microbial systems. For example, cloning and expression in non cellulolytic hosts of a cellulase gene that takes part in a complex aggregate, such as the cellulosome of *Clostridium thermocellum*, allows to separate that cellulase from all the other components of the system and to examine its catalytic properties [Gilkes et al., 1991]. Moreover, the expression of protein fused with suitable tags permits the separation of the chimeras also when they are expressed in a cellulolytic host.

Cellulase genes have been cloned from different microbial genera into various suitable hosts.

The genes encoding cellulases are chromosomal in both Bacteria and Fungi. In the Fungi, cellulase genes are randomly distributed over the genome and each gene has its own transcription regulatory elements. However, in two exceptional cases such as for *Phanerochaete chrysosphorium* and *Trichoderma reseei* was found a genes co-localization. In

P. chrysosphorium, a cluster of three cellobiohydrolase genes was found by restriction mapping and sequence analysis of library cosmid clones [Covert et al., 1992]

Recent studies revealed that also in *T. reseei*, a big number of the genes involved in cellulose and hemicellulose degradation is not randomly distributed over the genome but clustered in several areas located among chromosomal regions of synteny with the other groups of Sordariomycetes. Genes co-localization, that implies a coordinated gene regulation, is probably the reason of the *Trichoderma* elevated efficiency in cellulosic substrates degradation despite its smaller cellulolytic enzymes repertoire respect to other fungal species [Ouyang et al., 2006]. In Bacteria, cellulase genes are either scattered over the chromosome [Aubert et al., 1988] or clustered on the genome. In the species belonging to the *Clostridium* genus, the genes coding for the proteins which constitute the cellulosomal complexes are often co-localized in clusters. In *C. thermocellum* despite the fact that most of the cellulase and xylanase genes are randomly distributed, several clusters have been found, suggesting the presence of operons as units of gene regulations [Miettinen-Oinonen, 2004]. The cellulosomal genes cluster of *C. cellulolyticum* is composed of 12 genes which are transcribed in two large polycistronic mRNA of 14 and 12 kb [Abdou et al., 2008]. Similar arrangements have been found in *C. cellulovorans* where 9 genes which constitute the cbpA cellulosomal cluster are transcribed as polycistronic mRNAs of 8 and 12 Kb [Han et al., 2003].

9.1 Heterologous Cloning and Expression in Different Microbial Hosts

The cellulase genes isolated from different microbial genera have been initially cloned in *E. coli* [Beguin et al., 1987]. The expression of catalytically active cellulases in *E. coli* is generally achieved at low levels due to the absence of post-translational modifications such as the glycosylation and to the intracellular accumulation of the recombinant enzymes. Most of the cellulase genes bear a signal peptide that is not well recognized by the *E. coli* expression system. This often lowers the production level of the recombinant proteins or even impairs their expression.

The thermostable cellulase Cel12A from *Rhodothermus marinus*, which contains in its sequence a highly hydrophobic putative signal peptide, when cloned in *E. coli*, was produced at extremely low levels and was also cytotoxic causing an extensive cell lysis. A successful expression was achieved only cloning a deletion mutant of the cellulase gene lacking of the hydrophobic signal peptide region [Wicher et al., 2001]. A similar approach was utilized for the *sso1949* gene coding for an highly acid-stable and thermostable β-endoglucanase in *Sulfolobus solfataricus* that bears an N-terminal signal peptide linked to the catalytic domain by a serine and threonine-rich region sequence. The expression of the active enzyme was obtained only when the putative signal peptide of 24 amino acids was deleted. The attempt to improve the over-expression, using a N-terminal deletion mutant lacking also the serine and threonine-rich region was unsuccessful [Huang et al., 2005]. In the case of the cellulase gene *celB1* from *Bacillus* sp. (strain 186-1), the expression of the active enzyme and its exportation in the periplasm was obtained only after the substitution of its signal peptide with that of the *E. coli* periplasmic outer membrane protease (OmpT) [Sànchez-Torres et al., 1996].

Since the recombinant cellulases, as described before, are often accumulated in the cytoplasm, this results in improper protein folding, leading to the formation of insoluble aggregates known as inclusion bodies [Villaverde and Carrio, 2003]. To overcome this problem one strategy is to decrease the formation of the inclusion bodies *in vivo* by fusing the recombinant protein with a domain which can enhance the solubility of the chimera. This approach was adopted for the cellulosomal cellulase EngB from *Clostridium cellulovorans*, which has been expressed as insoluble proteins in *E. coli.* The expression of EngB in a soluble form was achieved by fusing its catalytic domain with the proline-threonine rich region (PT-linker) and the cellulose-binding domain (CBD) of the non cellulosomal cellulase EngD [Murashima et al., 2003]. In successive studies the CBDs of different protein from *Clostridium* species and the PT-linker from *C. cellulovorans* were studied for their capacities to improve the solubility of various recombinant cellulosomal proteins when fused to their N- or C-terminal ends. The better results were obtained with the EngD PT-linker that, fused at the C-terminal end of the recombinant cohesin domain Coh6, enhances of three folds its solubility [Xu and Foong, 2008].

Another method to avoid the intracellular accumulation of recombinant insoluble cellulases is to obtain their extracellular secretion. *E. coli* is very limited in its ability to secrete proteins into the extracellular environment [Pugsley, 1993]. Recombinant proteins such as endoglucanases, which are secreted by their source organisms, can be accumulated in the periplasmic space in *E. coli* [Missiakas and Raina, 1997]. For example, the recombinant cellulase from *Fibrobacter succinogenes* AR1, was successful expressed in *E. coli* under its own promoter and due to the signal peptide contained in its sequence, was, in the percentage of 80%, secreted in the periplasm and the remaining part was found in the extracellular medium [Cavicchioli and Watson, 1991].

A novel thermostable 1,4-β-endoglucanase CelI15 from *Bacillus subtilis* strain I15 was expressed in *E. coli* BL21(DE3) with a production level three times higher than that of the wild-type strain, and it was entirely found in extracellular medium [Yang et al., 2010]. Therefore, the CelI15 extracellular signal peptide could be functional in the heterologous host.

The studies of Zhou *et al.* [1999] were aimed to obtain the extracellular production in *E. coli* of the CelZ endoglucanase from *Erwinia chrysanthemi*. The production of CelZ that was previously expressed in *E. coli B* as a periplasmic product, was enhanced using a strong promoter derived from the *Zymomonas mobilis* genome; morever, its extracellular secretion was obtained reconstituting in *E. coli* the *E. chrysanthemi* secretion system II encoded by the *out* genes [He et al., 1991]. Several other recent studies are aimed to improve the ability of *E. coli* strains to secrete recombinant proteins that is very useful to eliminate the additional process steps for the release of enzymatic activity in the extracellular medium [Mergulhão et al., 2005].

E. coli is generally used as the initial host organism for the isolation and expression of bacterial genes. The choice of alternative hosts is motivated by the ability to secrete proteins into the extracellular medium, by the closer evolutionary kinship that allows more efficient expression, and by the capacity to effect post-translational modification such as glycosylation [Beguin, 1990].

To obtain cellulases that both are produced in high yield and secreted into the medium, several *Bacillus* species have been used as host cells. Zhang *et al.* [2010b] studied the expression of the non cellulosomal family 48 cellulase from *Clostridium phytofermentans* in *E. coli* and *B. subtilis*. CpCel 48 was expressed intracellularly in a soluble active form in *E. coli* with and without a histidine tag fused at its C-terminal end and in a secretory active form in *B. subtilis*. This was the first report on the expression of a secretory family 48 glycoside hydrolase in *B. subtilis* obtained by cloning the CpCel 48 gene into the pP43NMK *E. coli-B. subtilis* shuttle expression vector in-frame with the neutral protease B (NprB) signal peptide-encoding sequence.

In the above and the two successively mentioned reports is underlined the importance of the signal peptide sequences for the secretory production yields. In the case of the Cel9 endoglucanase from *Mixobacter* spA1, the signal peptide was functional either in *E. coli* and *B. subtilis* where the secretory expression of the active enzyme was achieved [Avitia et al., 2000]. Celdc from *Pyrocccus horikoshi,* instead, was expressed only at low levels in *E. coli* but a high extracellular production was obtained with the host-vector system of *Bacillus brevis*. Celdc lacking of the N-terminal 28 aminoacids signal peptide and C-terminal 12 aminoacids was cloned in the expression-secretion vector pNU226 downstream the promoter, the translation initiation region and the modified signal peptide sequences of the middle wall protein (MWP) gene of *B. brevis* 47 [Sagiya et al., 1994]. Several attempts altering the sequence of the signal peptide inserting different amino acids or short peptides at the cleavage site dramatically increased the cellulase production level [Kashima and Udaka, 2004].

While *B. subtilis* may suffice for many applications, *B. megaterium* and *B. stearothermophilus* are attractive alternative systems which offer the advantages of increased plasmid stability [Chen et al., 2008] and growth at a higher temperature [Soutschek-Bauer and Staudenbauer, 1987], respectively.

Since it is impossible to obtain through the prokaryotic biosynthetic machinery, post-translational modifications, several cellulase genes have been expressed in eukaryotic hosts. Among the several yeast species endowed with a broad range of properties useful to express recombinant cellulase genes at the correct level of post-translational maturation, *Saccharomyces cerevisiae* has received the most attention. Although most of the cellulases that have been successfully produced in *S. cerevisiae* were of fungal origin, there are reports of successful bacterial cellulase production [van Zyl et al., 2007; Van Rensburg et al., 1998; Parvez et al., 1994]

Most reports regarding the expression of cellulases and hemicellulases in yeast describes strong (or other constitutively expressed) promoters to drive expression of the heterologous genes. The promoter choice undoubtedly has a great influence on the expression levels but also leader sequences strongly affect the recombinant protein yields. In a recent report [Zhu et al., 2010], the expression of the endoglucanase Egl1 cloned from *Trichoderma viride CICC3038* in *S. cerevisiae* was achieved replacing the native signal sequence with the mating factor α prepro-leader sequence (MFα) signal peptide. A 61.5% enhancement of the specific endoglucanase activity was obtained accompanied by a faster substrate consumption and growth rate in presence of CMC as the sole carbon source.

Apart from the production of several saccharolytic enzymes [Oin et al., 2008; Hong et al., 2003], many efforts have been made for enabling *S. cerevisiae* to directly ferment cellulosic biomass to ethanol. The absence of a suitable technology for bioethanol production is due to the high costs required to obtain large amounts of cellulases for cellulose hydrolysis into fermentable sugars. A good solution to solve this problem is to develop a whole cell biocatalyst able to perform cellulase production, cellulose hydrolysis and sugars fermentation in a single consolidated bioprocessing (CBP) [Wen et al., 2010; Lynd et al., 2002]. In order to realize such CBP technology, engineering ethanologenic microorganisms to heterologously express a functional cellulase system is the most promising strategy. *S. cerevisiae* is one of the best candidates to realize a CBP mainly due to its superior traits, including high ethanol yield and tolerance, robustness in industrial fermentation, a wide variety of genetic engineering tools and safe status [Van Zyl et al., 2007]. Several research groups have expressed into *S. cerevisiae* different types of carbohydrate active enzymes in a free form mimicking a rudimentary cellulase system [Den Haan et al., 2007]. However, the best way to overcome the recalcitrant nature of cellulose and obtain the highest conversion yields into ethanol is to express the hydrolytic enzymes in a cellulosome fashion where the different enzymes work together synergistically [Mingardon et al., 2007]. Recent studies exploiting the cellulosomal modular nature are aimed to make a combination of recombinant chimeric components to construct artificial cellulosomes. Chimeric scaffoldins that contain cohesins from different species and an optional CBM are incubated with hybrid cellulases (either cellulosomal or noncellulosomal) from different origins expressed as fusion protein with suitable dockerin domains. In this way it is possible to obtain the most performant cellulosomal configuration for different substrate bioprocessing integrating suitable hydrolytic enzymes at specified positions. The new generation of designer cellulosomes in which three different enzymes has been integrated into the chimeric scaffoldin are found to be considerably more active than the corresponding free enzyme.

To further improve cellulose hydrolysis, also exploiting the effect of enzyme-substrate-microbe complex synergy [Lu et al., 2006], different researchers achieved the co-displaying of single cellulases on the yeast cell surface directly or after their insertion in a minicellulosome. In the first case, *Trichoderma reesei* endoglucanase II and cellobiohydrolase II, and *Aspergillus aculeatus* β-glucosidase I were simultaneously co-displayed as individual fusion proteins with the C-terminal-half region of α-agglutinin. The co-displaying of all three genes allowed an ethanol production of 3 g per liter after 40 h directly from the amorphous cellulose and this yield was impaired by the co-displaying of only β-glucosidase I and endoglucanase II [Fujita et al., 2004]. An improvement in both cellulose hydrolysis and ethanol production has been achieved when an entire minicellulosome has been assembled on the yeast cell surface. Tsai *et al.* [2009] showed the assemblage of an entire functional minicellulosome on the cell surface of *S. cerevisiae.* A trifunctional miniscaffoldin, consisting of an internal CBD flanked by three divergent cohesin domains from *C. thermocellum*, *C. cellulolyticum*, and *Ruminococcus flavefaciens,* was expressed and displayed on *S. cerevisiae* cell surface by using the GPI anchor linked at the N-terminal side of the scaffoldin [Boder and Wittrup, 1997]. Subsequently the incubation of the engineered *S cerevisiae* cells with *E. coli* lysates containing an endoglucanase (CelA) fused with a dockerin domain from *C. thermocellum*, an exoglucanase (CelE) from *C. cellulolyticum* fused with a dockerin domain from the same species and a β-glucosidase (BglA) from *C. thermocellum* tagged with the dockerin from *R. flavefaciens* resulted in the assembly of a functional minicellulosome on the yeast cell surface. The minicellulosome showed the synergistic effect for cellulose hydrolysis and the yeast produced ethanol directly from phosphoric acid-swollen cellulose (PASC) at a concentration of 3.5 g/l after 48 h of incubation.

The most recent studies on recombinant minicellulosomes have been made by Fei *et al.*, [2010]. They obtained the co-expresssion of all the cellulosomal components into a recombinant *S. cerevisiae* host and the *in vivo* assembling of a functional cellulosome on the cell surface. The cell surface display of the complete trifunctional cellulosome did not require the *in vitro* loading onto the scaffoldin of the enzymatic components, previously produced in *E. coli*, but the expression of the miniscaffoldin dictated the formation of the entire complex by the high-affinity interactions between cohesins and dockerins. In this study, were successfully displayed two miniscaffoldins, CipA3 and CipA1, based on the well-characterized scaffoldin protein CipA from *C. thermocellum* [Gerngross et al., 1993]. CipA3, containing a cellulose-binding domain (CBD) and three cohesin modules, was designed to assemble minicellulosomes with up to three enzymatic activities while CipA1, containing a CBD and one cohesin module was designed to assemble a spatially restricted unifunctional minicellulosomes on the yeast cell surface. The enzyme components used in this study, including *T. reesei* EGII and CBHII and *A. aculeatus* BGL1, had fungal origins, and all of them were functionally previously expressed in *S. cerevisiae* [Fujita et al., 2004]. The capability of synthesizing a trifunctional minicellulosomes gave to the yeast cells the ability to simultaneously break down and ferment PASC to ethanol with a titer of 1.8 g per liter. This yield was higher than those obtained when the single hydrolytic enzymes were displayed on cell surface through cohesin-dockerin interactions and were spatially distributed.

In addition, several cellulase genes have also been expressed efficiently in other microbial systems such as *Pichia pastoris*, *Humicola insolens*, *Streptomyces*, *Aspergillus oryzae* [Moriya et al., 2003; Rashid et al., 2008; Wonganu et al; 2008].

9.2 Cloning and Expression in Plant Systems

The expression of cellulolytic enzymes in transgenic plant offers a huge economic advantage over the more traditional production system from recombinant microorganisms. Transgenic plants can express different cellulases at high levels and the production scale up requires reduced capital investments compared to those needed for the purchase and maintenance of large fermentors and associated equipments. Several research groups have investigated on the practicality of producing various cellulases in crop plants [Ziegelhoffer et al., 2009; Jin et al., 2003; Hood et al., 2007]. Cellulolytic enzymes produced directly in biomass crops make possible the utilization of this resource firstly as a bioreactor to accumulate cellulase enzymes and, subsequently, as feedstock for fuel ethanol production. The heterologous enzymes extracted both from fresh or dry transgenic crop or produced in a dedicated crop such as alfalfa [Ullah et al., 2002] can be then added to the pretreated biomass [Sticklen, 2008 Oraby et al., 2007]. As an alternative, the expression of the cellulolytic activities in plant crops, not followed by the enzyme extraction, could potentially yield the biomass more favorable for bioprocessing [Sticklen, 2007] eliminating the pretreatment costs. Cellulose breakdown can be allowed to occur in the field, providing that the temperature optimum of the cellulase is suitable, or, alternatively, the residues can be harvested after the crop, and the sugars fermented off-site, to produce ethanol.

One of the most important requirements for an economically viable utilization of cellulases produced by transgenic plant is the high level of active enzyme expression. To accomplish this task the most important problem to solve is the improvement of plant transformation techniques. Stable nuclear genetic transformation consisting in the insertion of foreign gene(s) in the plant genomes through several transformation methods, such as *Agrobacterium tumefaciens* or polyethylene glycol-mediated systems or, more recently developed, microprojectile bombardment technique, results in low enzyme yields ranging about 1% of total soluble proteins (TSP) [Bogorad, 2008]. In this regard, chloroplast transformation offers a great advantage with recombinant protein yields over 10% TSP, and exceptional transformants reaching as high as >40%TSP [Gray et al., 2009]. In fact, differently from the nuclear, the plastid transformation results in thousands of transgene copies per cell that are actively transcribed and translated [McKenzie, 2008]. Moreover, the enzyme stability is enhanced by the reduced exposure to proteases. In such way, transplanctomic expression allows to reach the production of high levels of active, recoverable, and intact enzyme, the accumulation of which does not compromise plant growth and development. In this context, relevant studies have been reported on the expression of the highly thermotolerant endoglucanase E1 from *Acidothermus cellulolyticus*. E1 has a considerable potential for a successful production in plants due to its thermostability and reduced activity at ambient temperature that allows the enzyme accumulation in the cell with minimal effects on plant growth, and its easily recovery in an active form. Direct expression of the E1 protein as holoenzyme or as catalytic domain (CD) alone has been achieved in several plants with significantly varying levels of expression. In recombinant potato lines, the expression under the leaf-specific promoter allowed an accumulation of holoenzyme up to 2.6% of TSP [Sun et al., 2007]. Recent studies are in progress regarding the E1 production in transgenic tobacco. The E1 expression in tobacco chloroplast greatly increased the cellulase production, that was obtained at levels of 12% of chloroplast TSP [Ziegelhoffer et al., 2001]. However, implementation of this system, is not straightforward because it depends on plastid transformation which is not yet possible in most plant species such as the feedstock biomass crop.

A more direct and generally applicable strategy involves expression of a nuclear transgene and targeted secretion of the gene product into the apoplast. The highest levels of accumulation have been achieved when the E1CD was secreted into the apoplast of the leaves of primary *Arabidopsis thaliana* transformants, reaching levels up to 25.7% of the total soluble protein content [Ziegler et al., 2000]. Attempts to reach the expression of E1 in maize biomass crop are also reported [Biswas et al., 2006].

Two *Thermobifida fusca* thermostable cellulases, Cel6A and Cel6B, were also expressed in tobacco varieties following chloroplast transformation. In the first attempts Yu *et al.* [2007] inserted *cel6A* and *cel6B* in the chloroplast genome of a nicotine-free and a nicotine-containing tobacco variety.

Higher accumulation yield was obtained when the *cel6A* coding region was expressed in chloroplast of *Nicotiana tabacum* cv. Samsun with its start codon fused to a downstream box (DB) region [Gray et al., 2009]. The best results were obtained with tetanus toxin fragment C (TetC) DB region that allowed level of TetC-Cel6A accumulation of 10% of chloroplast TSP. These values have improved the accumulation of Cel6A over 100-fold respect to the yield of 0.1% TSP obtained upon the nuclear Cel6A expression [Ziegelhoffer et al., 1999].

The expression of the EgI endoglucanase from *R. albus* is an example of how the expression of cellulase activity could improve some specific characteristic of transgenic plants such as the digestibility level of silage plant. The *egI* gene that codes for one of the major cellulolytic enzyme from the rumen bacterium *R. albus*, has been expressed in tobacco cells BY2. These cells expressed an intracellular catalically active EgI at a level 30-folds higher than the wild type cells. Although the expression of *egI* in BY2 cells did not affect their growth, the enzyme was active toward the host cell wall after cell disruption. Transgenic tobacco plants transformed with the *Agrobacterium*-mediated method [Sakka et al., 2000] expressed also a strong CMC degrading activity. The transgenic tobacco plants were morphologically similar to the wild type plants grown in the same conditions but EgI was able to enhance the degradation of the plant tissue by macerating the plants. If this type of transgenic grass is fed to cattle, the cellulase released from the cells by mastication and disruption should degrade the cellulosic compounds enhancing the grass digestibility.

9.3 Cloning and Expression in Bombix Mori Cells and Larvae through the Baculovirus Expression System

The baculovirus vector system for heterologous gene expression in insect cells is the most suitable method to overcome problems such as the poor solubility and the overglycosylation of the recombinant protein produced in *E. coli* and yeasts hosts, respectively.

In recent reports [Zhou et al., 2010; Li et al., 2010], high cellulase expression levels of the endoglucanase EGII and EGI, from *T. reesei* and *T. viride* respectively, have been shown in the silkworm *Bombyx mori* cells and larvae using a baculovirus expression system. The cellulase genes have been introduced in bacmids, *E. coli* and *Bombyx mori* shuttle vectors, that consist in the baculovirus genome containing a bacterial origin of replication, a kanamycin resistance marker, a segment of DNA encoding the lacZα peptide and a targeting site for the bacterial transposon Tn7 (att-Tn7). To obtain the recombinant bacmid, the cellulase gene was firstly introduced in the multiple cloning site, flanked by the left and right bacterial transposon Tn7 sequences of a donor plasmid. After the introduction of the recombinant donor plasmid in the *E. coli* DH10β strain, that contain the bacmid and the helper plasmid coding for a transposase protein, the cellulose gene transpose into the att-Tn7 site in the bacmid genome. With this novel Bac-to-Bac system, the recombinant baculovirus, easily generated through gene transposition and previously propagated in *E. coli,* has been transfected in the *B. mori* BmN cells and larvae that produced high level of recombinant protein. In the case of EGI a further improvement of the cellulase yield was obtained utilizing mutant bacmid lacking the virus-encoded chitinase and cathepsin genes of *B. mori* nucleopolyhedrovirus. For EGII a putative yield of about 386 μg per larva (equal at a concentration of about 150 mg/l) of catalitically active cellulase was reached after the baculovirus infection.

10. Biotechnological Applications of Cellulases

10.1 Cellulases in Brewing and Wine Biotechnology

The macerating enzymes, comprising cellulases, hemicellulases and pectinases, hydrolyze the plant cell wall and, consequently, can be used in brewing and wine biotechnology to improve the quality of finished products and avoid the use of chemicals. Enzyme preparations are used in the brewing and distilling industries to reduce the viscosity of the mash and to improve the overall efficiency of the process. In fact, cellulolytic and hemicellulolytic enzymes allow the conversion of undigestible lignocellulosic biomass into fermentable sugars, with consequent increase of alcohol yield. The quality of the products results improved and, at the same time, the overall costs of production are reduced.

10.1.1 Beer Brewing Process

Barley is the most common cereal used for the production of beer although wheat, corn, and rice are also widely used. The main processes involved in beer production include milling to reduce the size of the dry malt in order to increase the availability of the carbohydrates; mashing where water is added to the malt; lautering where spent grains are removed from the wort, boiling of the wort with flavouring hops, fermentation of the wort liquor, maturation, conditioning, filtration and packaging of the final product. The high concentration of β-glucan in the brewing process, resulting from unsuitable brewing process or low quality barley, produces high viscosity of beer, formation of gelatinous precipitate, decrease of the extract yield, and lower run-off of wort [Bamforth, 1994; Guo et al., 2010; Bhat, 2000].

In brewing process, cellulases are used during the mashing stage in order to hydrolyze excess β-glucans and reduce the viscosity, thus improving the separation of the wort from the spent grains. Oksanen *et al.* [1985] observed that the endoglucanase and the cellobiohydrolase from the *Trichoderma* cellulase system produced a large reduction of the degree of polymerization of the β-glucans, and wort viscosity. Moreover, the increased addition of enzymes used resulted in improved filtering.

A. niger, *T. reesei,* and *P. funiculosum*, which are generally recognized as food grade microorganisms, are the major source of cellulases currently used in the mashing step, as these enzymes provide technological benefit to beer manufacture [de Castro et al., 2010; Karboune et al., 2008].

An alternative solution is that production of cellulolytic enzymes, enzymatic hydrolysis of the polysaccharidic fraction, and fermentation of the resulting sugars are all combined in a single step. *S. cerevisiae* is a promising candidate, as it produces ethanol at high concentrations, has GRAS status, and can be easily genetically manipulated. Unfortunately, *S. cerevisiae* completely lacks of a cellulose-degrading enzyme system but it can be employed industrially as host for expression of heterologous celluase genes. As example, the processive endoglucanase Cel9A of *T. fusca* was recently produced in *S. cerevisiae*. In addition, to improve the cellulolytic capability of the yeast and to investigate the level of synergy among cellulases produced by a recombinant host, the *cel9A* gene was co-expressed with *cel5A (egII)* and *cel7B (egI)* genes of *T. Reesei* [van Wyk et al., 2010]. Yeast strains with acquired ability to degrade barley β-glucans and to accumulate sulfite, can improve the quality of beer. In fact, sulfite is an important component of beer because it has antioxidant and antimicrobial activities and can also forms aldehyde adducts that stabilize the flavour of beer.

10.1.2 Wine Production

Wine manufacture is a biotechnological process in which yeast cells and enzymes are indispensable for ensuring a high quality product. The use of cellulases, hemicellulases and pectinases during wine making, allows a better skin maceration, and superior color extraction, particularly important in the production of red wine; in addition, it improves clarification, filtration, and the overall quality and stability of the wine [Galante et al., 1998a]. However, it is also important to recall that studies on the effect of enzymes on wine color and anthocyanin content have led to contradictory results [Sacchi et al., 2005].

The polysaccharidic fraction of wines comes from the pecto-cellulosic cell walls of grape berries [Pellerin et al., 1996; Visal et al., 2003; Ducasse et al., 2010] , and its composition and quantities depends on the wine making process that can be changed by using different enzymes [Ayestaran et al., 2004; Guadalupe et al., 2007].

Pectinase preparations, used in wine making, were lately modified by addition of cellulases and hemicellulases in small quantities to realize a more complete breakdown of the cells with consequent fruit liquefaction in a moderately short time period [Plank and Zent, 1993]. It was demonstrated that the mixture of macerating enzymes worked better than pectinases alone in grape processing [Haight and Gump, 1994].

Since 1980, the use of a β-glucanase from *Trichoderma* sp. has been proposed for wine making from grapes infected by *Botrytis cinerea* [Dubordieu et al., 1981; Villetaz et al., 1984]. This microorganism produces a soluble high molecular mass 1,3-β-glucan with short side chains linked through 1,6-β-glycosidic bonds, thus complicating wine filtration and clarification. To overcome this obstacle, a β-glucanase from *T. harzianum* was identified and patented to hydrolyze glucans for resolving undesirable effects generated by the presence of *B. cinerea*.

10.2 Cellulases in Animal Feed Biotechnology

Although cellulose is the main food resource for many animal species, most omnivores and herbivores are unable to produce the cellulases by themselves; contrariwise, the ruminants live in symbiosis with cellulolytic microorganisms (mixtures of highly specialized bacteria and protozoa localized in the digestive tract) that degrade cellulose under anaerobic conditions [Kobayashi et al., 2008]

Low forage digestibility limits the intake of accessible energy for animals, comprised ruminants; in addition, it contributes to increase nutrient excretion by livestock, and prevents the possibility of using low-quality feedstuffs. As plant polysaccharides are degraded relatively slowly and incompletely than other components of feedstuff, an efficient system for the complete enzymatic hydrolysis is required for improving the use of low-quality highly fibrous silage [Graham and Inborr, 1992; Chesson and Forsberg, 1997; Ozkose et al., 2009]. Research aimed to develop animal feed from different kinds of agro-industrial waste, in order to minimize feed costs, is under way. For this purpose, several Fungi, including some species of *Pleurotus*, are utilized to biodegrade the vegetable residues for their use as animal feed. *Pleurotus* can colonize different kinds of lignocellulosic residues, such as citric bagasse and rice straw, and increases nutritional values and digestibility of these raw materials thanks to its extracellular cellulolytic and hemicellulolytic enzymes.

The addition of lignocellulolytic digestive enzymes into animal diet is widespread lately, not only for ruminants [Bowman et al., 2002], but also for non-ruminant farm animals [Carneiro et al., 2008] and poultry [Woyengo et al., 2008].

In this context, cellulases have a wide range of potential applications in the animal feed industry; these hydrolytic enzymes allow to increase the nutritional quality of feed, through the improvement of cell wall digestion and efficiency of feed utilization, and also contribute to cut down excessive nutrient excretion by livestock.

They can be added to the fodder, also in the early step [Zhu et al., 1999], and several fibrolytic enzyme products, used at present as feed additives in ruminant diets, were originally developed as silage additives [Lewis et al., 1996]. Often, commercial enzymes utilized in the livestock feed industry, are obtained from microbial fermentation, and enzymatic products for animal diets are obtained from both Fungi (mostly *T. longibrachiatum, A. niger, A. oryzae*) and Bacteria (mostly *Bacillus* spp.) [Pendleton, 2000; Bhat and Hazlewood, 2001].

In the genus *Lactococcus, L. lactis* is one of the main species to be considered as a dairy-product-associated bacterium [Svec and Sledacek, 2008]. In addition, it could also be used as potential silage inoculant because it is recognized as a safe microorganism and farm environment is a natural habitat of this species. Moreover, the use of biological additives can control the amount and pattern of fermentation in forage-based silages by decreasing the populations of harmful microorganisms in the ensiled forage.

A gene encoding for a cellulase from the anaerobic rumen fungus *Neocallimastix* sp. was cloned and successfully expressed into two *L. lactis* strains (IL403 and MG1363). The transformed strains were then employed as silage additives for pre-biodegradation of the plant biomass to improve the fiber digestibility during the ensiling process [Ozkose et al., 2009].

Recently, a novel cellulase (CelA4) from the thermoacidophilic bacterium *Alicyclobacillus* sp. A4 has been purified and characterized [Bai et al., 2010]. This enzyme, highly acid stable and protease-resistant, hydrolyzes with high efficiency barley β-glucan, and under simulated gastric conditions, decreases the viscosity of barley-soybean feed to a greater extent. These properties make CelA4 a good candidate as a new commercial glucanase to improve the nutrient bioavailability of pig feed.

10.3 Cellulases in Pulp and Paper Biotechnology

10.3.1 *Biomechanical Pulping*

Mechanical pulping process is electrical energy intensive and results in low paper strength. Biomechanical pulping, defined as the enzymatic treatment of lignocellulosic materials before the mechanical pulping step, has shown at least 30% savings in electrical energy consumption, and significant improvements in paper strength properties.

The potential of enzymatic treatments has been assessed and the processes have proved successful [Gubitz et al., 1998; Bajpai, 1999]. Since biofibers were stronger than the conventional fibers, it was possible to reduce the amount of bleached softwood kraft pulp by at least 5% in the final product. Utilization of cellulases from fungal sources (*T. reesei, Aspergillus* sp.) [Buchert et al., 1998; Suurnakki et al., 2000] saves 33% electrical energy and significantly improves paper strength properties compared to the control. Fungal cellulases pretreatment reduced brightness, but brightness was restored to the level of bleached control with 60% more hydrogen peroxide.

A cellulase preparation produced by the ascomycete Fungus *Chrysosporium lucknowense* for using in the pulp and paper industry represents, at present, an attractive alternative to the well-known cellulases from Fungi like *Aspergillus* sp. and *T. reesei* for protein production on a commercial scale [Bukhtojarov et al., 2004; Hinz et al., 2009].

10.3.2 Biomodification of Fibers

In recent years the fiber biomodification has become more and more interesting because this process is environmentally friendly, consumes less energy and makes less damage to fiber than traditional process, also improving drainage, beatability and runnability of paper mills [Pellinen et al., 1989; Henriksson and Gatenholm, 2002; Yang et al., 2008]. For this purpose, cellulases together with other enzymes like hemicellulases can be used [Bhat, 2000; Noe et al., 1986]. As example, an endoglucanase from *T. reesei* provided with a dual activity on xylan and cellulose was utilized in fiber biomodification [Pere et al., 1995], and it showed a drainage improvement of 30% compared to the endoglucanase from same microorganism specific only for cellulose.

As cellulases used in the fiber biomodification can act on the surface and into the inner layers of cellulose fibers, a careful study on their mechanism of action has been done in the last years [Suurnakki et al., 2003]. The aim was to understand the changes produced on fibers in order to obtain a final product provided with better quality, namely improvement of fiber-fiber bonds with consequently better cohesion between the fibers in the finished product, and to lower the production costs. Particularly, Cadena *et al.* [2010] studied the endoglucanase cel9B from *Paenibacillus barcinonensis* in biopulping refining to investigate the ability of this multidomain enzyme to improve the paper strength property and reduce production costs.

10.3.3 Biodeinking

All over the world people offer more attention to the environment and so, the recycle of waste paper has to be considered also as a necessity for the protection of forest and economy. Paper mill will gain profit from the utilization of recycled fiber, since it is profitable to decrease pollution, cost, and investment.

Conventional deinking technology with alkali is characterized by a low efficiency on laser printed paper and is not retained environmentally friendly. Consequently, researchers have concentrated their attention to new deinking technologies [Moon and Nagarajan, 1998]. The principle of enzymatic deinking is based on the weakening of the connections between toner and fibers due to the enzyme attack with separation of toner particles from fibers [Yingjuan et al., 2005; Shufang et al., 2005]. The enzymatic deinking allows us to avoid the use of alkali; moreover, using enzymes at acidic pH it is possible to prevent the yellowing, modify the distribution of the ink particle size, improve fiber brightness strength, pulp freeness and cleanliness, reduce fine particles and reduce environmental pollution.

Until 2000, the use of enzymes to perform biodeinking was only investigated at the laboratory scale [Buchert et al., 1998; Bhat, 2000]. Subsequently, a mixture of cellulase, lipase, and amylase was employed in biodeinking process at industrial level [Morbak and Zimmermann, 1998].

The effect of combined deinking technology with ultrasounds, UV irradiation and enzyme on laser printed paper was investigated. The results confirmed that the dose of alkali can be reduced using biodeinking technology. Cellulases from different microorganisms such as *A. niger, T. reesei, Humicola insolens, Myceliophtora fergusii, Chrysosporium lucknowense, Fusarium* sp. were used for this purpose [Marques et al., 2003].

10.4 Cellulases in Food Biotechnology

10.4.1 Fruit and Vegetable Juices

The fruit and vegetable juices consumption in Europe, Australia, New Zealand and the USA has increased in recent years and therefore, have a significant importance from a commercial standpoint. Juices are consumed by a wide range of consumers throughout the year, for the availability of nutritious components from fruit and vegetables, but also for their perceived health benefits [Lampe, 1999; Kurowska et al., 2000]. For example, orange juice is rich in vitamin C, folic acid, potassium, and it is an excellent source of bioavailable antioxidant phytochemicals [Franke et al., 2005]. Juice is the liquid that is naturally contained in fruit or vegetable tissues. It is prepared by extraction, which involves maceration followed by pressing or decanting, to separate the juice from the solid, followed by clarification and stabilization. When fruit industries began to produce juice, the yields were low, and many difficulties were encountered in filtering the juice to an acceptable clarity. Enzymes can play a key role in this process improving yield, clarity and stability of the juice, and the addition of "macerating enzymes" is constantly increasing [Askar, 1998; Bhat, 2000; Dongowski and Sembries, 2001]. This mixture consists of a multi-enzyme system comprising proteases, amylases, pectinases, cellulases, hemicellulases and lysozyme from food-grade microorganisms (*A. niger* and *Trichoderma* sp.) useful in breaking the fruit tissues to release more juice. The enzymatic process of fruit juice production is claimed to offer a number of advantages over mechanical-thermal procedure. In particular, the use of cellulases and pectinases is an integral part of modern fruit processing technology involving treatment of fruit mashes as these enzymes not only facilitate easy pressing, but also increase juice recovery. In addition, they ensure the highest possible quality of end products such as aroma, phenolic components content and absence of cloudiness [Buchert et al., 2005; Ramadan and Thomas, 2007]. Kapasakalidis *et al.* [2009] tried to enhance cell wall degradation from black currant pomace by including a "cellulase-assisted" hydrolysis step as an essential treatment for the production of polyphenol-rich extracts that could be further processed for the manufacture of dietary supplements or food additives. For this purpose, a commercial preparation of cellulase from *T. reesei* was used. The enzyme treatment significantly increased plant cell wall polysaccharide degradation as well as enhanced the availability of phenols for subsequent methanolic extraction.

10.4.2 Olive Oil

Olive oil production is very important because it is an old tradition dating back a thousand years and represents one of the most interesting fields of Italian agriculture. It is important to note that the virgin olive oil is a healthy fat due to its high content of oleic acid and antioxidants, particularly phenolic compounds [Manna et al., 1999]. Nowadays, the olive oil extraction is carried out with technological industrial processes (continuous or discontinuous), although the quality and the quantity of the obtained oil are still to be improved. A way for trying to solve the problem could be the utilization of biotechnology in olive oil industry, also considering eco-sustainability and lower environmental impact of the enzymes [Voragen et al., 2001; Chiacchierini et al., 2007].

Extraction of olive oil involves: (1) crushing and grinding of olives in a stone or hammer mill; (2) passing the minced olive paste through a series of malaxeurs and horizontal decanters; (3) high-speed centrifugation to recuperate the oil. To obtain a product of high quality it is very important to utilize freshly picked, clean and not fully mature fruit, under cold pressing conditions. However, high amounts of oil have also been obtained with fully ripened fruits processed at temperatures higher than room temperature, but this leads to a worse quality. The oil has high acidity, rancidity and poor aroma [Galante et al., 1998a; Garcia et al., 2001 4, 5]. Specifically, during extraction the content of some components is significantly modified according to the extraction technique employed [Amirante et al.,

2001], while new components can be formed, as a result of chemical and/or enzymatic pathways [Ranalli et al., 1999a]. During the last decades, the enzyme preparations are used in the processing of fruits and vegetables to improve the yield and quality of the products. Since 1980, systematic research revealed that for the efficient maceration and extraction of oil from olives no single enzyme was sufficient but pectinases, cellulases and hemicellulases were found to be very important all together for this result. The commercial enzyme preparation Olivex, a pectinase preparation with low levels of cellulase and hemicellulase from *A. aculeatus*, was initially used for the extraction of olive oil [Fantozzi et al., 1977]. Afterwards, a commercially available combination of enzymes from different microorganisms (Cytolase 0) consisting of pectinases from *Aspergillus*, cellulases and hemicellulases from *Trichoderma*, proved to be superior than the enzymes from a single microorganism [Ranalli et al., 1999b]. More recently, the enzymatic complex Bioliva showed to have positive effects on colour pigments and chromatic parameters [Ranalli et al., 2005; Chiacchierini et al., 2007].

10.5 Cellulases in Textile and Laundry Biotechnology

Since the early part of the last century, enzymes such as the cellulases have been used for a wide range of applications in textile processing in replacement of the traditional methods.

10.5.1 Biostoning and Biopolishing

Jeans manufactured from denim are one of the world's most popular clothing items. In the late 1970s and early 1980s, industrial laundries developed methods for producing faded jeans by washing the garments with pumice stones, which partially removed the indigo dye revealing the white interior of the yarn, which leads to the faded, worn and aged appearance. This process was designated as “stone-washing”. The use of 1-2 kg stones per kg of jeans for 1 h during stone-washing met the market requirements, but caused several problems including rapid consumption of washing machines, and unsafe working conditions.

As an alternative to the stone-washing, biostoning is by far the most economical and environmental friendly way to treat denim. The cotton fabrics treated with the enzymes loose the indigo, which later is easily removed by mechanical abrasion in the wash cycle [Cavaco-Paulo, 1998; Yamada et al., 2005]. The substitution of pumice stones by an enzymatic treatment includes many advantages: washing machines lower consumption and elevated productivity, short treatment times, less intensive working conditions. Moreover, it is possible to operate in a more safe environment because pumice powder is not produced, and the process can be mechanized controlling, with the use of computer, the dosing devices of liquid cellulase preparations [Bhat, 2000].

Nevertheless, a very important problem during biostoning is the “back-staining”, namely the high propensity of the released dye to redeposit on the clothes. This process masks the overall blue/white contrast of the finished product; therefore, controlling the back-staining is essential.

Much interest of the researchers has been focused on the mechanism of cellulose adsorption of cellulases as the best cellulases to utilize for application in textile processing are those with sites on the surface of protein globule capable of binding indigo with low adsorption ability on cellulose [Galante et al., 1998b].

According to recent research, back-staining is dependent on pH value and type of enzyme [Muntazer and Sadeghian Maryan, 2010]. However, neutral and alkaline cellulases are preferred to acid cellulases due to the decrease of the staining intensity [Sinitsyn et al., 2001]. Among cellulases potentially useful in the textile industry, thermophilic cellulases have received great attention as additives in biostoning and biopolishing. The cellulase from the thermophilic fungus *T. emersonii* was used as additive in biopolish by treating the jute-cotton union fabrics in order to test its effectiveness [Gomes et al., 2007]. The enzyme enhanced whiteness, brightness and softness of the treated materials, and pilling and fuzziness of the treated samples were remarkably reduced without loss of tensile strength beyond acceptable limits.

In the textile wet processing, the biopolishing is usually carried out with desizing, scouring, bleaching, dyeing and finishing by utilization of cellulases. However, there are not clear indications about the best cellulase mixture to use also if, in general, less quantity of endoglucanases implies reduced loss of tissue weight [Miettinen-Oinonen and Suominen, 2002]. The use of these enzymes allow many improvements such as the removal of short fibers, surface fuzziness smooth, polished appearance, more color uniformity and brigthness, improved finishing, and fashionable effects. At last, due to increasing environmental concerns and constraints being imposed on textile industry, cellulase treatment of cotton fabrics is an environmentally friendly way of improving the property of the fabrics.

In 2007, Anish *et al.* [2006] isolated an endoglucanase from the alkalothermophilic bacterium *Thermomonospora* sp. The enzyme, used for denim biofinishing under alkaline conditions, was effective in removing hairiness with negligible weight loss and imparting softness to the fabric. Higher abrasive activity with lower back-staining was a preferred property for denim biofinishing exhibited by the *Thermomonospora* endoglucanase.

10.5.2 Laundry

The most important reason to use enzymes in detergents is that they are biodegradable and a very small quantity of these inexhaustible biocatalysts can replace very large quantity of chemicals. Since detergents hold ionic and anionic surfactants, and bleaching agents (oxidizing agents) that can partially or completely denature proteins, the enzymes for laundry must be resistant to anionic surfactants and oxidizing agents.

The accumulation of microfibrils on the surface of the fabrics makes the fabrics look hairy and scatters incident light, thereby lessening the brightness of the original colors. In detergent industry, cellulases are used to remove microfibrils from the surface of cellulosic fabrics, enhancing color brightness, hand feel and dirt removal from cotton garments that during repeated washings can become fluffy and dull. As consequence, the most promising candidates should have high defibrillation capacity, such as the endoglucanases from GH family 45 which are the most used in the detergent industry. Shimonaka *et al.* [2006] examined the properties of GH family 45 endoglucanases from Mucorales sp. The defibrillation activities of RCE1 and RCE2 from *Rhizopus oryzae*, MCE1 and MCE2 from *Mucor circinelloides*, and PCE1 from *Phycomyces nitens* were much higher than those of the other GH family 45 endoglucanases.

10.6 Cellulases in Bioethanol Production

The increasing concerns about environmental protection, the rising cost of fuels, the decrease of the world reserves of fossil energy and global weather change caused by increased carbon dioxide emissions, have directed scientific interest toward the production of bioethanol from renewable resources for a "greener" alternative energy which can respond to the high energy demand of the world [Li et al., 2009]. Yearly, photosynthesis produces more than 10^{11}tons of dry plant material worldwide, and cellulose constitutes almost half of this material [Leschine, 1995].

Ethanol, also called grain alcohol, is a clear colorless liquid, biodegradable, low in toxicity which produces little environmental pollution when burns to produce carbon dioxide and water. Bioethanol is the principle fuel used as a petrol substitute for road transport vehicles, and it is produced using biological renewable resource such as the"lignocellulosic biomass" materials [Hamelinck et al., 2005; Hill et al., 2006]. The advantage over fossil fuels is that bioethanol decrease the greenhouse gas emissions which are mainly produced by the road transport system. Another usefulness of bioethanol is represented by the possibility to further reduce the amount of carbon monoxide produced by the old engines, thus improving air quality without additional costs. Moreover, very important is the ease with which this biofuel can be simply integrated into the existing road transport system since it can be mixed with conventional fuel in quantities up to 5% without the need of engine modifications.

At first, the bioenergy industry was based on the fermentation of glucose derived from food crop using conventional technologies; however, starch raw materials are not sufficient enough to meet increasing demand, and are expensive.

However, it is possible to utilize biomass from different kinds of materials at lower price. These materials, like wood, municipal solid waste, waste paper, agricultural and industrial waste are already available to produce bioethanol and are not in competition with food sources [Kim and Dale, 2004; Lin and Tanaka, 2006].

The yield of fermentable sugars, for low cost fuel production, represents a principal test in global efforts to utilize renewable resources rather than fossil fuels. The lignocellulosic biomass refers to plant biomass which is composed of cellulose, hemicelluloses and lignin. In such kinds of biomass, the chains of cellulose and hemicelluloses are embedded in a lignin matrix, which hinders their efficient degradation. Cellulose and hemicelluloses can be hydrolyzed by enzymes or chemical methods into their sugars that can subsequently be converted into bioethanol by well established fermentation technologies. In general, the production of bioethanol from lignocellulosic biomass consists in three important steps:

1) pretreatment that allows delignification of the biomass to release cellulose and hemicellulose from their complex with lignin, making more accessible these polysaccharides to the enzymes so the hydrolysis could be much more effective [Mosier et al., 2005; Alvira et al., 2010];
2) saccharification: conversion of the polysaccharides into fermentable sugars. Enzymatic degradation of biomass has been extensively studied and requires the action of several enzymes acting in cooperation such as endoglucanase, β-glucosidase, xylanase, α-arabinosidase, β-xylosidase, and others;
3) fermentation by yeast or other appropriate microorganisms to obtain ethanol from the resulting mixture of hexose and pentose sugars.

Cellulases are essential for successful bioconversion of lignocellulosic biomass; thus, the search for cellulolytic enzymes is ongoing in last years, and various microorganisms of bacterial as well as fungal origin have been evaluated for their ability to degrade cellulosic substrates into glucose monomers [Kumar et al., 2008].

The primary interest in using Fungi comes from their capacity to produce significant amounts of cellulases which are secreted into the medium with following easy isolation and purification. The genera *Aspergillus* and *Trichoderma* are the most used for this purpose among the filamentous fungi genera. Already in 1950, a *Trichoderma* strain which produced a cellulase complex capable of degrading native cellulose was identified [Dashtban et al., 2009]. *T. reesei* RutC30 is known as an excellent cellulases producer, but the low content of β-glucosidase in its extract, which is required for total hydrolysis of cellulose to glucose, is pointed out as a disadvantage [Kim et al., 2003]. On the other hand, *A. niger* strains have been studied due to their ability to produce high levels of β-glucosidase, although the production of endoactive enzymes is deficient. In order to produce well-balanced extracts, mixed cultures from two genera are employed but, the synergism between the different groups of cellulases produced by pure cultures is often better than that observed from co-cultures. The search for new microorganisms and cellulases with potential use in lignocellulosic biomass degradation is incessantly in progress. As example, *A. niger* isolated from soil sampled from Ejura farms (Gahana) was used to hydrolyze corncobs, the main agrowaste from maize which accounts for 30% of its weight, into simple sugars for subsequent fermentation to bioethanol in a simultaneous saccharification and fermentation process. The highest ethanol concentration of 0.64 g per liter was recorded over the 24 h fermentation period [Zakpaa et al., 2009].

It is reported that in solid state cultivation *T. reesei* secretes a complex array of degradative enzymes. The production of cellulases by *T. reesei* F-418, cultivated on alkali treated rice straw, was recently investigated by Abd El-Zahern and Fadel, in order to produce bioethanol from rice straw, an abundant lignocellulosic waste by-product. The solid state fermentation technique was employed. After saccharification of the biomass obtained by cellulases produced from *T. reesei*, glucose fermentation step was conducted by *S. cerevisiae* SHF-5 under static condition giving 5.1% (v/v) ethanol after 24 h fermentation period [Abd El-Zaher and Fadel, 2010].

However, the isolation and characterization of glycoside hydrolases from Eubacteria are now becoming widely exploited. Bacteria often have a higher growth rate than Fungi allowing for higher recombinant production of enzymes. Moreover, bacterial glycoside hydrolases are often expressed in multi-enzyme complexes known as "cellulosome" providing better activity and synergy [Bayer et al., 2007].

It must be underlined that the drastic conditions required by many pretreatment methods such as high temperature, pressure, or low pH may generate problems when using mesophilic enzymes. In order to overcome this difficulty, microorganisms thriving in habitats characterized by extreme conditions can be taken under investigation as a source of polysaccharide-degrading enzymes, since they allow us to perform biotransformation processes at "non-conventional conditions" under which common enzymes are completely denatured.

Several extremophilic microorganisms belonging to Bacteria and Archaea domains produce cellulolytic strains which can be extremely resistant to environmental stresses.

Enzymes from these microorganisms can survive the harsh conditions found in the bioconversion processes, as they are resistant to high temperatures, low or high pH values, organic solvents and all common protein denaturing agents. These features make these biocatalysts a powerful tool in industrial biotransformation processes of lignocellulosic biomass degradation [Maki et al., 2009]. The exploitation of lignocellulosic biomass for the production of biofuel is potentially feasible; however, several biotechnological constraints must be overcome. One of the first requirements is the efficient production of a hydrolyzate rich in fermentable sugars; therefore, to obtain considerable cellulose degradation, a proficient enzyme blend containing all enzymes required for total hydrolysis of the polysaccharide is required. One example is represented by the enzyme extract from the hyperthermophilic and acidophilic archaeon *S. solfataricus*, which contains the main glycolytic activities (namely endoglucanase and β-glucosidase) required to hydrolyze cellulose into glucose. This extract was used to hydrolyze at high temperature agro-based raw materials such as brewer's spent grains after preliminary strong acid pretreatment. The enzyme saccharification produced high conversion of cellulose into fermentable glucose with a yield of 64% [Morana et al., 2009].

References

Abd El-Zaher, FH; Fadel, M. Production of bioethanol via enzymatic saccharification of rice straw by cellulase produced by Trichoderma reesei under solid state fermentation. *New York Sci J*, 2010, 3, 72-78.

Abdou, L; Boileau, C; De Philip, P; Pagés, S; Fiérobe, HP; Tardif, C. Transcriptional regulation of the Clostridium cellulolyticum cip-cel operon: a complex mechanism involving a catabolite-responsive element. *J Bacteriol*, 2008, 190, 1499-1506.

This article is not included in your organization's subscription. However, you may be able to access this article under your organization's agreement with Elsevier.Ahsan, M; Matsumoto, M; Karita, S; Kimura, T; Sakka, K; Ohmiya, K. Purification and characterization of the family J catalytic domain derived from the *Clostridium thermocellum endoglucanase*CelJ. *Biosci Biotechnol Biochem,* 1997, 61, 427-431.

Ahsan, MM; Kimura, T; Karita, S; Sakka, K; Ohmiya, K. Cloning, DNA sequencing, and expression of the gene encoding Clostridium thermocellum cellulase CelJ, the largest catalytic component of the cellulosome. *J Bacteriol*, 1996, 178, 5732-5740.

Aksenova, H; Rainey, FA; Janssen, PH; Morgan, HW; Zavarzin, GA. (1992) Spirochaeta thermophila sp. nov., an obligately anaerobic polysaccharolytic member of the genus Spirochaeta. *Int J Syst Bacteriol*, 1992, 42, 175-177.

Alfredsson, GA; Kristjansson, JK; Hjorleifsdottir, TS; Stetter, KO. *Rhodothermus marinus,* gen. nov., sp. nov., a thermophibic, halophilic bacterium from submarine hot springs in Iceland. *J Gen Microbiol*, 1988, 134, 299-306.

Ali, BRS; Zhou, L; Graves, FM; et al. Cellulases and hemicellulases of the anaerobic fungus Piromyces constitute a multiprotein cellulose-binding complex and are encoded by multigene families. *FEMS Microbiol Lett*, 1995, 125, 15-21.

Alvira, P; Tomás-Pejó, E; Ballesteros, M; Negro, MJ. Pretreatment technologies for an efficient bioethanol production process based on enzymatic hydrolysis: A review. *Bioresour Technol*, 2010, 101, 4851-4861.

Amirante, R; Cini, E; Montel, GL; Pasqualome, A. Influence of mixing and extraction parameters on virgin olive oil quality. *Grasas y Aceites*, 2001, 52, 198-201.

Ando, S; Ishida, H; Kosugi, Y; Ishikawa, K. Hyperthermostable endoglucanase from *Pyrococcus horikoshii. Appl Environ Microbiol*, 2002, 68, 430-433.

André, G; Kanchanawong, P; Palma, R; Cho, H; Deng, X; Irwin, D; Himmel, ME; Wilson, DB; Brady, JW. Computational and experimental studies of the catalytic mechanism of *Thermobifida fusca* cellulase Cel6A (E2*). Protein Eng*, 2003,16, 125-134.

Anish, R; Rahman, MS; Rao, M. Application of cellulases from an alkalothermophilic *Thermomonospora sp.* in biopolishing of denims. *Biotechnol Bioeng* , 2006, 96, 48-56.

Arai, T; Kosugi, A; Chan, H; Koukiekolo, R; Yukawa, H; Inui, M; Doi, RH. Properties of cellulosomal family 9 cellulases from *Clostridium cellulovorans. Appl Microbiol Biotechnol*, 2006, 71, 654-660.

Miettinen-Oinonen, A; Suominen, P. Enhanced production of *Trichoderma reesei* endoglucanases and use of the new cellulase preparations in producing the stonewashed effect on denim fabric. *Appl Environ Microbiol*, 2002, 68, 3956-3964.

Askar, A. Enzymes in fruit juice processing. *Fruit Proc*, 1998, 7, 273-276.

Atalla, RH; Vanderhart, DL. Native cellulose: a composite of two distinct crystalline forms. *Science*, 1984, 223, 283-285.

Aubert, JP; Béguin, R; Millet, J. Biochemistry and Genetics of Celhdose Degradation. *FEMS 5' Symposium 43*. Academic Press, London, United Kingdom, 1988.

Avitia, CI; Castellanos-Juàrez, FX; Sànchez, E; Téllez-Valencia, A; Fajardo-Cavazos, P; Nicholson, WL; Pedraza-Reyes, M. Temporal secretion of a multicellulolytic system in *Myxobacter sp. AL-1.* Molecular cloning and heterologous expression of cel 9 encoding a modular endocellulase clustered in an operon with cel 48, an exocellobiohydrolase gene. *Eur J Biochem*, 2000, 267, 7058-7064.

Ayestaran, B; Guadalupe, Z; Leon, D. Quantification of major grape polysaccharides *(Tempranillo v.)* released by maceration enzymes during the fermentation process. *Anal Chim Acta*, 2004,513, 29-39.

Bagnara, C; Toci, R; Gaudin, C; Bèlaìch, JP. Isolation and characterization of a cellulolytic microorganism, *Cellulomonas fermentans, sp. nov. Int J Syst Bacteriol*, 1985, 35, 502-507.

Bagnara-Tardif, C; Gaudin, C; Belaich, A; Hoest, P; Citard, T; Belaich, JP. Sequence analysis of a gene cluster encoding cellulases from *Clostridium cellulolyticum. Gene*, 1992, 119, 17-28.

Bai, Y; Wang, J; Zhang, Z; Shi, P; Luo; Huang, H; Feng, Y; Yao, B. Extremely acidic beta-1,4-glucanase, CelA4, from thermoacidophilic *Alicyclobacillus sp. A4* with high protease resistance and potential as a pig feed additive. *J Agric Food Chem*, 2010, 58, 1970-1975.

Baker, PW; Kennedy, J; Morrissey, J; O'Gara, F; Dobson, AD; Marchesi, JR. Endoglucanase activities and growth of marine-derived fungi isolated from the sponge *Haliclona simulans. J Appl Microbiol*, 2010, 108, 1668-1675.

Bamforth, C. β-glucan and β-glucanases in malting and brewing: practical aspects. *Brew Dig* 1994,69,12-16.

Barr, DJS; Kudo, H; Jakober, KD; Cheng, KJ. Morphology and development of rumen fungi*: Neocallimastix sp., Piromyces communis, and Orpinomyces bovis gen.nov., sp.nov. Can J Bot,* 1989, 67, 2815-2824.

Barras, F; van Gijsegem, F; Chatterjee, AK. Extracellular enzymes and pathogenesis of soft-rot *Erwinia. Ann Rev Phytopathol*, 1994, 32, 201-234.

Barros, RR; Oliveira, RA; Gottschalk, LM; Bon, EP. Production of cellulolytic enzymes by fungi *Acrophialophora nainiana* and *Ceratocystis paradoxa* using different carbon sources. *Appl Biochem Biotechnol*, 2010, 161, 448-454.

Bauer, MW; Driskill, LE; Calen, W; Snead, MA; Mathur, E; Kelly, RM. An endoglucanase, EglA, from the hyperthemophilic archeon *Pyrococcus furiosus hydrolyzes beta-1,4 bonds in mixed linkage (1–3), (1–4)-beta-D-glucans and cellulose. J Bacteriol*, 1999, 181, 284-290.

Bayer, E A; Chanzy, H; Lamed, R; Shoham, Y. Cellulose, cellulases and cellulosomes. *Curr Opin Struct Biol,* 1998, 8, 548–557.

Bayer, EA; Lamed, R; Himmel, ME. The potential of cellulases and cellulosomes for cellulosic waste management. *Curr Opin Biotechnol*, 2007, 18, 237-245.

Bayer, EA.; Lamed, R; White, BA; Flint, HJ. *From cellulosomes to cellulosomics. Chem Rec*, 2008, 8, 364-377.

Bayer, EA; Belaich, JP; Shoham, Y; Lamed, R. The cellulosomes: multienzyme machines for degradation of plant cell wall polysaccharides. *Annu Rev Microbiol*, 2004, 58, 521–554.

Bayer, EA; Kenig, R; Lamed, R. Adherence of *Chlostridium thermocellum* to cellulase. *J Bacteriol,* 1983, 156, 818-827.

Beguin, P. Molecular biology of cellulose degradation. *Ann Rev Microbiol*, 1990, 44, 219-248.

Beguin, P; Aubert, JP. The biological degradation of cellulose. *FEMS Microbiol Rev*, 1994, 13, 25-58.

Beguin, P; Gilkes, NR; Kilburn, DG; Miller, RC; O'Neil, GP; Antony, R; Warren, J. Cloning of cellulase genes. *Crit Rev Biotechnol*, 1987, 6, 129-162.

Béguin, P; Lemaire, M. The cellulosome: an exocellular, multiprotein complex specialized in cellulose degradation. *Crit Rev Biochem Mol Biol*, 1996, 31, 201-236.

Bergquist, PL; Gibbs, MD; Morris, DD; Te'o, VSJ; Saul, DJ; Morgan, HW. *Molecular diversity of thermophilic cellulolytic and hemicellulolytic bacteria. FEMS Microbiol Ecol*, 1999, 28, 99–110.

Beukes, NB; Pletschke, I. Effect of sulfur-containing compounds on *Bacillus cellulosome*-associated 'CMCase' and 'Avicelase' activities. *FEMS Microbiol Lett*, 2006, 264, 226-231.

Bhat, MK; McCrae, SI; Wood, TM .The endo-(1→4)-β-D-glucanase system of *Penicillium pinophilum* cellulase: Isolation, purification, and characterization of five major endoglucanase components. *Carbohydr Res,* 1989, 190, 279-297.

Bhat, MK. Cellulases and related enzymes in biotechnology. *Biotechnol Adv*, 2000, 18, 355-383.

Bhat, MK; Hazlewood, GP. Enzymology and other characteristics of cellulases and xylanases. Page 11 In: Bedford M, Partridge G editors. *Enzymes in farm animal nutrition.* Oxon, U.K: CABI Publishing; 2001; 1-11.

Bhat, MK; Bhat, S. Cellulose degrading enzymes and their potential industrial applications. *Biotechnol Adv,* 1997, 15, 583-620.

Bajpai, P. Application of enzymes in the pulp and paper industry. *Biotechnol Prog,* 1999, 15, 147-157.

Biely, P; Vršanska, M; Claeyssens, M. The endo-1,4-beta-glucanase I from *Trichoderma reesei*. Eur J Biochem, 1991, 200, 157-163.

Biswas, GCG; Ransom, C; Sticklen, M. Expression of biologically active *Acidothermus cellulolyticus* endoglucanase in transgenic maize plants. *Plant Sci*, 2006, 171, 617-623.

Blackall, LL; Hayward, AC; Sly, LI. Cellulolytic and dextranolytic gram-negative bacteria: revival of the genus *Cellvibrio. J Appl Bacteriol*, 1985, 59, 81-97.

Blake, CCF; Koenig, DF; Mair, GA; North, ACT; Phillips, DC; Sarma, VR. Structure of hen egg white lysozyme. A threedimensional Fourier synthesis at 2 A resolution. *Nature*, 1965,206, 757-763.

Boder, ET; Wittrup, KD. Yeast surface display for screening combinatorial polypeptide libraries. *Nat Biotechnol*, 1997, 15, 553–557.

Bogorad L. Engineering chloroplasts: an alternative site for foreign genes, proteins, reactions and products. *A trends Guide to Proteomics II*, 2000, 18, 257-263.

Bok, JD; Dinesh, A; Yernool, DA; Eveleigh, DE. Purification, characterization, and molecular analysis of thermostable cellulases CelA and CelB from *Thermotoga neapolitana*. *Appl Environ Microbiol*, 1998, 64, 4774-4781.

Bowman, GR; Beauchemin, K A; Shelford, J A. The proportion of the diet to which fibrolytic enzymes are added affects nutrient digestion by lactating dairy cows. *J Dairy Sci*, 2002, 85, 3420-3429.

Broda, P; Birch, PRJ; Brooks, PR; Sims, PFG. Lignocellulose degradation by *Phanerochaete chrysosporium:* gene families and gene expression for a complex process. *Mol Microbiol*, 1996, 19, 923-932.

Bronnenmeier, K; Kern, A; Liebl, W; Staudenbauer, WL. Purification of *Thermotoga maritima* enzymes for the degradation of cellulosic materials. *Appl Environ Microbiol*, 1995, 61, 1399-1407.

Bryant, MP; Bacterial species of the rumen. *Bacteriol Rev,* 1959, 23, 125-153.

Buchert, J; Koponen, JM; Suutarinen, M; Mustranta, A; Lille, M; Törrönen, R; Poutanen, K. Effect of enzyme aided pressing on anthocyanin yield and profiles in bilberry and blackcurrant juices. *J Sci Food Agric*, 2005, 85, 2548-2556.

Buchert, J; Oksanen, J; Pere, J; Siika-Aho, M; Suurnäkki, A; Viikari, L. Applications of *Trichoderma reesei* enzymes in the pulp and paper industry.In: Harman GE, Kubicek CP editors. *Trichoderma and Gliocladium*. London: Taylor & Francis. 1998;343-364.

Bukhtojarov, FE; Ustinov, BB; Salanovich, TN; Antonov, AI; Gusakov, AV; Okunev, ON; Sinitsyn, AP. Cellulase complex of the fungus *Chrysosporium lucknowense*: isolation and characterization of endoglucanases and cellobiohydrolases. *Biochemistry* (Mosc), 2004, 69, 542-551.

Cadena, EM; Vidal, T; Torres, AL. Influence of the hexenuronic acid content on refining and ageing in eucalyptus TCF pulp, *Biores Technol*, 010, 101, 3554-3560.

Cannio, R; Di Prizito, N; Rossi, M; Morana A. A xylan-degrading strain of *Sulfolobus solfataricus:* isolation and characterization of the xylanase activity. *Extremophiles*, 2004, 8, 117-124.

Carneiro, MSC; Lordelo, MM; Cunha, LF; Freire, JPB. Effects of dietary fibre source and enzyme supplementation on faecal apparent digestibility, short chain fatty acid production and activity of bacterial enzymes in the gut of piglets. *Anim Feed Sci Technol*, 2008, 146, 124-136.

Cavaco-Paulo, A. Mechanism of cellulase action in textile processes. *Carbohydr Polym*, 1998, 37, 273-277.

Cavicchioli, R; Watson, K. Molecular cloning, expression, and characterization of endoglucanase genes from *Fibrobacter succinogenes AR1. Appl Environ Microbiol*, 1991, 57, 359-365.

Chabra, SR; Kelly, RM. Biochemical characterization of *Thermotoga maritima* endoglucanase Cel74 with and without a carbohydrate binding module (CBM). *FEBS Lett,* 2002, 531, 375-380.

Chen, BY; Wang, HT. Utility of enzymes from *Fibrobacter succinogenes and Prevotella ruminicola* as detergent additives. *J Ind Microbiol Biotechnol*, 2008, 35, 923-930.

Chen, H; Li, D; Blum, L; Ljungdahl, LG. Two genes of the anaerobic fungus *Orpinomyces sp.* strain PC-2 encoding cellulases with endoglucanase activities may have arisen by gene duplication. *FEMS Microbiol Lett,* 1998, 159, 63-68.

Chen, H; Xu, B; Liao, JH; Guan, XY.; Wu, Q. Expression of endoglucanase gene in *Bacillus Megaterium* and characterization of recombinant enzyme. *Yi Chuan*, 2008, 30, 649-654.

Chen, PJ; Wei, TC; Chang, YT; Lin, LP. Purification and characterization of carboxymethyl cellulase from *Sinorhizobium fredii. Bot Bull Acad Sinica*, 2004, 45, 111-118.

Chesson, A; Forsberg, CW. Polysaccharide degradation by rumen microorganisms. In: Hobson PN editors. *The Rumen Microbial Ecosystem.* London :Elsevier Applied Science;1997; 251-284.

Chiacchierini, E; Mele, G; Restuccia, D; Vinci, G. Impact evaluation of innovative and sustainable extraction technologies on olive oil quality. *Trends Food Sci Technol*, 2007, 18, 299-305.

Christakopoulos, P; Kekos, D; Macris, BJ; Claeyssens, M; Bhat, MK. Purification and mode of action of a low molecular mass endo-1,4-β-D-glucanase from *Fusarium oxysporum. J Biotechnol,* 1995, 39, 85-93.

Claeyssens, M; Nerinckx, W; Piens, K. Carbohydrases from *Trichoderma reesei and other microorganisms: structures, biochemistry, genetics and applications.* Cambridge: The Royal Society of Chemistry, 1998.

ClaeyssensM ; Tomme P. Structure-activity relationships in cellulolytic enzymes. In: Coughlan MP, editor. *Enzyme systems in biomass conversion.* London: Elsevier Applied Science; 1989; 37–49.

Clausen, CA. Dissociation of the multienzyme complex of the brown-rot fungus *Postia placenta. FEMS Microbiol Lett*, 1995, 127, 73-78.

Cohen, GN; Barbe, V; Flament, D; Galperin, M; Heilig, R; Lecompte, O; Poch, O; Prieur, D; Querellou, J; Ripp, R et al. An integrated analysis of the genome of the hyperthermophilic archaeon *Pyrococcus abyssi. Mol Microbiol*, 2003, 47, 1495-1512.

Cohen, R; Suzuki, MR; Hammel, KE. Processive endoglucanase active in crystalline cellulose hydrolysis by the brown rot basidiomycete *Gloeophyllum trabeum. Appl Environ Microbio*l, 2005, 71, 2412-2417.

Conesa, A; Punt, PJ; van Luijk, N; van den Hondel, CAMJJ. The secretion pathway in filamentous fungi: a biotechnological view. *Fung Genet Biol*, 2001, 33, 155-171.

Cooney, DG; Emerson, R. *Thermophilic fungi: an account of their biology, activities and classification.* San Francisco: Freeman WH; 1964.

Cornillot, E; Croux, C; Soucaille, P. Physical and genetic map of the *Clostridium acetobutylicum* ATCC 824 chromosome. *J Bacteriol,* 1997, 179, 7426-7434.

Coughlan, MP. The properties of fungal and bacterial cellulases with comment on their production and application. *Biotechnol Genet Eng Rev*, 1985b, 3, 39-109.

Coughlan, MP. Mechanisms of cellulose degradation by fungi and bacteria *J Anim Feed Sci* 1991, 32, 77-100.

Coutinho, PM; Henrissat, B. (1999) The modular structure of cellulases and other carbohydrate-active enzymes: an integrated database approach. In: Ohmiya K, Hayashi K, Sakka K, Kobayashi Y, Karita S, Kimura T editors. *Genetics, biochemistry and ecology of cellulose degradation*, Uni Publishers Co., Tokyo, Japan: 1999; 15–23.

Coutinho, PM; Henrissat, B. In: Gilbert HJ, Davies G, Henrissat B, Svensson B. *Recent advances in carbohydrate bioengineering.* Cambridge The Royal Society of Chemistry; 1999.

Covert, SF; Wymelenberg, AV; Cullen, D. Structure, organization, and transcription of a cellobiohydrolase gene cluster from *Phanerochaete chrysosporium. Appl Environ Microbiol*, 1992, 58, 2168-2175.

Darland, G; Brock, TD. *Bacillus acidocaldarius sp.nov.,* an acidophilic thermophilic spore-forming bacterium. *J Gen Microbiol*, 1971, 67, 9-15.

Dashtban, M ; Schraft, H; Qin, W. Fungal bioconversion of lignocellulosic residues; opportunities & perspectives. *Review. Int J Biol Sci*, 2009, 5, 578-595

Davies, GJ; Brzozowski, AM; Dauter, M; Varrot, A; Schülein, M. Structure and function of *Humicola insolens* family 6 cellulases: structure of the endoglucanase, Cel6B, at 1.6 A resolution. *Biochem J,* 2000, 348, 201-207.

de Castro, AM; de Albuquerque de Carvalho, ML; Leite, SG; Pereira, N Jr. Cellulases from *Penicillium funiculosum:* production, properties and application to cellulose hydrolysis. *J Ind Microbiol Biotechnol*, 2010, 37, 151-158.

De Rosa, M; Gambacorta, A; Bu'lock, JD. Extremely thermophilic acidophilic bacteria convergent with *Sulfolobus acidocaldarius. J Gen Microbiol*, 1975, 86, 156-164.

Deckert, G; Warren, PV; Gaasterland, T; Young, WG; Lenox, AL; Graham, DE; Overbeek, R; Snead, MA; Keller, M; Aujay, M; Huber, R; Feldman, RA; Short, JM; Olsen, GJ; Swanson, RV. The complete genome of the hyperthermophilic bacterium *Aquifex aeolicus. Nature*, 1998, 392, 353-358.

Demain, AL; Newcomb, M.; Wu, JH. *Cellulase, clostridia, and ethanol. Microbiol Mol Biol Rev*, 2005, 69, 124–154.

Den Haan, R; Rose, SH; Lynd, LR.; van Zyl, WH. Hydrolysis and fermentation of amorphous cellulose by recombinant *Saccharomyces cerevisiae. Metab Eng*, 2007, 9, 87–94.

Denman, S; Xue, XL; Patel, B. Characterization of a *Neocallimastix patriciarum* cellulase cDNA (celA) homologous to *Trichoderma reesei* cellobiohydrolase II. *Appl Environ Microbiol*, 1996, 62, 1889-1896.

Deobald, LA; Crawford, DL. Lignocellulose biodegradation. In: Hurst CJ, Knudsen GR, Stetzenbach LD, Walter MV editors. *Manual of Environmental Microbiology.* Washington DC: ASM Press, 1997; 730–737.

Dijkerman, R; Op den Camp, HJM; van der Drift, C; Vogels GD. The role of the cellulolytic highmolecular mass (HMM) complex of the anaerobic fungus *Piromyces sp.* strain E2 in the hydrolysis of microcrystalline cellulose. *Arch Microbiol*, 1997, 167, 137-142.

Doi, RH. Cellulases of mesophilic microorganisms: cellulosome and noncellulosome producers. *Ann N Y Acad Sci*, 2008, 1125, 267-279.

Doi, RH; Goldstein, M; Hashida, S; Park, JS; Takagi, M. The *Clostridium cellulovorans* cellulosome. *Crit Rev Microbiol*, 1994, 20, 87-93.

Dongowski, G; Sembries, S. Effects of Commercial Pectolytic and Cellulolytic Enzyme Preparations on the Apple Cell Wall. *J Agric Food Chem*, 2001, 49, 4236- 4242.

Dubordieu, D; Ribereau-Gayon, P; Fournet, B. Structure of the exocellular ß-D-glucan from *Botrytis cinerea*. *Carbohydr Res*, 1981, 93,294-9.

Ducasse, MA; Canal-Llauberes, RM; de Lumley, M; Williams, P; Souquet, JM; Fulcrand, H; Doco, T; Cheynier, V. Effect of macerating enzyme treatment on the polyphenol and polysaccharide composition of red wines. *Food Chem*, 2010, 118, 369-376.

Eberhardt, RY; Gilbert, HJ; Hazlewood, GP. Primary sequence and enzymic properties of two modular endoglucanases, Cel5A and Cel45A, from the anaerobic fungus *Piromyces equi*. *Microbiology*, 2000, 146, 1999-2008.

Eckert, K; Schneider, EA. Thermoacidophilic endoglucanase (CelB) from *Alicyclobacillus acidocaldarius* displays high sequence similarity to arabinofuranosidases belonging to family 51 of glycoside hydrolases. *Eur J Biochem*, 2003, 270, 3593-3602.

Eckert, K; Zielinski, F; Lo Leggio, L; Schneider, E. Gene cloning, sequencing, and characterization of a family 9 endoglucanase (CelA) with an unusual pattern of activity from the thermoacidophile *Alicyclobacillus acidocaldarius ATCC27009*. *Appl Microbiol Biotechnol*, 2002, 60, 428-436.

Elberson, MA; Malekzadeh, F; Yazdi, MT et al. *Cellulomonas persica sp. nov.* and *Cellulomonas iranensis sp. nov.*, mesophilic cellulose-degrading bacteria isolated from forest soils. *Int J Syst Evol Microbiol*, 2000, 50, 993-996.

Elshafei, AM; Hassan, MM; Haroun, BM; Abdel-Fatah, OM; Atta, HM; Othman, AM. Purification and properties of an endoglucanase of *Aspergillus terreus DSM 826*. *J Basic Microbiol*, 2009, 49, 426-432.

Endo, K; Hakamada, Y; Takizawa, S; Kubota, H; Sumitomo, N; Kobayashi, T; Ito, S. A novel alkaline endoglucanase from an alkaliphilic Bacillus isolate: enzymatic properties, and nucleotide and deduced amino acid sequences. *Appl Microbiol Biotechnol*, 2001, 57, 109-116.

Enger, MD; Sleeper, BP. Multiple cellulase system from *Streptomyces antibioticus*. *J Bacteriol*, 1965, 89, 23-27.

Eriksen, J; Goksoyr, J. Cellulases from *Chaetomium thermophile var. dissitum*. *Eur J Biochem*, 1977, 445-450.

Eriksson, KE; Pettersson, B. Extracellular enzyme system utilized by the fungus *Sporotrichum pulverulentum (Chrysosporium lignorum)* for the breakdown of cellulose. 1. Separation, purification, and physico-chemical characterization of five endo-1,4-beta-glucanases. Eur J Biochem, 1975a, 51, 193-206.

Eriksson, KE; Pettersson, B. Extracellular enzyme system used by the fungus *Sporotrichum pulverulentum (Chrysosporium lignorum)* for the breakdown of cellulose. 3. Purification, physicochemical characterization of an exo-1,4-beta-glucanases. *Eur J Biochem*, 1975b, 51, 213-218.

Fantozzi, P; Petruccioli, G; Montedoro, G. *Trattamenti con additivi enzimatici alle paste di oliva sottoposte ad estrazione, per pressione unica: Influenze delle cultivars, dell'epoca di raccolta e della conservazione.* Riv Ital Sost Grasse, 1977, 54, 381–388.

Fauth, U; Romaniec, MP; Kobayashi, T; Demain, AL. Purification and characterization of endoglucanase Ss from *Clostridium thermocellum*. *Biochem J*, 1991, 279, 67-73.

Fei, W; Sun, J; Zhao, H. Yeast Surface Display of Trifunctional Minicellulosomes for Simultaneous Saccharification and Fermentation of Cellulose to Ethanol. *Appl Environ Microbiol,* 2010, 76, 1251-1260.

Fengel, D; Wegener, G.1984. *Wood: Chemistry, Ultrastructure, Reactions.* De Gruyter, Berlin.

Fiala, G; Stetter, KO. *Pyrococcus furiosus sp. nov.* represents a novel genus of marine heterotrophic archaebacteria growing optimally at 100°C. *Arch Microbiol*, 1986, 145, 56-61.

Fierobe, HP; Mingardon, F; Mechaly, A; Bélaïch, A; Rincon, MT; Pagés, S; Lamed, R; Tardif, C; Bélaïch, JP; Bayer, EA. Action of designer cellulosomes on homogeneous versus complex substrates: controlled incorporation of three distinct enzymes into a defined trifunctional caffoldin. *J Biol Chem*, 2005, 280, 16325-16334.

Fontes, CMGA.; Gilbert, HJ. Efficient Nanomachines Designed to Deconstruct Plant Cell Wall Complex Carbohydrates. *Annu Rev Biochem*, 2010, 79, 23.1–23.27.

Foreman, PK; Brown, D; Dankmeyer, L; Dean, R; Diener, S; Dunn-Coleman, NS; Goedegebuur, F; Houfek, TD; England, GJ; Kelley, AS; Meerman, HJ; Mitchell, T; Mitchinson, C; Olivares, HA; Teunissen, PJM; Yao, J; Ward, M. Transcriptional regulation of biomass-degrading enzymes in the filamentous fungus *Trichoderma reesei. J Biol Chem*, 2003, 278, 31988-31997.

Franke, AA; Cooney, RV; Henning, SM; Custer, LJ. Bioavailability and antioxidant effects of orange juice components in humans. *J Agric Food Chem*, 2005, 53, 5170-5178.

Fujii, T; Fang, X; Inoue, H; Murakami, K; Sawayama, S. Enzymatic hydrolyzing performance of *Acremonium cellulolyticus* and *Trichoderma reesei* against three lignocellulosic materials. *Biotechnol Biofuels*, 2009, 2, 1-24.

Fujino, T; Béguin, P; Aubert, JP. Organization of a *Clostridium thermocellum* gene cluster encoding the cellulosomal scaffolding protein CipA and a protein possibly involved in attachment of the cellulosome to the cell surface. *J Bacteriol*, 1993, 175, 1891-1899.

Fujino, T; Sasaki, T; Ohmiya, K; Shimizu, S. Purification and properties of an endo-1,4-beta-glucanase translated from a *Clostridium* josui gene in *Escherichia coli. Appl Environ Microbiol*, 1990, 56, 1175-1178.

Fujita, Y; Ito, J; Ueda, M; Fukuda, H; Kondo, A. Synergistic saccharification, and direct fermentation to ethanol, of amorphous cellulose by use of an engineered yeast strain codisplaying three types of cellulolytic enzyme. *Appl Environ Microbiol*, 2004, 70, 1207-1212.

Fukui, T; Atomi, H; Kanai, T; Matsumi, R; Fujiwara, S; Imanaka, T. Complete genome sequence of the hyperthermophilic archaeon *Thermococcus kodakaraensis KOD1* and comparison with *Pyrococcus* genomes. *Genome Res*, 2005, 15, 352-363.

Fukumori, F; Sashihara, N; Kudo, T; Horikoshi, K. Nucleotide sequence of two cellulase genes from alkalophilic *Bacillus sp.* strain N-4 and their strong homology. *J Bacteriol*, 1986, 168, 479-485.

Galante, YM; De Conti, A; Monteverdi, R. *Application of Trichoderma* enzymes in food and feed industries. In:Harman GF, Kubicek CP, editors. *Trichoderma and Gliocladium-Enzymes, biological control and commercial applications.* London: Taylor & Francis; 1998a; 327-342.

Galante, YM; De Conti, A; Monteverdi, R. Application of *Trichoderma* enzymes in textile industry. In: Harman GF, Kubicek CP, editors. *Trichoderma & Gliocladium-Enzymes, biological control and commercial applications*. Vol. 2. London: Taylor & Francis, 1998b. pp. 311–326.

Ganju, RK; Murthy, SK; Vithayathil, PJ. Purification and functional characteristics of an endocellulase from *Chaetomium thermophile var. coprophile*. *Carbohydr Res,* 1990, 197, 245-255.

Gao, Z; Ruan, L; Chen, X; Zhang, Y; Xu, X. A novel salt-tolerant endo-beta-1,4-glucanase Cel5A in *Vibrio sp.* G21 isolated from mangrove soil. *Appl Microbiol Biotechnol*, 2010, 87, 1373-1382.

Garcia, A; Brenes, M; Moyano, MJ; Alba, J; Garcia, P; Garrido, A. Improvement of phenolic compound content in virgin olive oils by using enzymes during malaxation. *Journal of Food Engineering*, 2001, 48, 189-194.

Gardner, KH; Blackwell, J. T*he structure of native cellulose Biopolymers*, 2004,13, 1975-2001.

Gerbi, C; Bata, J; Breton, A; Prensier, G. Glycoside and polysaccharide hydrolase activity of the rumen anaerobic fungus *Caecomyces communis (Sphaeromonas communis SENSU ORPIN)* at early and final stages of the developmental cycle. *Curr Microbiol*, 1996, 32, 256–259.

Gerngross, UT; Romaniec, MP; Kobayashi, T; Huskisson, NS; Demain, AL. Sequencing of a *Clostridium thermocellum* gene (cipA) encoding the cellulosomal SL-protein reveals an unusual degree of internal homology. *Mol Microbiol,* 1993, 8, 325–334.

Ghangas, GS; Wilson DB. Cloning of the *Thermomonospora fusca endoglucanase* E2 gene in *Streptomyces lividans:* affinity purification and functional domains of cloned product. *Appl Environ Microbiol*, 1988, 54, 2521-2526.

Gilad, R; Rabinovich, L; Yaron, S; Bayer, EA; Lamed, R; Gilbert, HJ; Shoham, Y. CelI, a noncellulosomal family 9 enzyme from *Clostridium thermocellum,* is a processive endoglucanase that degrades crystalline cellulose. *J Bacteriol*, 2003, 185, 391-398.

Gilbert, HJ. Cellulosomes: microbial nanomachines that display plasticità in quaternari structure. *Molec Microbiol*, 2007, 63, 1568-1576.

Gilkes, NR; Kilburn, DG; Miller, RC; Warren, RAJ. Bacterial cellulases. *Bioresour Technol*, 1991, 36, 21-35.

Gilkes, NR; Warren, RAJ; Miller, RC Jr.; Kilburn, DG. Precise excision of the cellulose binding domains from two *Cellulomonas fimi* cellulases by a homologous protease and the effect on catalysis. *J Biol Chem*, 1988, 263, 10401–10407.

Gomes, I; Sarkar, PK; Rahman, SR; Rahim, MA; Gomes, DJ. Production of Cellulase from *Talaromyces emersonii* and evaluation of its application in eco-friendly functional finishing of jute-based fabrics. *Bangladesh J Microbiol*, 2007, 24, 109-114.

Gonzalez, JM; Masuchi, Y; Robb, FT; Ammerman, JW; Maeder, DL; et al. *Pyrococcus horikoshii sp. nov., a hyperthermophilic archaeon isolated from a hydrothermal vent at the Okinawa Trough. Extremophiles*, 1998, 2, 123-130.

Graham, H; Inborr, J. Application of xylanase-based enzymes in commercial pig and poultry production. In: Visser J, Beldman G, Kusters-van Someren MA, Voragen AGJ editors. *Xylan and Xylanases; Progress in Biotechnology*. Amsterdam: Elsevier; 1992; 535-538.

Gray B; Ahner, BA; Hanson, MR. High-level bacterial cellulase accumulation in chloroplast-transformed tobacco mediated by downstream box fusions. *Biotechnol Bioeng*, 2009, 4, 1045-1054.

Guadalupe, Z; Palacios, A; Ayestaran, B. Maceration enzymes and mannoproteins: A possible strategy to increase colloidal stability and color extraction in red wines. *J Agric Food Chem*, 2007, 55, 4854-4862.

Gubitz, GM; Mansfield, S.D; Bohm, D; Saddler, J.N.Effect of endoglucanases and hemicellulases in magnetic and flotation deinking of xerographic and laser-printed papers. 1998, *J Biotechnol*, 65, 209-215.

Guo, Q; Zhang, W; Ma, L; Chen, Q; Chen, J; Zhang, H; Ruan, H; He, G. A food-grade industrial arming yeast expressing β-1,3-1,4-glucanase with enhanced thermal stability. *J Zhejiang Univ Sci B*, 2010, 11,41-51.

Guo, R; Ding, M; Zhang, SL; Xu, GJ; Zhao, FK. Molecular cloning and characterization of two novel cellulase genes from the mollusk *Ampullaria crossean*. *J Comp Physiol [B]*, 2008,178, 209-215.

Haight, KG; Gump, BH. The use of macerating enzymes in grape juice processing. *Am J Enol Vitic*, 1994, 45, 113-116.

Hakamada, Y; Endo, K; Takizawa, S; Kobayashi, T; Shirai, T; Yamane, T et al. Enzymatic properties, crystallization, and deduced amino acid sequence of an alkaline endoglucanase from *Bacillus circulans*. *Biochim Biophys Acta*, 2002, 1570, 174-180.

Halldórsdóttir, S; Thórólfsdóttir, ET; Spilliaert, R; Johansson, M; Thorbjarnardóttir, SH; Palsdottir, A et al. Cloning, sequencing and overexpression of a *Rhodothermus marinus* gene encoding a thermostable cellulase of glycosyl hydrolase family 12. *Appl Microbiol Biotechnol*, 1998, 49, 277-284.

Hamelinck, CN; van Hooijdonk, G; Faaij, APC. Ethanol from lignocellulosic biomass: techno-economic performance in short-, middle- and long-term. *Biomass Bioen*, 2005, 28, 384-410.

Han, SO; Yukawa, H; Inui, M; Doi RH. Transcription of Clostridium cellulovorans cellulosomal cellulase and hemicellulase genes. *J Bacteriol*, 2003, 185, 2520-2527.

Han, SO; Yukawa, H; Inui, M; Doi, RH. Effect of carbon source on the cellulosomal subpopulations of Clostridium cellulovorans. *Microbiology*, 2005, 151, 1491–1497.

Hasper, AA; Dekkers, E; van Mil, M; van de Vondervoort, PJ; de Graaff, LH. *EglC, a new endoglucanase from Aspergillus niger with major activity towards xyloglucan*. *Appl Environ Microbiol*, 2002, 68, 1556-1560.

Hassan, MM; Elshafei, AM; Haroun, BM; Atta, HM; Abdel-Fatah, OM; Othman, AM. Screening of different fungal strains for cellulose formation using different agricultural wastes. *N Egypt J Microbiol*, 2008, 19, 210-225.

Hastrup, ACS; Jensen, B; Clausen , C; Green, F. The effect of CaCl2 on growth rate, wood decay and oxalic acid accumulaion in Serpula lacrymans and related brown-rot fungi. *Holzforshung*, 2006, 60, 339-345.

Hazlewood, GP; Laurie, JL; Ferreira, LM; Gilbert, HJ. *Pseudomonas fluorescens subsp.* cellulosa: an alternative model for bacterial cellulase. *J Appl Bacteriol*, 1992, 72, 244-251.

He, SY; Schoedel, C; Chatterjee, AK; Collmer, A. Cloned Erwinia chrysanthemi out genes enable Escherichia coli to selectively secrete a diverse family of heterologous proteins to its milieu. *Proc Natl Acad Sci USA*, 1991, 88, 1079–1083.

He, YL; Ding, YF; Long, YQ. Two cellulolytic Clostridium species: Clostridium cellulosi sp. nov. and Clostridium cellulofermentans sp. nov. *Int J Syst Bacteriol*, 1991, 41, 306-309.

Henriksson, G; Nutt, A; Henriksson, H; Pettersson, B; Stahlberg, J; Johansson, G; Pettersson, G. Endoglucanase 28 (Cel12A), a new Phanerochaete chrysosporium cellulase. *Eur J Biochem,* 1999, 259, 88-95.

Henriksson, A;Gatenholm, P. Suface properties of CTMP fibers modified with xylans, *Cellulose*, 2002, 9, 55-64.

Henrissat, B. A classification of glycosyl hydrolases based on amino acid sequence similarities. *Biochem J*, 1991, 280, 309-316.

Henrissat, B; Bairoch, A. *New families in the classification of glycosyl hydrolases based on amino acid sequence similarities. Biochem J*, 1993, 293, 781-788.

Henrissat, B; Bairoch, A. Updating the sequence-based classification of glycosyl hydrolases. *Biochem J,* 1996, 316, 695-696.

Henrissat, B; Teeri, TT; Warren, RAJ. A scheme for designating enzymes that hydrolyse the polysaccharides in the cell walls of plants. *FEBS Lett,* 1998, 425, 352-354.

Hill, J; Nelson, E; Tilman, D; Polasky, S; Tiffany, D. Environmental, economic, and energetic costs and benefits of biodiesel and ethanol biofuels. *Proc Natl Acad Sci U S A,* 2006, 103, 11206-11210.

Hinz, SWA; Pouvreau, L; Joosten, R; Bartels, J; Jonathan, M C; Wery, J; Schols, HA. *Hemicellulase production in Chrysosporium lucknowense C1, J Cereal Sci*, 2009, 50, 1-6.

Hirvonen, M; Papageorgiou, AC. Crystallization and preliminary crystallographic analysis of a family 45 endoglucanase from the thermophilic fungus Melanocarpus albomyces. *Acta Crystallogr D Biol Crystallogr*, 2002, 58, 336-338.

Hong, J; Tamaki, H; Yamamoto, K; Kumagai, H. Cloning of a gene encoding a thermostable endo-beta-1,4-glucanase from Thermoascus aurantiacus and its expression in yeast. *Biotechnol Lett,* 2003, 25, 657-661.

Hongpattarakere, T. Hyperthermostable cellulolytic and hemicellulolytic enzymes and their biotechnological applications. *J Sci Technol*, 2002, 24, 481-491.

Hood, EE; Love, R; Lane, A; Bray, J; Clough, R; Pappu, K; Drees, C; Hood, KR; Yoon, S; Ahmad, A; Howard, JA. Subcellular targeting is a key condition for high-level accumulation of cellulase protein in transgenic maize seed. *Plant Biotechnol J*, 2007, 5, 709–719.

Horikoshi, K; Nakao, M; Kurono, Y; Sashihara, N. Cellulase of an alkalophilic Bacillus strain isolated from soil. *Can J Microbiol*, 1984, 30, 774-779.

Hreggvidsson, GO; Kaiste E; Holst, O; Eggertsson, G; Palsdottir, A; Kristjansson, JK. An extremely thermostable cellulase from the thermophilic eubacterium Rhodothermus marinus. *Appl Environ Microbiol,* 1996, 62, 3047-3049.

Huang, XP; Monk, C. Purification and characterization of a cellulase (CMCase) from a newly isolated thermophilic aerobic bacterium Caldibacillus cellulovorans gen. nov., sp. nov. *World J Microbiol Biotechnol*, 2004, 20, 85-92.

Huang, Y; Krauss, G; Cottaz, S; Driguez, H; Lipps G. A highly acid-stable and thermostable endo-beta-glucanase from the thermoacidophilic archaeon Sulfolobus solfataricus. *Biochem J*, 2005, 385, 581-588.

Huber, R; Langworthy TA; Konig, H; Thomm, M; Woese, CR; Sleyter UB; Stetter, CO. Thermotoga maritima sp. nov. represents a new genus of unique extremely thermophilic eubacteria growing up to 90°C. *Arch Microbiol*, 1986, 144, 324-333.

Hui, JPM; White, TC; Thibault, P. Identification of glycan structure and glycosylation sites in cellobiohydrolase II and endoglucanase I and II from Trichoderma reesei. *Glycobiol*, 2002, 12, 837-849.

Igarashi, K; Ishida, T; Hori, C; Samejima, M. Characterization of an endoglucanase belonging to a new subfamily of glycoside hydrolase family 45 of the basidiomycete Phanerochaete chrysosporium. *Appl Environ Microbiol*, 2008, 74, 5628-5634.

Ikeda, Y; Hayashi, H; Okuda, N; Park, EY. Efficient cellulase production by the filamentous fungus Acremonium cellulolyticus. *Biotechnol Prog,* 2007, 23, 333-338.

Irwin, DC; Spezio, M; Walker, LP; Wilson, DB. Activity studies of eight purified cellulases: specificity, synergism, and binding domain effects. *Biotechnol Bioeng*, 1993, 42, 1002–1013.

Irwin, DC; Zhang, S; Wilson, DB. Cloning, expression and characterization of a family 48 exocellulase, Cel48A, from Thermobifida fusca. *Eur J Biochem*, 2000, 267, 4988-4997.

Ito, S. Alkaline cellulase from alkalophilic Bacillus: enzymatic properties, genetics, and application to detergents. *Extremophiles*, 1997, 1, 61-66.

Jagtap, S; Rao, M. Purification and properties of a low molecular weight 1,4-beta-d-glucan glucohydrolase having one active site for carboxymethyl cellulose and xylan from an alkalothermophilic Thermomonospora sp. *Biochem Biophys Res Commun*, 2005, 329, 111-116.

Jannasch, HW; Huber, R; Belkin, S; Stetter, KO. Thermotoga neapolitana sp. nov. of the extremely thermophilic, eubacterial genus Thermotoga. *Arch Microbiol*, 1988, 150, 103-110.

Jatinder, K; Chadha, BS; Saini, HS. Optimization of medium components for production of cellulases by Melanocarpus sp. MTCC 3922 under solid-state fermentation. *World J Microbiol Biotechnol*, 2006, 22, 15-22.

Jensen, BR. *Mixed β-glucanase, xylanase enzyme preparation, produced by a strain of Humicola insolens.* Ultraflo^{TM}L. Unpublished report No. 2002-37749-01 from Novozymes A/S, 2002, Bagsvaerd, Denmark.

Jeya, M; Joo, AR; Lee, KM; Sim, WI; Oh, DK; Kim, YS et al. Characterization of endo-beta-1,4-glucanase from a novel strain of Penicillium pinophilum KMJ601. *Appl Microbiol Biotechnol,* 2010, 85, 1005-1014.

Jin, R; Richter, S; Zhong, R; Lamppa, GK. Expression and import of an active cellulase from a thermophilic bacterium into the chloroplast both in vitro and in vivo. *Plant Mol Biol*, 2003, 51, 493–507.

Jindou, S; Brulc, JM.; Levy-Assaraf, M.; Rincon, MT.; Flint, HJ; Berg, ME et al. Cellulosome gene cluster analysis for gauging the diversity of the ruminal cellulolytic bacterium Ruminococcus flavefaciens. *FEMS Microbiol Lett,* 2008, 285, 188–194.

Jorgensen, H; Kristensen, JB; Felby, C. Enzymatic conversion of lignocellulose into fermentable sugars: Challenges and opportunities. *Biofuels Bioprod Bioref*, 2007, 1, 119-134.

Jun, CW; Min, MZ; Sel, KM. Purification of the cellulase complex produced by Penicillium camemberti and its partial characterization. *Folia Microbiol* (Praha), 1992, 37, 199-204.

Kakiuchi, M; Isui, A; Suzuki, K; Fujino, T; Fujino, E; Kimura, T et al. Cloning and DNA sequencing of the genes encoding Clostridium josui scaffolding protein CipA and cellulase CelD and identification of their gene products as major components of the cellulosome. *J Bacteriol*, 1998, 180, 4303–4308.

Kang, MS; Im, WT; Jung, HM; Kim, MK; Goodfellow, M; Kim, KK et al. Cellulomonas composti sp. nov., a cellulolytic bacterium isolated from cattle farm compost. *Int J Syst Evol Microbiol,* 2007, 57, 1256-1260.

Kapasakalidis, PG; Rastall, RA; Gordon, MH. Effect of a cellulase treatment on extraction of antioxidant phenols from black currant (Ribes nigrum L.) pomace. *J Agric Food Chem*, 2009, 57, 4342-4351.

Karboune, S; Geraert, PA; Kermasha, S. Characterization of selected cellulolytic activities of multi-enzymatic complex system from Penicillium funiculosum. *J Agric Food Chem*,2008, 56, 903-909.

Karita, S; Nakayama, K; Goto, M; Sakka, K; Kim, WJ; Ogawa S. A novel cellulolytic, anaerobic, and thermophilic bacterium, Moorella sp. strain F21. *Biosci Biotechnol Biochem,* 2003, 67, 183-185.

Karlsson, J; Momcilovic, D; Wittgren, B; Schülein, M; Tjerneld, F; Brinkmalm, G. Enzymatic degradation of carboxymethyl cellulose hydrolyzed by the endoglucanases Cel5A, Cel7B, and Cel45A from Humicola insolens and Cel7B, Cel12A and Cel45Acore from Trichoderma reesei. *Biopolymer*, 2002, 63, 32-40.

Karlsson, J; Siika-aho, M; Tenkanen, M; Tjerneld, F. Enzymatic properties of the low molecular mass endoglucanases Cel12A (EG III) and Cel45A (EG V) of Trichoderma reesei. *J Biotechnol*, 2002, 99, 63-78.

Karlsson, J; Saloheimo, M; Siika-aho, M; Tenkanen, M; Penttilä, M; Tjerneld, F. Homologous expression and characterization of Cel61A (EG IV) of Trichoderma reesei. *FEBS J,* 2001, 268, 6498-6507.

Karnchanatat, A; Petsom, A; Sangvanich, P; Piapukiew, J; Whalley, JSA; Reynolds, DC et al. A novel thermostable endoglucanase from the wood-decaying fungus Daldinia eschscholzii (Ehrenb.:Fr.) Rehm. *Enzyme Microb Technol*, 2008, 42, 404-413.

Kashima, Y; Udaka, S. High-level production of hyperthermophilic cellulase in the Bacillus brevis expression and secretion system. *Biosci Biotechnol Biochem*, 2004, 68, 235-237.

Kaur, J; Chadha, BS; Kumar, BA; Saini, HS. Purification and characterization of two endoglucanases from Melanocarpus sp. MTCC 3922. *Bioresour Technol*, 2007, 98, 74-81.

Kelly, JA, Sielecki, AR; Sykes, BD; Philips, DC. X-ray crystallography of the binding of the bacterial cell wall trisaccharide NAM-NAG-NAM to lysozyme. *Nature*, 1979, 282, 875-878.

Khandke, KM; Vithayathil, PJ; Murthy, SK. Purification of xylanase, beta-glucosidase, endocellulase, and exocellulase from a thermophilic fungus, Thermoascus aurantiacus. *Arch Biochem Biophys*, 1989, 274, 491-500.

Kim, KC; Seung-Soo, Y; Oh Young, A; Seong-Jun, K. Isolation andcharacteristics of Trichoderma harzianum FJ1 producing cellulases and xylanase. *J Microbiol Biotechnol*, 2003, 13, 1-8.

Kim, S; Dale, BE. Global potential bioethanol production from wasted crops and crop residues. *Biomass Bioen,* 2004, 26, 361-375.

Kim, CH. Characterization and substrate specificity of an endo-beta-1,4-D-glucanase I (Avicelase I) from an extracellular multienzyme complex of Bacillus circulans. *Appl Environ Microbiol*, 1995, 61, 959-965.

Kim, HW; Ishikawa, K. Complete saccharification of cellulose at high temperature using endocellulase and beta-glucosidase from Pyrococcus sp. *J Microbiol Biotechnol*, 2010, 20, 889-892.

Kim, HW; Ishikawa, K. Structure of hyperthermophilic endocellulase from Pyrococcus horikoshii. *Proteins*, 2010, 78, 496-500.

Kim, JO; Park, SR; Lim, WJ; Ryu, SK; Kim, MK; An, CL et al. Cloning and characterization of thermostable endoglucanase (Cel8Y) from the hyperthermophilic Aquifex aeolicus VF5. *Biochem Biophys Res Commun*, 2000, 279, 420-426.

Kim, JY; Hur, SH; Hong, JH. Purification and characterization of an alkaline cellulase from a newly isolated alkalophilic Bacillus sp. HSH-810. *Biotechnol Lett*, 2005, 27, 313-316.

Kleman-Leyer, K; Kirk, TK. Three native cellulose depolymerizing endoglucanases from solid-substrate cultures of the brown-rot fungus Meruliporia (Serpula) incrassata. *Appl Environ Microbiol*, 1994, 60, 2839-2845.

Klemm, D; Schmauder, HP; Heinze, T. *In: Biopolymers:Polysaccharides II;* Vandamme E, De Baets A , Steinbu¨chel A, editors. Weinheim, Germany; Wiley-VCH; 2002; 275-319

Klemm, D; Heublein, B; Fink, HP; Bohn, A. Cellulose: Fascinating Biopolymer and Sustainable Raw Material. Review. *Polym Sci*, 2005, 44, 3358-3393.

Kleywegt, GJ; Zou, JY, Divne, C; Davies, GC; Sinning, I; Stahlberg, J et al. The crystal structure of the catalytic core domain of endoglucanase I from Trichoderma reesei at 3.6 A° resolution, and a comparison with related enzymes. *J Mol Biol*, 1997, 272, 383-397.

Kobayashi, Y; Shinkai, IT; Koike, S. Ecological and physiological characterization shows that Fibrobacter succinogenes is important in rumen fiber digestion. *Folia Microbiol*, 2008, 53, 195-200.

Kochavi, D; Videback, T; Cedronii, D. Optimization processing conditions in enzymatic stone-washing. *Ann Dyest Rep*, 1990, 9, 24-28.

Koshland, DE. Stereochemistry and the mechanism of enzymatic reactions. *Biol Rev*, 1953, 28,416-436.

Kosugi, A; Murashima, K; Doi, R. H. Xylanase and acetyl xylan esterase activities of XynA, a key subunit of the Clostridium cellulovorans cellulosome for xylan degradation. *Appl Environ Microbiol,* 2002, 68, 6399–6402.

Kosugi, A; Murashima, K; Doi, RH. Characterization of two noncellulosomal subunits, ArfA and BgaA, from Clostridium cellulovorans that cooperate with the cellulosome in plant cell wall degradation. *J Bacteriol*, 2002, 184, 6859–6865.

Kotchoni, OS; Shonukan, OO; Gachomo, WE. Bacillus pumilus BpCRI6, a promising candidate for cellulase production under conditions of catabolite repression. *Afr J Biotechnol,* 2003, 2, 40-146.

Kuhad, RC; Singh, A; Eriksson, KEL. Microorganisms and enzymes involved in the degradation of plant fiber cell walls. *Adv Biochem Eng Biotechnol*, 1997, 57, 45-125.

Kuhls, K; Lieckfeldt, E; SamuelsDagger, GJ et al. Molecular evidence that the asexual industrial fungus Trichoderma reesei is a clonal derivative of the ascomycete Hypocrea jecorina. *Proc Natl Acad Sci USA,* 1996, 93, 7755–7760.

Kumar, R; Singh, S; Singh, OV. Bioconversion of lignocellulosic biomass: biochemical and molecular perspectives. *J Ind Microbiol Biotechnol*, 2008, 35, 377–391.

Kurokawa, J; Hemjinda, E; Arai, T; Kimura, T; Sakka, K; Ohmiya, K. Clostridium thermocellum cellulase CelT, a family 9 endoglucanase without an Ig-like domain or family 3c carbohydrate-binding module. *Appl Microbiol Biotechnol*, 2002, 59, 455-461.

Kurowska, EM; Spence, JD; Jordan, J; Wetmore, S; Freeman, DJ; Piché, LA et al. HDL-cholesterol-raising effect of orange juice in subjects with hypercholesterolemia. *Am J Clin Nutr*, 2000, 72, 1095-1100.

Lamed, R; Setter, E; Kenig, R; Bayer, EA. The cellulosome - a discrete cell surface organelle of Clostridium thermocellum which exhibits separate antigenic, cellulose-binding and various cellulolytic activities. *Biotechnol Bioeng,* 1983, 13, 163-181.

Lampe, JW. Health effects of vegetables and fruit: assessing mechanisms of action in human experimental studies. *Am J Clinl Nutr,* 1999, 70, 475S-490S.

Langsford, ML; Gilkes, NR; Wakarchuk, WW et al. The cellulase system of Cellulomonas fimi. *J Gen Microbiol,* 1984, 130, 1367-1376.

Lao, G; Ghangas, GS; Jung, ED; Wilson, DB. DNA sequences of three β-1,4-endoglucanase genes from Thermomonospora fusca. *J Bacteriol*, 1991, 173, 3397-3407.

Larsson, AM; Bergfors, T; Dultz, E; Irwin, DC; Roos, A; Driguez, H et al. *Crystal structure of Thermobifida fusca endoglucanase Cel6A in complex with substrate and inhibitor.Biochemistry*, 2005, 44, 12915-12922.

Lee, SM ; Koo, YM.. Pilot-scale production of cellulose using Trichoderma reesei Rut C-30 in fed-batch mode, *J Microbiol Biotechnol*, 2001, 11, 229-233.

Lee, SS; Ha, JK; Cheng, KJ. The effects of sequential inoculation of mixed rumen protozoa on the degradation of orchard grass cell walls by anaerobic fungus Anaeromyces mucronatus 543. *Can J Microbiol*, 2001, 47, 754–760.

Lee, YJ; Kim, BK; Lee, BH; Jo, KI; Lee, NK; Chung, CH et al. Purification and characterization of cellulase produced by Bacillus amyoliquefaciens DL-3 utilizing rice hull. *Bioresour Technol*, 2008, 99, 378-386.

Lehninger, AL. *Principles of biochemistry.*Worth, New York. 1982

Leibovitz, E; Béguin, P. A new type of cohesin domain that specifically binds the dockerin domain of the Clostridium thermocellum cellulosome-integrating protein CipA. *J Bacteriol*, 1996, 178, 3077–3084.

Lemaire, M; Béguin, P. Nucleotide sequence of the celG gene of Clostridium thermocellum and characterization of its product, endoglucanase CelG. *J Bacteriol,* 1993, 175, 3353-3360.

Leschine, SB. Cellulose degradation in anaerobic environments. *Annu Rev Microbiol*, 1995, 49, 399-426.

Lewis, GE; Hunt, CW; Sanchez, WK; Treacher, R; Pritchard, GT; Feng, P. Effect of direct-fed fibrolytic enzymes on the digestive characteristics of a forage-based diet fed to beef steers. *J Anim Sci,* 1996, 74, 3020-3028.

Li, XH; Yang, HJ; Bhaskar, R; Park, EY; Jiang, LJ; Wang, D et al. Enhanced cellulose production of the Trichoderma viride mutated by microwave and ultraviolet. *Microbiol Res,* 2010, 165, 190-198.

Li, X; Yang, H; Roy, B; Wang, D; Yue, W; Jiang, L et al. The most stirring technology in future: Cellulase enzyme and biomass utilization. *Afr J Biotechnol*, 2009, 8 , 2418-2422

Li, S; Sauer, W.; Huang, S.; Gabert, V. Effect of β-glucanase supplementation to hulless barley- or wheat-soybean meal diets on the digestibilities of energy, protein, β-glucans, and amino acids in young pigs. *J Anim Sci*, 1996, 74, 1649–1656.

Li, X; Calza, R. Fractionation of cellulases from the ruminal fungus Neocallimastix frontalis EB188. *Appl Environ Microbiol*, 1991, 57, 3331-3336.

Li, X; Gao, P. Isolation and partial properties of cellulose decomposing strain of Cytophaga sp. LX-7 from soil. *J Appl Microbiol*, 1997, 82, 73-80.

Li, X; Wang, D; Zhou, F; Yang, H; Bhaskar, R; Hu, J; Sun, C; Miao, Y. Cloning and expression of a cellulase gene in the silkworm, Bombyx mori by improved Bac-to-Bac/BmNPV baculovirus expression system. *Mol Biol Rep*, 2010, DOI 10.1007/s11033-010-0025-2

Li, XZ. Streptomyces cellulolyticus sp. nov., a new cellulolytic member of the genus Streptomyces. *Int J Syst Bacteriol*, 1997, 47, 443-445.

Li, YN; Shi, GY; Wang, W; Wang, ZX. A newly isolated Rhizopus microsporus var. chinensis capable to secrete amyloltic enzymes with raw-starch-digesting activity. *J Microbiol Biotechnol*, 2010, 20, 383-390.

Liebl, W. Cellulolytic enzymes from Thermotoga species. *Methods Enzymol*, 2001, 330, 290-300.

Liebl, W; Ruile, P; Bronnenmeier, K; Riedel, K; Lottspeich, F; Greif, I. Analysis of a Thermotoga maritima DNA fragment encoding two similar thermostable cellulases, CelA and CelB, and characterization of the recombinant enzymes. *Microbiology*, 1996, 142, 2533-2542.

Lima, AO; Quecine, MC; Fungaro, MH; Andreote, FD; Maccheroni, W Jr; Araújo, WL et al. Molecular characterization of a beta-1,4-endoglucanase from an endophytic Bacillus pumilus strain. *Appl Microbiol Biotechnol*, 2005, 68, 57-65.

Limauro, D; Cannio, R; Fiorentino, G; Rossi, M; Bartolucci, S. Identification and molecular characterization of an endoglucanase gene, celS, from the extremely thermophilic archaeon Sulfolobus solfataricus. *Extremophiles*, 2001, 5, 213-219.

Lin, Y; Tanaka, S. Ethanol fermentation from biomass resources: current state and prospects. *Appl Microbiol Biotechnol*, 2006, 69, 627-642.

Lin, SB; Stutzenberger, FJ. Purification and characterization of the major β-1,4-endoglucanase from Thermomonospora curvata. *J Appl Bacteriol*, 1995, 79, 447-453.

Liu, JH; Selinger, LB; Hu, YJ; Moloney, MM; Cheng, KJ; Beauchemin, KA. An endoglucanase from the anaerobic fungus Orpinomyces joyonii: characterization of the gene and its product. *Can J Microbiol*, 1997, 43, 477-485.

Ljungdahl, LG. The cellulase/hemicellulase system of the anaerobic fungus Orpinomyces PC-2 and aspects of its applied use. *Ann N Y Acad Sci,* 2008, 1125, 308-321.

Lo Leggio, L; Parry, NJ; van Beeumen, J; Claeyssens, M; Bhat, MK; Pickersgill, RW. Crystallization and preliminary X-ray analysis of the major endoglucanase from Thermoascus aurantiacus. *Acta Cryst*, 1997, D 53, 599-604.

Loginova, LG; Beliakova, LA; Guzhova, EP; Iusupova, IKh; Burdenko, LG. Thermophilic Myceliophthora thermophila decomposes cellulose. *Mikrobiologyia*, 1983, 52, 605-608.

Lu, M; Li, D; Zhang, C. Purification and properties of an endocellulase from the thermophilic fungus Chaetomium thermophile. *Wei Sheng Wu Xue Bao*, 2002, 42, 471-477.

Lu, Y; Zhang, YH; Lynd, LR. 2006. Enzyme-microbe synergy during cellulose hydrolysis by Clostridium thermocellum. *Proc Natl Acad Sci USA*, 2006, 103, 16165–16169.

Lucas, R; Robles, A; García, MT; Alvarez De Cienfuegos, G; Gálvez, A. Production, purification, and properties of an endoglucanase produced by the hyphomycete Chalara (Syn. Thielaviopsis) paradoxa CH32. *J Agric Food Chem*, 2001, 49, 79-85.

Lynd, LR; Weimer, PJ; van Zyl, WH; Pretorius, IS. Microbial cellulose utilization: fundamentals and biotechnology. *Microbiol Mol Biol Rev*, 2002, 66, 506-577.

Machado de Castro, A; de Albuquerque de Carvalho, ML; Ferreira Leite, SG; Pereira, N Jr. Cellulases from Penicillium funiculosum: production, properties and application to cellulose hydrolysis. *J Ind Microbiol Biotechnol*, 2010, 37, 151–158.

Madden, RH. Isolation and Characterization of Clostridium stercorarium sp. nov., cellulolytic thermophile. *Int J Syst Bacteriol*, 1983, 33, 837-840.

Madden, RH; Bryder, MJ; Poole, NJ. Isolation and characterization of an anaerobic, cellulolytic bacterium, Clostridium papyrosolvens sp. nov. *Int J Syst Bacteriol*, 1982, 32, 87-91.

Maki, M; Leung, KT; Qin, W. The prospects of cellulase-producing bacteria for the bioconversion of lignocellulosic biomass. *Int J Biol Sci*, 2009, 5, 500-516.

Mandels, M. Applications of cellulases. *Biochem Soc Trans*, 1985, 13, 414-415.

Manna, C; Galletti, P; Cucciolla, V; Montedoro, GF; Zappia, V. Olive oil hydroxytyrosol protects human erythrocytes against oxidative damages. *J Nutr Biochem*, 1999, 10, 159-165.

Mansfield, SD; Saddler, JN; Gubitz, GM. Characterization of endoglucanases from the brown rot fungi Gloeophyllum sepiarium and Gloeophyllum trabeum. *Enzyme Microb Technol*, 1998, 23, 133-140.

Marques, S; Pala, H; Alves, L; Amaral-Collaco, MT; Gama, FM; Girio, FM. Characterisation and application of glycanases secreted by Aspergillus terreus CCMI 498 and Trichoderma viride CCMI 84 for enzymatic deinking of mixed office waste papers, *J Biotechnol*, 2003,100, 209-219.

Martinez, D; Berka, RM; Henrissat, B; Saloheimo, M; Arvas, M; Baker, SE et al. Genome sequencing and analysis of the biomass-degrading fungus Trichoderma reesei (syn. Hypocrea jecorina) *Nat Biotechnol*, 2008, 26, 553-560.

Matsui, H; Ban-Tokuda, T. Studies on carboxymethyl cellulase and xylanase activities of anaerobic fungal isolate CR4 from the bovine rumen. *Curr Microbiol*, 2008, 57, 615-619.

Maurelli, L; Giovane, A; Esposito, A; Moracci, M; Fiume, I; Rossi, M et al. Evidence that the xylanase activity from Sulfolobus solfataricus Oalpha is encoded by the endoglucanase precursor gene (sso1354) and characterization of the associated cellulase activity. *Extremophiles*, 2008, 12, 689-700.

Mayer, F; Coughlan, MP; Mori, Y; Ljungdahl, LG. Macromolecular organization of the cellulolytic complex of Clostridium thermocellum as revealed by electron microscopy. *Appl Environ Microbiol*, 1987, 53, 2785–2792.

McCarthy, T; Hanniffy, O; Savage, AV; Tuohy, MG. Catalytic properties and mode of action of three endo-beta-glucanases from Talaromyces emersonii on soluble beta-1,4- and beta-1,3;1,4-linked glucans. *Int J Biol Macromol*, 2003, 33, 141-148.

McGavin, M; Forsberg, CW. Isolation and characterization of endoglucanases 1 and 2 from Bacteroides succinogenes S85. *J Bacteriol,* 1988, 70, 2914-2922.

McHale, A; Coughlan, MP. Synergistic hydrolysis of cellulose by components of the extracellular system of Talaromyces emersonii. *FEBS Lett*, 1980, 117, 319-322.

McHale, A; Coughlan, MP. Identification of various components produced during growth on cellulosic media by Talaromyces emersonii. *Biochim Biophys Acta*, 1981, 662, 145-151.

McKenzie, B. *Heterologous expression of cellulase enzymes in transplastidic Nicotiana tabacum cv. Petit Havana*. School of Applied Sciences, Department of Biotechnology and Environmental Biology, RMIT University, Melbourne, Australia, 2008.

Mergulhão, FJ; Summers, DK; Monteiro, GA. Recombinant protein secretion in Escherichia coli. *Biotechnol Adv*, 2005, 23, 177-202.

Miettinen-Oinonen, A. *Trichoderma reesei strains for production of cellulases for the textile industry.* VTT Publications, Finland, Espoo 2004,

Miettinen-Oinonen, A; Londesborough, J; Joutsjoki, V; Lantto, R; Vehmaanperä, J; Primalco Ltd. Biotec. Three cellulases from Melanocarpus albomyces for textile treatment at neutral pH. *Enzyme Microb Technol*, 2004, 34, 332-341.

Miettinen-Oinonen A; Suominen, P. Enhanced production of Trichoderma reesei endoglucanases and use of the new cellulase preparations in producing the stonewashed effect on denim fabric. *Appl Environ Microbiol*, 2002, 68, 3956-3964.

Mingardon, F; Chanal, A; Lopez-Contreras, AM; Dray, C; Bayer, EA; Fierobe, HP. Incorporation of fungal cellulases in bacterial minicellulosomes yields viable, synergistically acting cellulolytic complexes. *Appl Environ Microbiol*, 2007, 73, 3822–3832.

Missiakas, D; Raina, S. Protein folding in the bacterial periplasm. *J Bacteriol*, 1997, 179, 2465–2471.

Mohagheghi, A; Grohmann, K; Himmel, M; Leighton, L; Updegraff DM. Isolation and characterization of Acidothermus cellulolyticus gen. nov., sp. nov., a new genus of thermophilic, acidophilic, cellulolytic bacteria. *Int J Syst Bacteriol*, 1986, 36, 435-443.

Moloney, AP; McCrae, SI; Wood, TM; Coughlan. MP. Isolation and characterization of the 1,4-beta-D-glucan glucanohydrolases of Talaromyces emersonii. *Biochem J*, 1985, 225, 365-374.

Monserrate, E; Leschine, SB; Canale-Parola, E. Clostridium hungatei sp. nov., a mesophilic, N2-fixing cellulolytic bacterium isolated from soil. *Int J Syst Evol Microbiol*, 2001, 51, 123-132.

Moon, T; Nagarajan, R. Deinking of xerographic and laser-printed paper using block copolymers. *Colloids Surf. A: Physiochem Eng Aspects*, 1998, 132, 275-288.

Morana, A; Esposito, A; Maurelli, L; Ruggiero, G; Ionata, E; Rossi, M et al. A novel thermoacidophilic cellulase from Alicyclobacillus acidocaldarius. *Protein Pept Lett*, 2008, 5, 1017-1021.

Morana, A; Maurelli, L; La Cara, F; Rossi, M. Extremely thermophilic enzymes: their utilization in agricultural waste conversion for bioethanol production. In: *Bioethanol: production, benefits and economics*. New York: Nova Science Publishers Inc.; 2009; 93-106.

Morbak, AL; Zimmermann, W. Deinking of mixed office paper, old newspaper and vegetable oil based ink printed paper using cellulase, xylanases and lipases. *Progr Paper Recycling*, 1998, 7, 14-27.

Moriya, T; Murashima, K; Nakane, A; Yanai, K; Sumida, N; Koga, J et al. Molecular cloning of endo-beta-D-1,4-glucanase genes, rce1, rce2, and rce3, from Rhizopus oryzae. *J Bacteriol,* 2003, 185, 1749-1756.

Moriya, T; Watanabe, M; Sumida, N; Okakura, K; Murakami, T. Cloning and overexpression of the avi2 gene encoding a major cellulase produced by Humicola insolens FERM BP-5977. *Biosci Biotechnol Biochem*, 2003, 67, 1434-1437.

Moriya, T; Watanabe, M; Sumida, N; Okakura, K; Murakami, T. Cloning and overexpression af the avi2 gene encoding a major cellulase produced by Humicola insolens FERM BP-5977. *Biosci Biotechnol Biochem*, 2003, 67, 1434-1437.

Mosier, N; Wyman, C; Dale, B; Elander, R; Lee, YY; Holtzapple, M et al. Features of promising technologies for pretreatment of lignocellulosic biomass. *Bioresour Technol*, 2005, 96,673-686.

Mountford DO, Orpin CG. *Anaerobic fungi: biology, ecology and function*. New York, NY Marcel Dekker, Inc; 1994.

Muntazer M; Sadeghian Maryan, A. Influences of different enzymatic treatment on denim garment. *Appl Biochem Biotechnol*, 2010, 160, 2114-2128.

Murashima, K; Kosugi, A; Doi, RH. Solubilization of cellulosomal cellulases by fusion with cellulose-binding domain of noncellulosomal cellulase EngD from Clostridium cellulovorans. *Prot Struct Funct Genet*, 2003, 50 620–628.

Murashima, K; Nishimura, T; Nakamura, Y; Koga, J; Moriya, T; Sumida, N et al. Purification and characterization of new endo-1,4-beta-D-glucanses from Rhizopus oryzae. *Enzyme Microb Technol,* 2002, 30, 319-326.

Murray, WD; Sowden, LC; Colvin, JR. Bacteroides cellulosolvens sp. nov., a cellulolytic species from sewage sludge. *Int J Syst Bacteriol*, 1984, 34, 185-187.

Nakari-Seta La, T; Penttila, M. Production of Trichoderma reesei cellulases on glucose-containing media. *Appl Environ Microbiol*, 1995, 61, 3650-3655.

Nishiyama, Y; Sugiyama, J ; Chanzy, H; Langan, P. Crystal structure and hydrogen bonding system in cellulose Ir from synchrotron X-ray and neutron fiber diffraction, *J Am Chem Soc*, 2002, 124, 9074–9082.

Noe, P; Chevalier, J; Mora, F; Comtat, J. Action of enzymes in chemical pulp fibres. Part II: enzymatic beating, *J Wood Chem Technol*, 1986, 6, 167-184.

Ogawa, A; Suzumatsu, A; Takizawa, S; Kubota, H; Sawada, K; Hakamada, Y et al. Endoglucanases from Paenibacillus spp. form a new clan in glycoside hydrolase family 5. *J Biotechnol*, 2007, 129, 406-414.

Ohara, H; Noguchi, J; Karita, S; Kimura, T; Sakka, K; Ohmiya, K. Sequence of egV and properties of EgV, a Ruminococcus albus endoglucanase containing a dockerin domain. *Biosci Biotechnol Biochem,* 2000, 64, 80-88.

Okada, G. Cellulase of Aspergillus niger. *Methods Enzymol*, 1988, 160, 259-264.

Oksanen, J; Ahvenainen, J; Home, S. Microbial cellulase for improving filtrability of wort and beer. *Proc Eur Brew Chem Helsinki*, 1985, 419-25.

Olutiola, PO. Cellulase enzymes in culture filtrates of Ceratocystis paradoxa. *Mycologia*, 1976, 68, 1083-1092.

Olutiola, PO. Characterization of cellulase from Humicola lanuginosa. *Cell Mol Life Sci*, 1982, 38,1332-1333.

Oraby, H; Venkatesh, B; Dale, B; Ahmad, R; Ransom, C; Oehmke, J et al. Enhanced conversion of plant biomass into glucose using transgenic rice-produced endoglucanase for cellulosic ethanol. *Transgenic Res*, 2007, 16, 739–749.

Orpin, CG. Isolation of cellulolytic phycomycete fungi from the caecum of the horse. *J Gen Microbiol*, 1981,123, 287-296.

Orpin, CG. Studies on the rumen flagellate Neocallimastix frontalis. *J Gen Microbiol*, 1975, 91, 249-262.

Orpin, CG. *Studies on the rumen flagellate Sphaeromonas communis*. *J Gen Microbiol*, 1976, 94, 270-280.

O'Sullivan, AC. Cellulose: the structure slowly unravels. *Cellulose*, 1997, 4, 173-207.

Ouyang, J; Yan, M; Kong, D; Xu, L. A complete protein pattern of cellulase and hemicellulase genes in the filamentous fungus Trichoderma reesei. *Biotechnol J*, 2006, 1, 1266-1274.

Ozkose, E; Thomas, BJ; Davies, DR; Griffith, GW; Theodorou, M.K. Cyllamyces aberensis gen.nov. sp.nov., a new anaerobic gut fungus with branched sporangiophores isolated from cattle. *Can J Bot*, 2001, 79, 666-673.

Ozkose, E; Akyol, I.; Kar, B; Comlekcioglu, U; Ekinci, MS. Expression of fungal cellulase gene in Lactococcus lactis to construct novel recombinant silage inoculants. *Folia Microbiol*, 2009,54, 335-342.

Paakkari, T; Serimaa, R; Fink, HP. Structure of amorphous cellulose. *Acta polimerica*, 40,731-734 34 by Saddler JM, MH. Washington, DC: American Chemical Society

Pages, S; Belaich, A; Fierobe, HP; Tardif C; Gaudin, C; Belaich, JP. Sequence analysis of scaffolding protein CipC and ORFXp, a new cohesin-containing protein in Clostridium cellulolyticum: comparison of various cohesin domains and subcellular localization of ORFXp. *J Bacteriol*, 1999, 181, 1801–1810.

Pages, S; Gal, L; Belaich, A; Gaudin, C; Tardif, C; Belaich, J. Role of scaffolding protein CipC of Clostridium cellulolyticum in cellulose degradation. *J Bacteriol*, 1997, 179, 2810–2816.

Park, SR; Kim, MK; Kim, JO; Cho, SJ; Cho, YU; Yun, HD. Cloning and sequencing of cel5Z gene from Erwinia chrysanthemi PY35. *Mol Cells,* 2000, 10, 269-274.

Parry, NJ; Beever, DE; Owen, E; Nerinckx, W; Claeyssens, M; Van Beeumen, J et al. Biochemical characterization and mode of action of a thermostable endoglucanase purified from Thermoascus aurantiacus. *Arch Biochem Biophys*, 2002, 404, 243-253.

Parvez, S; Rajoka, MI; Fariha, F; Malik, KA. Cloning of endoglucanase genes from Cellulomonas biazotea into E. coli and S. cerevisiae using shuttle vector YEp24. *Folia Microbiol,* 1994, 39, 251-254.

Patel, GB; Khan, AW; Agnew, BJ; Colvin, JR. Isolation and characterization of an anaerobic, cellulolytic microorganism, Acetivibrio cellulolyticus gen. nov., sp. nov. *Int J Syst Bacteriol*, 1980, 30, 179-185.

Payen, A. Mémoire sur la composition du tissu propre des plantes et du ligneux. C R Hebd Seances Acad Sci, 1938, 7, 1052–1056.

Pellerin, P; Doco, T, Vidal, S; Williams, P; Brillouet, J M; Oneill, MA. Structural characterization of red wine RhamnoGalacturonan II. *Carbohydr Res*, 1996,290,183–197.

Pellinen, J; Abuhasan, J; Chang, H M. Biological delignification of pulp by Phanerochaete chrysosporium, *J Biotechnol*, 1989, 10, 161-170.

Pendleton, B. The regulatory environment. In: Muirhea S, editor. *Direct-fed microbial, enzyme, and forage additive compendium*. Minnetonka,MN: The Miller Publishing Co; 2000.

Pere, J; Siika-Aho, M; Buchert, J; Viikari, L. Effects of purified T. reesei cellulases on the fibre properties of kraft pulp. *Tappi J,* 1995, 78, 71-78.

Petitdemange, E; Caillet, F; Giallo, J; Gaudin, C. Clostridium cellulolyticum sp. nov., a cellulolyltic mesophilic species from decayed grass. *Int J Syst Bacteriol*, 1984, 34, 155-159.

Picart, P; Diaz, P; Pastor, FI. Cellulases from two Penicillium sp. strains olated from subtropical forest soil: production and characterization. *Lett Appl Microbiol*, 2007, 45, 108-113.

Pitulle, C; Yang, Y; Marchiani, M; Moore, ER; Siefert, JL; Aragno, M; Jurtshuk, PJ;Fox, GE. Phylogenetic position of the genus Hydrogenobacter. *Int J Syst Bacteriol*, 1994, 44, 620-626.

Plank, PFH; Zent, JR. Use of enzymes in wine making and grape processing. In: Gump BH , Pruett DJ editors. *Beer and wine production analysis, characterisation, and technological advances*. Washington : American Chemical Society DC; 1993; 191-196.

Pohlschröder, M; Leschine, SB; Canale-Parola, E. Multicomplex cellulase-xylanase system of Clostridium papyrosolvens C7. *J Bacteriol*, 1994, 176, 70-76.

Posta, K; Béki, E; Wilson, DB; Kukolya, J; Hornok, L. *Cloning, characterization and phylogenetic relationships of cel5B, a new endoglucanase encoding. J Basic Microbiol*, 2004, 44, 383-399.

Pothiraj, C; Balaji, P; Eyini, M. Enhanced production of cellulases by various fungal cultures in solid state fermentation of cassava waste. *Afr J Biotechnol*, 2006, 5, 1882-1885.

Pugsley, AP. The complete general secretory pathway in gram-negative bacteria. *Microbiol Rev,* 1993, 57, 50–108

Qin,Y; Wei, X; Liu, X; Wang, T; Qu, Y. Purification and characterization of recombinant endoglucanase of Trichoderma reesei expressed in Saccharomyces cerevisiae with higher glycosylation and stability. *Prot Express Purific*, 2008, 58, 162-167.

Rainey, FA; Donnison, AM; Janssen, PH; Saul, D; Rodrigo, A; Bergquist, PL et al. Description of Caldicellulosiruptor saccharolyticus gen. nov., sp. nov: an obligately anaerobic, extremely thermophilic, cellulolytic bacterium. *FEMS Microbiol Lett*, 1994, 120, 263–266.

Rajnish, KN; Choudhary, GM; Gunasekaran, P. Functional characterization of a putative endoglucanase gene in the genome of Zymomonas mobilis. *Biotechnol Lett*, 2008, 30, 1461-1467.

Ramadan, M.F; Thomas, MJ. Impact of enzymatic treatment on chemical composition, physicochemical properties and radical scavenging activity of goldenberry (Physalis peruviana L.) juice. *J Sci Food Agric*, 2007, 87, 52-460.

Ranalli, A; Ferrante, ML; De Mattia, G; Costantini, N. Analytical evaluation of virgin olive oil of first and second extraction. *J Agric Food Chem*, 1999a, 47, 417-424.

Ranalli, A; Sgaramella, A; Surricchio, G. The new ``Cytolase 0" enzyme processing aid improves quality and yields of virgin olive oil. *Food Chem*, 1999b, 66, 443-454.

Ranalli, A; Malfatti, A; Lucera, L; Contento, S; Sotiriou, E. Effect of processing techniques on the natural colourings and the other functional constituents in virgin olive oil. *Food Res Int*, 2005, 38, 873-878.

Rashid, MH; Javed, MR; Kawaguchi, T; Sumitani, J; Arai, M. Improvement of Aspergillus oryzae for hyperproduction of endoglucanase: expression cloning of cmc-1 gene of Aspergillus aculeatus. *Biotechnol Lett*, 2008, 30, 2165-2172.

Reese, ET; Mandels, M. Enzymatic degradation. In: Bikales NM, Segal L editors. *Cellulose and cellulose derivatives*. New York, N.Y: Wiley Interscience; 1971; 1079-1094.

Reese, ET; Sui, RGH; Levinson, HS. The biological degradation of soluble cellulose derivatives and its relationship to the mechanism of cellulose hydrolysis. *J Bacteriol*, 1950, 59, 485-497. 64

Rincon, MT; Cepeljnik, T; Martin, JC; Lamed, R; Barak, Y; Bayer, EA et al. Unconventional mode of attachment of the Ruminococcus flavefaciens cellulosome to the cell surface. *J Bacteriol*, 2005, 187, 7569–7578.

Rincon, MT; Ding, SY; McCrae, SI.; Martin, JC; Aurilia, V; Lamed, R et al. Novel organization and divergent dockerin specificities in the cellulosome system of Ruminococcus flavefaciens. *J Bacteriol*, 2003, 185, 703–713.

Roy, SK; Dey, SK; Raha, SK; Chakrabarty, SL. Purification and properties of an extracellular endoglucanase from Myceliophthora thermophila D-14 (ATCC 48104). *J Gen Microbiol*, 1990, 136, 1967-1971.

Sabathé, F; Bélaïch, A; Soucaille, P. Characterization of the cellulolytic complex (cellulosome) of Clostridium acetobutylicum. *FEMS Microbiol Lett*, 2002, 217, 15-22.

Sacchi, KL; Bisson, L F; Adams, DO.A review of the effect of winemaking techniques on phenolic extraction in red wines. *American Journal of Enology and Viticulture*, 2005, 56, 197-206.

Saiki, T; Kobayashi, Y; Kawagoe, K; Beppi, T. Dictyoglomus thermophilum gen. nov., sp. nov., a chemoorganotrophic, anaerobic, thermophilic bacterium. *Int J Syst Bacteriol*, 1985, 35, 253-259.

Sakka, K; Kimura, T; Karita, S; Ohmiya, K. Molecular breeding of cellulolytic microbes, plants and animals for biomass utilization. *J Biosci Bioeng*, 2000, 90, 227-233.

Sakon, J; Adney, WS; Himmel, ME; Thomas, SR; Karplus, PA. Crystal structure of thermostable family 5 endocellulase E1 from Acidothermus cellulolyticus in complex with cellotetraose. *Biochemistry*, 1996, 35, 10648–10660.

Salamitou, S; Raynaud, O; Lemaire, M; Coughlan, M; Béguin, P; Aubert, JP. Recognition specificity of the duplicated segments present in Clostridium thermocellum endoglucanase CelD and in the cellulosome-integrating protein CipA. *J Bacteriol*, 1994, 176, 2822–2827.

Saloheimo, M; Lehtovaara, P; Penttilä, M; Teeri, TT; Ståhlberg, J; Johansson, G et al. EGIII, a new endoglucanase from Trichoderma reesei: the characterization of both gene and enzyme. *Gene*, 1988, 63, 11-22.

Sami, AJ; Akhtar, MW. Purification and characterization of two low molecular-weight endoglucanases of Cellulomonas flavigena. *Enz Microbiol Technol*, 1993, 15, 586-592.

Sanchez, CR; Schvartz, PC; Ramos, BH. Growth and endoglucanase activity of Acetivibrio cellulolyticus grown in three different cellulosic substrates. *Rev Microbiol*, 1999, 30, 310-314.

Sànchez-Torres, J; Pérez, P; Santamarìa, RIA Cellulase gene from a new alkalophilic Bacillus sp. (strain N186-1). Its cloning, nucleotide sequence and expression in Escherichia coli. *Appl Microbiol Biotechnol*, 1996, 46, 149-155.

Sandgren, M; Shaw, A; Ropp, TH; Wu, S; Bott, R; Cameron, AD et al. The X-ray crystal structure of the Trichoderma reesei family 12 endoglucanase 3, Cel12A, at 1.9 A° resolution. *J Mol Biol,* 2000, 308, 295-310.

Saraswathy, VN; Roberts, MF; Demain, AL. True cellulase production by Clostridium thermocellum grown on different carbon sources. *Biotechnol Lett*, 1993, 15, 641-646.

Schülein, M. Enzymatic properties of cellulases from Humicola insolens. *J Biotechnol*, 1997, 57, 71-81.

Schuster, A; Schmoll, M. Biology and biotechnology of Trichoderma. *Appl Microbiol Biotechnol*, 2010, 87, 787-799.

Schwarz, WH. The cellulosome and cellulose degradation by anaerobic bacteria. *Appl Microbiol Biotechnol*, 2001, 56, 634-649. 34

Schwarz, WH; Schimming, S; Ruecknagel, KP; Burgschweiger, S; Kreil, G; Staudenbauer, WL. Nucleotide sequence of celC gene encoding endoglucanase C of Clostridium thermocellum. *Gene*, 1988, 63, 23–30.

Shanmughapriya, S; Kiran, GS; Selvin, J; Thomas, TA; Rani, C. Optimization, purification, and characterization of extracellular mesophilic alkaline cellulase from sponge-associated Marinobacter sp. MSI032. *Appl Biochem Biotechnol*, 2009, 162, 625-640.

She, Q; Sing, RH; Confalonieri, F; Zivanovic, Y; Allard, G; Awayez, MJ et al. The complete genome of the crenarchaeon Sulfolobus solfataricus P2. *Proc Natl Acad Sci USA*, 2001, 98, 7835-7840.

Shimokawa, T; Shibuya, H; Nojiri, M; Yoshida, S; Ishihara, M. Purification, molecular cloning, and enzymatic properties of a family 12 endoglucanase (EG-II) from Fomitopsis palustris: role of EG-II in larch holocellulose hydrolysis. *Appl Environ Microbiol,* 2008, 74, 5857-5861.

Shimonaka, A; Baba, Y; Koga, J; Nakane, A; Kubota, H; Kono, T. Molecular cloning of a gene encoding endo-beta-D-1,4-glucanase PCE1 from Phycomyces nitens. *Biosci Biotechnol Biochem,* 2004, 68, 2299-2305.

Shimonaka, A; Koga, J; Baba, Y; Nishimura, T; Murashima, K; Kubota, H et al. Specific characteristics of family 45 endoglucanases from Mucorales in the use of textiles and laundry. *Biosci Biotechnol Biochem*, 2006, 70, 1013-1016.

Shoseyov, O; Takagi, M; Goldstein, MA.; Doi, RH. Primary sequence analysis of Clostridium cellulovorans cellulose binding protein A. *Proc Natl Acad Sci USA*, 1992, 89, 3483–3487.

Shufang, W; Shaojun, D; Zhongzheng, L. Function of endocellulase in the deinking process of mixed office waste paper. *Chem Ind Forest Prod*, 2005, 25, 87-90.

Sigoillot, C, Lomascolo, A; Record, E et al. Lignocellulolytic and hemicellulolytic system of Pycnoporus cinnabarinus: isolation and characterization of a cellobiose dehydrogenase and a new xylanase. *Enzyme Microb Technol*, 2002, 31, 876–883.

Simankova, MV; Chernych, NA; Osipov, GA; Zavarzin, GA. Halocella cellulolytica, gen. nov. sp. nov., a new obligately anaerobic, halophilic, cellulolytic bacterium. *Syst Appl Microbiol*, 1993, 16, 385-389.

Singh, J; Batra, N; Sobti, RC. Purification and characterisation of alkaline cellulase produced by a novel isolate, Bacillus sphaericus JS1. *J Ind Microbiol Biotechnol*, 2004, 31, 51-56.

Sinitsyn, A P; Gusakov, AV; Grishutin, SG; Sinitsyna, O A; Ankudimova,N V. Application of microassays for investigation of cellulose abrasive activity and backstaining. *J Biotechnol*, 2001, 89, 233-238.

Sleat, R; Mah, RA; Robinson, R. Isolation and characterization of an anaerobic, cellulolytic bacterium, Clostridium cellulovorans sp. nov., *Appl Environ Microbiol*, 1984, 48, 88-93.

Soutschek-Bauer, E; Staudenbauer, WL. Synthesis and secretion of a heat-stable carboxymethylcellulose from Clostridium thermocellum in Bacillus subtilis and Bacillus stearothermophilus. *Mol Gen Genet*, 1987, 208, 537-541.

Stackebrandt, E; Kandler, O. Taxonomy of the genus Cellulomonas, based on phenotypic characters and deoxyribonucleic acid-deoxyribonucleic acid homology, and proposal of seven neotype strains. *Int J Syst Bacteriol*, 1979, 29, 273-282.

Steenbakkers, PJ; Freelove, A; Van Cranenbroek, B; Sweegers, BM; Harhangi, HR; Vogels, GD et al. The major component of the cellulosomes of anaerobic fungi from the genus Piromyces is a family 48 glycoside hydrolase. *DNA*, 2002, 13, 313-320.

Stetter, KO; Konig, H; Stackebrandt, E. Pyrodictium gen. nov., a new genus of submarine disc-shaped sulphur reducing archaebacteria growing optimally at 105°C. *Syst Appl Microbiol*, 1983, 4, 535-551.

Stewart, CS; Flint, HJ. Bacteroides (Fibrobacter) succinogenes, a cellulolytic anaerobic bacterium from the gastrointestinal tract. *Appl Microbiol Biotechnol*, 1989, 30, 433-439.

Sticklen, MB Plant genetic engineering for biofuel production: towards affordable cellulosic ethanol. *Nat Rev Genet*, 2008, 9, 433-443.

Sticklen, MB. Feedstock crop genetic engineering for alcohol fuels. *Crop Sci,* 2007, 47, 2238–2248.

Stolk, AC. Thermophilic species of Talaromyces and Thermoascus. *Antonie van Leeuwenhoek*, 1965, 31, 262-276.

Stoppok, W; Rapp, P; Wagner, F. Formation, location, and regulation of endo-1,4-β-glucanases and β-glucosidases from Cellulomonas uda. *Appl Environ Microbiol*, 1982, 44, 44-53.

Sukhumavasi, J; Ohmiya, K; Shimizu, S; Ueno, K. Clostridium josui sp. nov., a cellulolytic, moderate thermophilic species from Thai compost. *Int J Syst Bacteriol*, 1988, 38, 179-182.

Sun, Y; Cheng, JJ; Himmel, M.E; Skory, CD; Adney, WS; Thomas, SR et al. Expression and characterization of Acidothermus cellulolyticus E1 endoglucanase in transgenic duckweed Lemna minor 8627. *Bioresour Technol*, 2007, 98, 2866-2872.

Suominen, P; Mäntylä, A; Karhunen, T; Hakola, S; Nevalainen, H. High frequency one-step gene replacement in Trichoderma reesei II. Effect of deletion of individual cellulase genes. *Mol Gen Genetics*, 1993, 241, 523-530.

Suurnäkki, A; Tenkanen, M; Siika-aho, M; Niku-Paavola, ML; Viikari, L; Buchert, J. Trichoderma reesei cellulases and their core domains in the hydrolysis and modification of chemical pulp. *Cellulose*, 2000, 7, 189-209.

Suurnakki, A; Oksanen, T; Kettunen, H; Buchert, J. The effect of mannan on physical properties of ECF bleached softwood kraft fibre handsheets. *Nord Pulp Paper Res J*, 2003,18, 429-435.

Svec, P; Sledacek, I.. Characterization of Lactococcus lactis subsp. lactis isolated from surface waters. *Folia Microbiol*, 2008, 53, 53-56.

Svetlichny, VA; Svetlichnya, TP. Dictyoglomus turgidus, sp. nov., a new extreme thermophilic eubacterium isolated from hot springs in the Uzon Volcano Crater. *Mikrobiologiya*, 1988, 57, 435-441.

Svetlichnyi, VA; Svetlichnaya, TP; Chernykh, NA; Zavarzin, GA. Anaerocellum thermophilum gen. nov sp. nov: an extremely thermophilic cellulolytic eubacterium isolated from hot springs in the Valley of Geysers. *Microbiology*, 1990, 59, 598-604.

Takahata, Y; Nishijima, M; Hoaki, T; Maruyama, T. Thermotoga petrophila sp. nov. and Thermotoga naphthophila sp. nov., two hyperthermophilic bacteria from the Kubiki oil reservoir in Niigata, Japan. *Int J Syst Evol Microbiol*, 2001, 51, 1901-1909.

Takai, K; Nakamura, K; Toki, T; Tsunogai, U; Miyazaki, M; Miyazaki, J et al. Cell proliferation at 122°C and isotopically heavy CH4 production by a hyperthermophilic methanogen under high-pressure cultivation. *Proc Natl Acad Sci USA*, 2008, 105, 10949–10954.

Takashima, S; Nakamura, A; Masaki, H; Uozumi, T. Purification and characterization of cellulases from Humicola grisea. *Biosci Biotech Biochem*, 1996, 60, 77-82.

Takashima, S; Nakamura, A; Hidaka, M; Masaki, H; Uozumi, T. Cloning, sequencing, and expression of the cellulase genes of Humicola grisea var. thermoidea. *J Biotechnol*, 1996, 50, 137-147.

Takashima, S; Nakamura, A; Masaki, H; Uozumi, T. Cloning, sequencing, and expression of a thermostable cellulase gene of Humicola grisea. *Biosci Biotechnol Biochem*, 1997, 61, 245-250.

Tamaru, Y; Doi, RH. Pectate lyase A, an enzymatic subunit of the Clostridium cellulovorans cellulosome. *Proc Natl Acad Sci USA*, 2001, 98, 4125–4129.

Tamoi, M; Kuritaki, H; Fukamizo, T. β-1,4-Glucanase-like protein from the cyanobacterium Synechocystis PCC6803 is a β-1,3-1,4-glucanase and functions in salt stress tolerance. *Biochem J,* 2007, 405, 139-146.

Tang, B; Pan, H; Zhang, Q; Ding, L. Cloning and expression of cellulase gene EG1 from Rhizopus stolonifer var. reflexus TP-02 in Escherichia coli. *Bioresour Technol*, 2009, 100, 6129-6132.

Tao, YM; Zhu, XZ; Huang, JZ; Ma, SJ; Wu, B; Long, MN; Chen, QX. Purification and properties of endoglucanase from a sugar cane bagasse hydrolyzing strain, Aspergillus glaucus XC9. *J Agric Food Chem*, 2010, 58, 6126-6130.

Taya, M; Hinoki, H; Yagi, T; Kobayashi, T. Isolation and characterization of an extremely thermophilic, cellulolytic, anaerobic bacterium. *Appl Microbiol Biotechnol*, 1988, 29, 474-479.

Tomme, P; Warren, RA; Gilkes, NR. Cellulose hydrolysis by bacteria and fungi. *Adv Microb Physiol*, 1995, 37, 1-81.

Tong, CC; Cole, AL; Shepherd, MG. Purification and properties of the cellulases from the thermophilic fungus Thermoascus aurantiacus. *Biochem J,* 1980, 191, 83-94.

Trinci, APJ; Davies, DR; Gull, K; Lawrence, MI; Nielsen, BB; Rickers, A; Theodorou, MK. Anaerobic fungi in herbivorous animals. *Mycol Res*, 1994, 98, 129-152.

Tsai, SL; Oh, J; Singh, S; Chen, R; Chen, W. Functional Assembly of Minicellulosomes on the Saccharomyces cerevisiae Cell Surface for Cellulose Hydrolysis and Ethanol Production. *Appl Environ Microbiol*, 2009, 75, 6087-6093.

Tuohy, MG; Murray, PG; Gilleran, CT; Collins, CM; Reen, FJ; McLoughlin, LP; Lydon, AGS; Moloney, AP; Heneghan, MN; O'Donoghue, AJ; Mahon, CS. *Talaromyces emersonii enzyme systems*. WO/2007/091231, Aug 16, 2007.

Udaka, S; Yamagata, H. High level secretion of heterologous proteins by Bacillus brevis. *Methods Enzymol*, 1993, 217, 23-33.

Ullah, AHJ; Sethumadhavan, K; Mullaney, EJ; Ziegelhoffer, T; Austin-Phillips, S. Cloned and expressed fungal phyA gene in alfalfa produces a stable phytase. *Biochem Biophys Res Commun*, 2002, 290, 1343–1348.

Uzcategui, E; Johansson, G; Ek, B; Pettersson, G. The 1,4-beta-D-glucan glucanohydrolases from Phanerochaete chrysosporium. Re-assessment of their significance in cellulose degradation mechanisms. *J Biotechnol*, 1991, 21, 143-160.

Valaskova, V; Baldrian, P. Degradation of cellulose and hemicelluloses by the brown rot fungus Piptoporus betulinus: production of extracellular enzymes and characterization of the major cellulases. *Microbiology*, 2006, 152, 3613-3622.

Van Leeuwenhoek, A. Free and cell-bound cellulase activity of Cellulomonas flavigena. *Antheunisse J*, 1984, 50, 7-15.

van Peij, NN; Gielkens, MM; de Vries, RP; Visser, J; de Graaff, LH. The transcriptional activator XlnR regulates both xylanolytic and endoglucanase gene expression in Aspergillus niger. *Appl Environ Microbiol*, 1998, 64, 3615-3619.

Van Rensburg, P; Van Zyl, WH; Pretorius, IS. Engineering yeast for efficient cellulose degradation. *Yeast*, 1998, 14, 67-76.

van Wyk, N; den Haan, R; van Zyl, WH. Heterologous co-production of Thermobifida fusca Cel9A with other cellulases in Saccharomyces cerevisiae. *Appl Microbiol Biotechnol*, 2010, 87, 1813-1820.

van Zyl, WH; Lynd, LR; den Haan, R; McBride, JE. Consolidated bioprocessing for bioethanol production using Saccharomyces cerevisiae. *Adv Biochem Engin Biotechnol*, 2007, 108, 205–235.

Vanden Wymelenberg, A; Denman, S; Dietrich, D; Bassett, J; Yu, X; Atalla, R et al. Transcript analysis of genes encoding a family 61 endoglucanase and a putative membrane-anchored family 9 glycosyl hydrolase from Phanerochaete chrysosporium. *Appl Environ Microbiol,* 2002, 68, 5765–5768.

Varel, VH; Tanner, RS; Woese, CR. Clostridium herbivorans sp. nov., a cellulolytic anaerobe from the pig intestine. *Int J Syst Bacteriol*, 1995, 45, 90-94.

Vasella, A; Davies, GJ; Bohm, M. Glycosidase mechanisms.*Curr Opi. Che. Bio*, 2002, 6, 619-629.

Vidal, S;Williams, P; Doco, T; Moutounet, M; Pellerin, P. The polysaccharides of red wine: total fractionation and characterization. *Carbohydr Polym*, 2003, 54,439-447.

Villaverde, A; Carrio, MM. Protein aggregation in recombinant bacteria: biological role of inclusion bodies. *Biotechnol Lett*, 2003, 25, 1385-1395.

Villetaz, JC; Steiner, D; Trogus, H. The use of a ß-glucanase as an enzyme in wine clarification and filtration. *Am J Enol Vitic*, 1984, 35, 253-6.

Vocadlo,DJ; Davies, GJ. Mechanistic insights into glycosidase chemistry *Curr Opin Chem Biol* 2008, 12, 539-555.

Voragen, A G J; Beldman, G; Schols, HA. Chemistry and enzymology of pectins. In: McCleary BV, Prosky L Editors. *Advanced dietary fibre technology*. Oxford: Blackwell; 2001; 379-398.

Wachinger, G; Bronnenmeier, K; Staudenbauer, WL; Schrempf, H. Identification of mycelium-associated cellulase from Streptomyces reticuli. *Appl Environ Microbiol*, 1989, 55, 2653-2657.

Wang, JL; Ruan, H; Zhang, HF; Zhang, Q; Zhang, HB; He, GQ et al. Characterization of a thermostable and acidic tolerable β-glucanase from aerobic fungi Trichoderma koningii ZJU-T. *J Food Sci*, 2007, 72, C452-456.

Warner, CD; Hoy, JA; Shilling, TC; Linnen, MJ; Ginder, ND; Ford et al. Tertiary structure and characterization of a glycoside hydrolase family 44 endoglucanase from Clostridium acetobutylicum. *Appl Environ Microbiol*, 2010, 76, 338-346.

Warnick, TA; Methé, BA; Leschine, SB. Clostridium phytofermentans sp. nov., a cellulolytic mesophile from forest soil. *Int J Syst Evol Microbiol*, 2002, 52, 1155-1160.

Watanabe, H; Tokuda, G. Animal cellulases. *Cell Mol Life Sci*, 2001, 58, 1167-1178.

Waters, DM; Murray, PG; Ryan, LA; Arendt, EK; Tuohy, MG. Talaromyces emersonii thermostable enzyme systems and their applications in wheat baking systems. *J Agric Food Chem*, 2010, 58, 7415-7422.

Wen, F; Sun, J; Zhao, H. Yeast surface display of trifunctional minicellulosomes for simultaneous saccharification and fermentation of cellulose to ethanol. *Appl Environ Microbiol,* 2010, 76, 1251-1260.

Wicher, KB; Abou-Hachem, M; Halldórsdóttir, S; Thorbjarnadóttir, SH; Eggertsson, G; Hreggvidsson, GO et al. Deletion of a cytotoxic, N-terminal putative signal peptide results in a significant increase in production yields in Escherichia coli and improved specific activity of Cel12A from Rhodothermus marinus. *Appl Microbial Biotechnol*, 2001, 55, 578-584.

Wilson, CA; Wood, TM. The anaerobic fungus Neocallimastix frontalis: isolation and properties of a cellulosome-type enzyme fraction with the capacity to solubilize hydrogen-bond-ordered cellulose. *Appl Microbiol Biotechnol*, 1992, 37, 125-129.

Windberger, E; Huber, R; Trincone, A; Fricke, H; Stetter, KO. Thermotoga thermarum sp. nov. and Thermotoga neapolitana occurring in African continental solfataric springs. *Arch Microbiol*, 1989, 151, 506-512.

Wisotzkey, JD; Jurtshuk, PJr; Fox, GE; Deinhard, G; Poralla, K. Comparative sequence analyses on the 16S rRNA (rDNA) of Bacillus acidocaldarius, Bacillus acidoterrestris, and Bacillus cycloheptanicus and proposal for creation of a new genus, Alicyclobacillus gen. nov. *Int J Syst Bacteriol*, 1992, 42, 263-269.

Withers, SG;. Dombroski, D; Berven, LA; Kilburn, DG; Miller, RC; Warren, RA et al. Direct H-NMR determination of the stereochemical course of hydrolyses catalyzed by glucanase components of the cellulase complex. *Biochem Biophys Res Comm* 1986, 139, 487–494.

Wittmann, S; Shareck, F; Kluepfel, D; Morosoli, R. Purification and characterization of the CelB endoglucanase from Streptomyces lividans 66 and DNA sequence of the encoding gene. *Appl Environ Microbiol*, 1994, 60, 1701-1703.

Wonganu, B; Pootanakit, K; Boonyapakron, K; Champreda, V; Tanapongpipat, S; Eurwilaichitr, L. Cloning, expression and characterization of a thermotolerant endoglucanase from Syncephalastrum racemosum (BCC18080) in Pichia pastoris. *Prot Express purific*, 2008, 58, 78-86.

Wood, TM; McCrae, SI. Cellulase from Fusarium solani: purification and properties of the C1 component. *Carbohydr Res*, 1977, 57, 117-133.

Wood, TM; Wilson, CA; McCrae, SI; Joblin, KN. A highly active extracellular cellulase from the anaerobic rumen fungus Neocallimastix frontalis. *FEMS Microbiol Lett,* 1985, 34, 37-40.

Woyengo, TA; Sands, J S;Guenter, W; Nyachoti, CM. Nutrient digestibility and performance responses of growing pigs fed phytase- and xylanase-supplemented wheat-based diets. *J Anim Sci,* 2008, 86, 848-857.

Xia, Z ; Beaudry, AR; Bourbonnais, R. Effects of cellulases on the surfactant assisted acidic deinking of ONP and OMG. *Prig Pap Recycl*, 1996, 5, 46-58.

Xu, Y; Foong, FC. Characterization of a cellulose binding domain from Clostridium cellulovorans endoglucanase-xylanase D and its use as a fusion partner for soluble protein expression in Escherichia coli. *J Biotechnol*, 2008, 135, 319-325.

Yablonsky, MD; Bartley, T; Elliston, KO; Kahrs, SK; Shalita, ZP; Eveleigh, DE. Characterization and cloning of the cellulase complex of Microbispora bispora. In: Aubert JP; Béguin P; Millet J. editors. *FEMSD Symposioum No. 43. Biochemistry and genetics of cellulose degradation*. New York: Academic Press; 1988; 249–264.

Yamanobe, T; Mitsuishi, Y; Takasaki, Y. Isolation of a cellulolytic enzyme producing microorganism, culture conditions and some properties of the enzyme. *Agric Biol Chem*, 1987, 51, 65-74. 81

Yamada, M; Amano, Y; Horikawa, E; Nozaki, K; Kanda, T. Mode of action of cellulases on dyed cotton with a reactive dye. *Biosci Biotechnol Biochem*, 2005, 69, 45-50.

Yang, D; Weng, H; Wang, M; Xu, W; Li, Y; Yang, H. Cloning and expression of a novel thermostable cellulose from newly isolated Bacillus subtilis strain I15. *Mol Biol Rep*, 2010, 37, 1923-1929.

Yang, Q; Zhan, H; Wang, S; Li, K; Fu, S. Bio-modification of eucalyptus chemithermomechanical pulp, *Front. Chem Eng China*, 2008, 2, 28-33.

Yaron, S; Morag, E; Bayer, EA; Lamed, R; Shoham, Y. Expression, purification and subunit-binding properties of cohesins 2 and 3 of the Clostridium thermocellum cellulosome. *FEBS Lett*, 1995, 360, 121–124.

Yingjuan, F; Menghua, Q; Huiren, H. Effect of types and properties on the deinkability of wastepaper. *Trans Chin Pulp Paper*, 2005, 20, 155-159.

Yoon, JJ; Cha, CJ; Kim, YS; Kim, W. Degradation of cellulose by the major endoglucanase produced from the brown-rot fungus Fomitopsis pinicola. *Biotechnol Lett*, 2008, 30, 1373-1378.

Yu, LX; Gray, BN; Rutzke, CJ; Walker, LP; Wilson, DB; Hanson, MR. Expression of thermostable microbial cellulases in the chloroplasts of nicotine-free tobacco. *J Biotechnol,* 2007, 131, 362-369.

Zakpaa, HD; Mak-Mensah, EE; Johnson, FS. Production of bio-ethanol from corncobs using Aspergillus niger and Saccharomyces cerevisae in simultaneous saccharification and fermentation. *Afr J Biotechnol*, 2009, 8, 3018-3022.

Zechel, DL; Withers, SG. Glycosidase mechanisms: anatomy of a finely tuned catalyst. *Acc. Chem Res* 2000, 333, 11-18.

Zechel,DL; Withers,SG. Dissection of nucleophilic and acid–base catalysis in glycosidases. *Curr Opin Chem Biol*. 2001, 5, 643-649.

Zhang ,YH; Lynd, LR. Towards an aggregated understanding of enzymatic hydrolysis of cellulose: non complexed cellulase systems. *Biotechnol Bioeng*, 2004, 88, 797-824.

Zhang, XZ; Sathitsuksanoh, N; Zhang, YH. Glycoside hydrolase family 9 processive endoglucanase from Clostridium phytofermentans: heterologous expression, characterization, and synergy with family 48 cellobiohydrolase. *Bioresour Technol*, 2010a, 101, 5534-5538.

Zhang, XZ; Zhang, Z; Zhu, Z; Sathitsuksanoh, N; Yang, Y; Zhang, YHP. The noncellulosomal family 48 cellobiohydrolase from Clostridium phytofermentans ISDg: heterologous expression, characterization, and processivity. *Appl Microbiol Biotechnol*, 2010b, 86, 525-533.

Zhang, Z; Wang, Y; Ruan, J. Reclassification of Thermomonospora and Microtetraspora. *Int J Syst Bacteriol*, 1998, 48, 411-422.

Zhou, L; Wu, X; Lan, L; Liu, J. Expression of Trichoderma reesei endo-beta-glucanase II in silkworm, Bombyx mori L. by using BmNPV/Bac-to-Bac expression system and its bioactivity assay. *Biotechnol Lett*, 2010, 32, 67-72.

Zhou, S; Yomano, LP; Saleh, AZ; Davis, FC; Aldrich, HC; Ingram, LO. Enhancement of expression and apparent secretion of Erwinia chrysanthemi endoglucanase (encoded by celZ) in Escherichia coli B. *Appl Environ Microbiol*, 1999, 65, 2439-2445.

Zhu, H; Yao, S; Wang, S. MFα Signal peptide enhances the expression of cellulase eg1 gene in yeast. *Appl Biochem Biotechnol*, 2010, 162, 617-624.

Zhu, Y; Nishino, N; Kishida, Y;Uchida, S. Ensiling characteristics and ruminal degradation of Italian ryegrass and lucerne silages treated with cell wall degrading enzymes. *J Sci Food Agric*, 1999,79, 1987-1992.

Ziegelhoffer, T; Raasch, JA; Austin-Phillips, S. Dramatic effects of truncation and sub-cellular targeting on the accumulation of recombinant microbial cellulase in tobacco. *Mol Breeding*, 2001, 8, 147–158.

Ziegelhoffer, T; Raasch, JA; Austin-Phillips, S. Expression of Acidothermus cellulolyticus E1 endo-β-1,4-glucanase catalytic domain in transplastomic tobacco. *Plant Biotechnol J*, 2009, 7, 527-536.

Ziegelhoffer, T; Will, J; Austin-Phillips, S. Expression of bacterial cellulase gene in transgenic alfalfa (Medicago sativa L.), potato (Solanum tuberosum L.) and tobacco (Nicotiana tabacum L.). *Mol Breeding*, 1999, 5, 309–318.

Ziegler, MT; Thomas, SR; Danna, KJ. Accumulation of a thermostable endo-1,4-□-D-glucanase in the apoplast of Arabidopsis thaliana leaves. *Mol Breeding*, 2000, 6, 37–46.

Zverlov, V; Mahr, S; Riedel, K; Bronnenmeier, K. Properties and gene structure of a bifunctional cellulolytic enzyme (CelA) from the extreme thermophile Anaerocellum thermophilum with separate glycosyl hydrolase family 9 and 48 catalytic domains. *Microbiology*, 1998, 144, 457-565.

Zverlov, VV; Kellermann, J; Schwarz, WH. Functional subgenomics of Clostridium thermocellum cellulosomal genes: identification of the major catalytic components in the extracellular complex and detection of three new enzymes. *Proteomics*, 2005, 5, 3646-3653.

In: Cellulase: Types and Action, Mechanism, and Uses
Editor: Adam E. Golan
ISBN: 978-1-61761-983-0

Chapter 2

BIOTECHNOLOGICAL APPLICATIONS OF MICROBIAL CELLULASES

***Sunil Kumar*[1*], *Brijesh Kumar Mishra*[2] *and P. Subramanian*[3]**
AICRP on Post Harvest Technology, College of Technology & Engineering, Maharana Pratap University of Agriculture & Technology, Udaipur, India[1]
Department of Molecular Biology & Biotechnology, Rajasthan College of Agriculture, Maharana Pratap University of Agriculture & Technology, Udaipur, India[2]
Department of Dairy & Food Micrbiology, College of Dairy & Food Science Technology, Maharana Pratap University of Agriculture & Technology, Udaipur, India[13]

ABSTRACT

Cellulases, responsible for the hydrolytic cleavage of cellulose, are composed of a complex mixture of enzymes with different specificities to hydrolyse glycosidic bonds. Cellulases can be grouped into three major enzyme classes viz. endoglucanase, exoglucanase and β-glucosidase. Endoglucanases, often called carboxy methyl cellulases (CMCase), are proposed to initiate random attack at multiple internal sites in the amorphous regions of the cellulose fiber to open up sites for subsequent attack of cellobiohydrolases. Exoglucanase, better known as cellobiohydrolase, is the major component of the microbial cellulase system accounting for 40-70% of the total cellulase proteins and can hydrolyse highly crystalline cellulose. It removes mono-and dimers from the end of the glucose chain. β-glucosidase hydrolyse glucose dimers and in some cases cello-oligosaccharides to release glucose units. Generally, the endo- and exoglucanase work synergistically in cellulose hydrolysis but the underlined mechanism is still unclear. Microorganisms generally appear to have multiple distinct variants of endo- and exoglucanases. A diverse spectrum of cellulolytic microorganisms have been isolated and identified over the years and this list still continues to grow. Cellulases play a paramount role in natural carbon cycle by hydrolysing the lignocellulosic structures. Besides their

* Corresponding author's email: sunilsaini2007@rediffmail.com
Edted by: Dr Ajay Pal, Scientist C, Defence Food Research Laboratory, Mysore, India (Email: ajaydrdo @rediffmail.com)

applications in pharmaceutical industry, cellulases are also widely used in textile industry, in laundry detergents and in pulp and paper industry for various purposes. The cost of enzyme preparation is a major impediment in its commercial application. Recombinant DNA technology (RDT) and protein engineering have great potential for making significant improvements in increased production and higher specific activity of cellulases. The role of cellulases holds the key for transformation of organic wastes especially agricultural residue into biofuels through fermentation. Although the process is at its infant stage, this is an important aspect for sustainable development.

INTRODUCTION

Biomass can be defined as the mass of organic material from any biological entity. A wide range of biomass resources are available on this planet for conversion into useful bioproducts. These may include whole plants or their parts, plant constituents, processing byproducts, materials of marine origin and animal byproducts, municipal and industrial wastes, etc. [Smith et al., 1987]. These resources can be utilized to produce new biomaterials like bioethanol, biogas, food and animal feed and in industries like laundary, pulp and paper and textile industry, but a comprehensive understanding of raw materials' composition is essential before the desired functional elements can be obtained [Howard et al., 2003]. Lignocellulosic biomass holds the reputation of most abundant renewable source of organic matter on the earth as estimated by the Food and Agriculture Organization [FAOSTAT, 2006]. Around 2.9 x 10^3 million tons from cereal crops, 1.6 x 10^2 million tons from pulse crops, 14 million tons from oil seed crops and 5.4 x 10^2 million tons from plantation crops are produced worldwide, annually [Rajaram and Verma, 1990]. The fraction of cellulose varies in different biomasses and their parts [Table 1]. It is estimated that the yearly biomass production of cellulose alone is 1.5 trillion tons, making it an essentially inexhaustible source of raw material for ecofriendly products [Kim and Yun, 2006]. Therefore, the bioconversion of large amounts of lignocellulosic biomass into fermentable sugars has many potential biotechnological applications for sustainable development. In this domain, cellulases have a pivotal role to perform. Cellulases, which hydrolyze cellulose and other commodity chemicals to produce glucose, can be classified into three types: endoglucanase (endo-1,4-β-D-glucanase, EG, EC 3.2.1.4); cellobiohydrolase (exo-1,4-β-D-glucanase, CBH, EC 3.2.1.91) and β-glucosidase (1,4-β-D-glucosidase, BG, EC 3.2.1.21) [Hong et al., 2001; Li et al., 2006]. The endoglucanases catalyse random cleavage of internal bonds of the cellulose chain, while cellobiohydrolases attack the chain ends, releasing cellobiose. β-glucosidases are only active on cello-oligosaccharides and cellobiose, and release glucose monomers units from the cellobiose [Kumar et al. 2008]. Scientific fraternity has exploited their applications in various industries such as starch processing, animal feed production, grain alcohol fermentation, malting and brewing, extraction of fruit and vegetable juices, pulp and paper industry, and textile industry [Adsul et al., 2007; Kaur et al., 2007]. Various efforts are underway to enhance production and application in the area of cellulose biotechnology.

Table 1. Lignocellulose contents of common agricultural residues and wastes [Source: Howard et al., 2003].

Lignocellulosic materials	Cellulose (%)	Hemicellulose (%)	Lignin (%)
Hardwood stems	40-55	24-40	18-25
Softwood stems	45-50	25-35	25-35
Nut shells	25-30	25-30	30-40
Corn cobs	45	35	15
Paper	85-99	0	0-15
Wheat straw	30	50	15
Rice straw	32.1	24	18
Sorted refuse	60	20	20
Leaves	15-20	80-85	0
Cotton seeds hairs	80-95	5-20	0
Newspaper	40-55	25-40	18-30
Waste paper from chemical pulps	60-70	10-20	5-10
Primary wastewater solids	8-15	NA	24-29
Fresh bagasse	33.4	30	18.9
Solid cattle manure	1.6-4.7	1.4-3.3	2.7-5.7
Grasses	25-40	25-50	10-30

NA: Data not available

CELLULOSE BIOTECHNOLOGY

In nature, the cellulose fibers are generally embedded in a matrix of other structural biopolymers, primarily hemicelluloses and lignin. An important feature of this crystalline array is the relative impermeability to not only macromolecules like enzymes but in some cases to micromolecules even water also. At the molecular level, cellulose is a homopolysaccharide of β -D-glucopyranose units, linked by β (1, 4)-glycosidic bonds. Cellobiose is the smallest repetitive unit of cellulose and can be converted into glucose residues. The existence of several types of surface irregularities along with the crystalline and amorphous regions in the polymeric structure of cellulose provides heterogeneity to the system. This heterogeneity makes the cellulose fibers to swell when partially hydrated, with the result that the micro-pores and cavities become sufficiently large enough to allow penetration of larger molecules including enzymes [Sukumaran et al., 2005].

Microbial degradation of lignocellulosic waste is accomplished by a consortium of several substrate specific enzymes, the most prominent of which are the cellulases [Sukumaran et al., 2005]. Microbial cellulase composition may consist of one or more CBH components, one or more EG components and possibly β-glucosidases. The complete cellulase system comprising CBH, EG and BG components synergistically act together to convert crystalline cellulose into glucose units. Cellulases hydrolyze cellulose (β-1, 4-D-glucan linkages) and produce glucose as primary products along with cellobiose and cellooligosaccharides. The exocellobiohydrolases and the endoglucanases act together to hydrolyze cellulose to small cellooligosaccharides. The oligosaccharides (mainly cellobiose) are subsequently hydrolyzed to glucose by a major β-glucosidase [Sukumaran et al., 2005].

Bioconversion of cellulose into fermentable sugars is an emerging aspect of biotechnology that has invested enormous research efforts, as it is a prerequisite for the subsequent production of bioenergy. Although sugars and starch comprise the feedstock for 90% of the produced bio-ethanol today, but the most prevalent forms of sugar in nature are cellulose and hemicellulose. Cellulosic biomass can be converted to bio-ethanol by hydrolysis followed by fermentation. In hydrolysis, the biomass is converted into sugar, and the resulting sugar is converted to ethanol in fermentation [Kumar et al. 2008]. This process is enormously complicated than just fermentation of hexose sugars [De Ruyck et al., 1996] and still far from being cost effective as compared to the production of bioethanol from starch or sugar crops.

Potential Sources of Cellulolytic Enzymes

Cellulases are inducible enzymes and the most problematic and expensive aspect of its industry scale production is use of an appropriate inducer. Cellulase production on a commercial scale is induced by growing the fungus on solid cellulose or by culturing the organism in the presence of a disaccharide inducer such as lactose. However, on an industrial scale, both methods of induction results in high costs. Since the enzymes are inducible by cellulose, it is possible to use cellulose containing media for production but here again the process is controlled by the dynamics of induction and repression [Sukumaran et al., 2005].

While several bacteria, fungi and mushrooms can metabolize cellulosic biomass as an energy source [Maheshwari et al. 2009], only few strains are capable of secreting a complex of cellulase enzymes, which could have practical application in the enzymatic hydrolysis of cellulose. Besides *Trichoderma reesei,* other fungi such as *Humicola, Penicillium* and *Aspergillus* have the ability to yield high levels of extracellular cellulases. Aerobic bacteria such as *Cellulomonas, Cellovibrio* and *Cytophaga* are capable of cellulosic biomass degradation during solid state as well as submerged fermentation [Mishra et al. 2007; Mishra and Nain, 2010]. However, the microbes commercially exploited for cellulase preparations are mostly limited to *T. reesei, H. insolens, A. niger, Thermomonospora fusca, Bacillus* sp, and a few other organisms (Table 2).

The search for potential microbial strains of cellulolytic enzymes is continuing in the interest of cellulose biotechnology. Although various microorganisms of bacterial as well as fungal origin have been evaluated for their ability to degrade cellulosic substrates into glucose monomers, relatively very few microorganisms have been screened for their cellulase production potential [Das et al., 2007; Yu et al., 2007]. Generally, microorganisms are reported to secrete either endoglucanase or β-glucosidase (components of cellulase complex). Only those organisms, which produce appropriate levels of endoglucanase, exoglucanase and β-glucosidase, would effectively be capable of degrading native lignocellulosic biomass. Wojtczak et al. [1987] first reported that several strains of *Trichoderma* produce sufficient levels of extracellular cellulase complex capable of degrading native cellulose. Since then, many microorganisms have been isolated but only a few have been exploited for commercial utilization [Demain et al., 2005; Lynd et al., 2002]. Every component of the cellulase enzyme complex (endogucanase, exoglucanase and β-glucosidase) is essential for complete cellulose hydrolysis and generally in most of the microorganisms, β-glucosidase is either lacking or present in relatively small amounts. As a result, sugars, the end product of hydrolysis, do not accumulate quickly since cellobiose inhibits the endo and exoglucanases synthesis through feedback inhibition [Bisaria and Ghose, 1981].

Table 2. Major microorganisms employed in cellulase production [Source: Sukumaran et al. 2005].

Microorganism		References
Genus	Species	
Aspergillus	A. niger	Ong et al., 2004
	A. nidulans	Kwon et al., 1992
	A. orynze (recombinant)	Takashima et al., 1998
Fusarium	F. solani	Wood and McCrae, 1977
	F. oxysporum	Ortega, 1990
Humicola	H. insolens	Schulein, 1997
	H. grisea	Takashima et al., 1996
Melanocarpus	M. albomyces	Oinonen et al., 2004
Penicillium	P. brasilianum	Jorgensen et al., 2003
	P. occitanis	Chaabouni et al., 1995
	P. decumbans	Mo et al., 2004
Trichoderma	T. reesei	Schulein, 1988
	T. longibrachiatum	Fowler et al., 1999
	T. harzianum	Galante et al., 1998
Acidothermus	A. cellulolyticus	Tucker et al., 1989
Bacillus	Bacillus sp	Mawadza et al., 2000
	Bacillus subtilis	Heck et al., 2002
Clostridium	C. acetobutylicum	Lopez-Contreras et al., 2004
	C. thremocellum	Nochure et al., 1993
Pseudomonas	P. cellulosa	Yamane et al., 1970
Rhodothemus	R. marinus	Hreggvidsson et al., 1996
Cellulomonas	C. fimi	Shen et al., 1996
	C. bioazotea	Rajoka and Malik, 1997
	C. uda	Nakamura and Kitamura, 1983
Streptomyces	S. drozdowiczii	Grigorevvski de-Limaa et al., 2005
	S. sp	Okeke and Paterson, 1992
	S. lividans	Theberge et al., 1992
Thermononospora	T. fusca	Wilson, 1988
	T. curvata	Fennington et al., 1982

One way to meet this deficiency is to add β-glucosidase to the reaction mixture containing other cellulase components. Another approach might be the design of a suitable bioreactor in which cellobiose is removed continuously from the reaction mixture and treated in a separate reactor to yield glucose. The decay of lignocellulosic material catalyzed by enzymes from cellulolytic fungi is of great significance for carbon cycling in ecosystem. The primary interest in fungal cellulases stems from the fact that several fungi produce extracellular cellulases in significant amounts. Like bacterial cellulases, fungal cellulases act synergistically with endoglucanases, exoglucanases and β-glucosidases for cellulosic hydrolysis [Zhou and Ingram, 2000]. Apart from the cellulolytic fungus *Trichoderma viride*, many other fungi produce cellulases and degrade treated cellulosic material or soluble cellulose derivatives such as carboxymethylcellulose [Mishra et al., 2009]. However, they are not very effective on crystalline cellulosic substrates. Besides *Trichoderma viride*, the other cellulase producing mesophilic strains are *Agaricus sp.*, *Ganoderma sp.*, *Fusarium oxysporium*, *Piptoporus betulinus*, *Penicillium echinulatum*, *P. purpurogenum*, *Aspergillus niger* and *A. fumigatus* etc. [Singh et al., 1989; Sharma et al., 2001; Szijarto et al., 2004; Valaskova and Baldrian, 2006; Martins et al., 2008]. The cellulases from *Aspergillus* usually have high β-glucosidase activity but lower endoglucanase levels, whereas, *Trichoderma* has high endo and exoglucanase components but lower β-glucosidase levels, and hence has limited efficiency in cellulose hydrolysis. Thermophillic fungi such as *Sporotrichum thermophile*, *Scytalidium thermophillum, Clostridium straminisolvens* and *Thermonospora curvata* also produce the cellulase complex and can degrade native cellulose [Hutnan et al., 2000; Kato et al., 2004; Kaur et al., 2004]. Such thermophilic organisms may be valuable sources of thermostable cellulases. Similarly, various bacterial strains can produce cellulase complexes aerobically as well as anaerobically. Some of the bacterial strains producing cellulases are *Rhodospirillum rubrum*, *Cellulomonas fimi*, *Clostridium stercorarium*, *Bacillus polymyxa*, *Pyrococcus furiosus*, *Acidothermus cellulolyticus*, *Saccharophagus degradans* and *Cytophaga hutchinsonii* [Kato et al., 2005; Taylor at al., 2006; Das et al., 2007; Mishra et al., 2007]. However, bacterial cellulases exist as discrete multi-enzyme complexes, called cellulosomes that consist of multiple subunits that interact with each other synergistically and degrade cellulosic substrates efficiently [Bayer et al., 2004]. The cellulosome is believed to allow concerted enzyme activity in close proximity to the bacterial cell, enabling optimum synergism among the different units of cellulase complex. Concomitantly, the cellulosome also minimizes the distance over which cellulose hydrolysis products must diffuse, allowing efficient uptake of these oligosaccharides by the host cell [Schwarz, 2001]. Cellulosome preparations from *C. thermocellum* are very efficient in hydrolyzing microcrystalline cellulose [Lamed and Bayer, 1988].

APPLICATION OF CELLULASES

Cellulases have enormous biotechnological potential for various industries including chemicals, fuel, food, brewing and wine, animal feed, textile and laundry, pulp and paper as well as in agriculture sector [Bhat, 2000; Sun and Cheng, 2002; Beauchemin et al., 2003]. It is estimated that approximately 20% of the world's sale of industrial enzymes consists of cellulases [Bhat, 2000]. Some of the emerging areas of biotechnological applications of microbial cellulases have been discussed here:

Bioethanol

Over-utilization of Earth's available fossil energy (hydrocarbons) is a major challenge for the twenty-first century. Alternative energy sources based on sustainable, regenerative and eco-friendly processes are important resources to address this challenge. Bioconverted energy products including ethanol, methane, hydrogen etc. are being considered as integral constituents of biofuels. Ethanol presently has the largest market due to its use as a chemical feedstock or as a fuel additive or primary fuel [Kerr and Service, 2005]. The production of ethanol from sugars or starch has negative impact on the economics of the process, thus making ethanol more expensive compared with fossil fuels. Hence, several attempts are being made for the production of ethanol using lignocellulosic biomass to lower the production costs [Farrell et al., 2006]. Various lignocellulosic rich crop residues like wheat straw, rice straw, corn cob, sunflower stalks, sunflower hulls and water-hyacinth have been exploited for ethanol production [Roberto et al., 2003; Sharma et al., 2004]. However, rapid and efficient fermentation of hydrolysates is limited because of the simultaneous generation of a range of inhibitory compounds during the hydrolysis of lignocellulosic materials.

Global crude oil production is predicted to decline from 25 billion barrels to approximately 5 billion barrels in 2050 [Campbell and Laherrere, 1998]. Brazil produces ethanol through cane juice fermentation whereas in the USA corn is used. In the US, fuel ethanol has been used in gasohol or oxygenated fuels since the 1980s. These gasoline fuels contain up to 10% ethanol by volume [Sun and Cheng, 2002]. It is estimated that 4540 million litres of ethanol is used by the US transportation sector and this number will rise phenomenally since the US automobile manufacturers plan to manufacture a significant number of flexi-fueled engines which can use an ethanol blend of 85% ethanol and 15% gasoline by volume [Sun and Cheng, 2002]. In India, 5% ethanol blending in petrol had been made mandatory but due to high cost of ethanol from sugar molasses, now the limit has been changed to voluntary blending. Therefore, the bioconversion of lignocellulosic biomass into bioethanol is pertinent for the developing countries like India.

Perhaps, the most important application of cellulases currently being actively investigated is in the utilization of lignocellulosic wastes for the production of biofuel. The lignocellulosic residues represent the most abundant renewable resource available to mankind but their use is limited only due to lack of cost effective technologies. A potential application of cellulase is the conversion of cellulosic materials to glucose and other fermentable sugars, which in turn can be used as microbial substrates for the production of single cell proteins or a variety of fermentation products like ethanol [Kundu et al., 1983]. The strategy employed currently in bioethanol production from lignocellulosic residues is a multi-step process involving pre-treatment of the residue to remove lignin and hemicellulose fraction, cellulase treatment at 50 ^{o}C to hydrolyze the cellulosic residue to generate fermentable sugars, and finally use of a fermentative microorganism to produce alcohol. The cellulase preparation needed for the bio-ethanol plant is prepared in the premises using same lignocellulosic residue as substrate, and the organism employed is almost always *T. ressei.* To develop efficient technologies for biofuel production, significant research has been directed towards the identification of efficient cellulase systems and process conditions. In addition, studies have also been directed at the biochemical and genetic improvement of the existing organisms utilized in the process. The use of pure enzymes in the conversion of biomass to ethanol or to fermentation products is currently uneconomical due to the high cost of commercial cellulases. Effective strategies are yet to resolve and active research has to be taken up in this direction. Overall, cellulosic biomass is an attractive resource that can serve as substrate for the production of many value added metabolites and different cellulolytic enzymes that can be used at commercial level [Sukumaran et al., 2005].

Biogas Production

Biogas (methane) has the potential to yield more energy than any other current type of bio-fuel (e.g. bio-diesel, bio-ethanol). Biogas can be produced from a wide range of conventional lignocellulosic biomass [Antony et al., 2007; Levin et al., 2007]. The experimental evidences suggested that maize, wheat, rye, sunflower and other variety of lignocellulosic biomass can be utilized efficiently for biogas production [Amon et al., 2007]. The yield of methane was observed to be 1,500-2,000 metric tons per hectare per year when maize was used as a lignocellulosic substrate while yield in the range of 3,200-4,500 metric ton per hectare per year was achieved using cereal crop wastes. Apart from these, other lignocellulosic materials derived from sunflowers and alpine grass have also been reported as potential substrate for biogas production (2,600–4,550 metric ton per hectare per year) [Amon et al., 2007]. Hydrogen has also been regarded as a viable energy option. It has been demonstrated that the indigenous microbes were capable of producing significant amounts of hydrogen by fermentation of aqueous hydrolysates of the steam-pretreated hemicellulosic fraction of corn stover [Rohit et al., 2007]. Application of cellulosic enzymes/microorganisms to the lignocellulosic biomass may hasten the process of decomposition which in turn will increase the biogas production [Khandelwal, 2004].

Textile Industry

Cellulases have become the third largest group of enzymes used in the industry after a decade of their introduction. They are used in the biostoning of denim garments for producing softness and the faded look of denim garments replacing the use of pumica stones which were traditionally employed in the industry [Bhat, 2000]. They act on the cellulose fiber to release the indigo dye used for coloring the fabric, producing the faded or rugged look of denim. *H. insolens* cellulase is most commonly employed in the biostoning, though use of acidic cellulase from *Trichoderma* along with proteases is also found equally good. Cellulases are utilized for digesting off the small fiber ends protruding from the fabric resulting in a better finish. Cellulases have also been used in softening, defibrillation, and in processes for providing localized variation in the color density of fibers [Cortez et al., 2001; Sukumaran et al., 2005].

Laundry and Detergents

Cellulases, in particular endoglucanase and cellobiohydrolase, are commonly used in detergents for cleaning textiles. Several reports showed that endoglucanase variants, in particular from *T. reesei,* are suitable for use in detergents [Galante and Formantici, 2003]. *T. viride, T. harzianum* and *A. niger* are other industrially utilized natural sources of cellulases. Alkaliphilic and thermophilic cellulase preparations, mainly from *Humicola* species *(H. insolens and H. grisea* var. *thermoidea)* are commonly added in washing powders and detergents [Uhlig, 1998].

Food and Animal Feed

In food industry, cellulases are used in extraction and clarification of fruit and vegetable juices, production of fruit nectars and purees, and in the extraction of olive oil [Galante et al., 1998]. Glucanases are added to improve the malting of barley in beer manufacturing, and in wine industry, better maceration and color extraction is achieved by use of exogenous hemicellulases and glucanases. Cellulases are also used in carotenoid extraction to be used as food colourant. Enzyme preparations containing hemicellulase and pectinase in addition to cellulases are used to improve the nutritive quality of forages. Improvements in feed digestibility and animal performance have been reported using cellulases in feed processing [Graham and Balnave, 1995]. Bedford et. al. [2003] described the feed additive use of *Trichoderma* cellulases in improving the feed conversion ratio and/or increasing the digestibility of a cereal-based feed.

Keeping in view the huge market potential of fibre-degrading enzymes in animal feed industry, a number of commercial preparations have been produced [Beauchemin et al., 2001, 2003]. The use of fibre-degrading enzymes for ruminants such as cattle and sheep for improving feed utilization, milk yield and body weight gain have attracted considerable interest. Steers fed with an enzyme mixture containing xylanase and cellulase showed an increased live-weight gain of approximately 30-36% [Beauchemin et al., 1995]. In dairy cows, the milk yield increased in the range of 4-16% on various commercial cellulolytic enzyme treated forages [Beauchemin et al., 2001].

Pulp and Paper Industry

In the pulp and paper industry, cellulases and hemicellulases have been employed for biomechanical pulping for modification of the coarse mechanical pulp and hand sheet strength properties, de-inking of recycled fibers and for improving drainage and runnability of paper mills [Akhtar, 1994]. Cellulases are employed in ink removal as well as coating and toners for paper. Bio-characterization of pulp fibers is another application where microbial cellulases are employed. Cellulases are also used in preparation of easily biodegradable cardboard. The enzyme is employed in the manufacture of soft paper including paper towels and sanitary paper, and the same enzymic preparation is used to remove adhered paper [Hsu and Lakhani, 2002].

Apart from above mentioned applications, cellulases are also employed in formulations for removal of industrial slime, in protoplast research, and for generation of antibacterial chitooligosaccharides to be used in food preservation, immuno-modulation and as a potent antitumor agent [Tsai et al., 2000; Qin et al., 2004].

Factors Affecting Cellulase Enzyme Production

Chemical Factors

Effect of Carbon Sources

Since any cellulose biotechnological process is likely to base on crude enzymes, it is important to increase their activities in the culture supernatants by selecting the best carbon and nitrogen sources and optimizing their concentrations [Gomes et al., 2000]. Cellulase production was found to be dependent on the nature of the carbon source used in the culture medium. Various lignocellulose carbon sources have been tested for their ability to induce cellulase production. The impact of various carbon sources on cellulase biosynthesis by *A. terreus* M11 is summarised in Table 3 [Gao et al., 2008]. Corn stover proved to be the best carbon source for CMCase, FPase and β-glucosidase production among the tested lignocellulosic biomass. The result showed that corn stover is composed of cellulose (39.54%), hemicellulose (25.76%), Klason lignin (17.49%) and ash (5.04%). Besides, the efficiency of enzyme production also depends on the bare chemical composition of the raw material, accessibility of various components and their chemical and physical associations. Wheat straw, rice straw and corn stover have been known as an ideal substrate for cellulose production [Panagiotou et al., 2003; Mishra and Nain, 2010].

Several investigations done till date have indicated that cellulases are inducible enzymes, and different carbon sources have been analysed to find their role in effecting the enzymatic levels. Cellobiose (2.95 mM) may act as an effective inducer of cellulases synthesis in *Nectria catalinensis* [Pardo and Forchiassin, 1999]. An increased rate of endoglucanase biosynthesis in *Bacillus sp.* was reported in the presence of cellobiose or glucose (0.2%) in the culture medium [Paul and Verma, 1990]. Yeoh et al. [1986] had reported the inhibition of β-glucosidase activity at higher concentrations of cellobiose and gentibiose to an extant of 80%; similarly, laminaribiose and glucose also led to a 55–60% reduction in the enzymatic activity. Later, Shiang et al. [1991] described a possible regulation mechanism of cellulose biosynthesis and proposed that sugar alcohols, sugar analogues, xylose, glucose, sucrose, sorbose, cellobiose, methylglucoside etc. at a particular concentration may induce a cellulose regulatory protein called cellulase activator molecule (CAM). The level and yield of CAM get affected possibly due to substrate concentration and some unknown factors imparted by moderators. Many different agro-industrial wastes, synthetic or natural, have been examined as the carbon source for the process. Among the cellulosic materials, sulfate pulp, printed papers, mixed waste paper, wheat straw, paddy straw, sugarcane bagasse, jute stick, carboxymethylcellulose, corncobs, groundnut shells, cotton, ball milled barley straw, delignified ball milled oat spelt xylan, larch wood xylan, etc. have been used as the substrates for cellulase production [Doppelbauer et al., 1987; Singh et al., 1990; Gunju et al., 1990;

Mishra and Nain, 2010]. The observations indicated that the production of cellulases increased with increase in substrate concentration up to 12% during solid-state-fermentation using *Aspergillus niger*. Further increase in substrate concentration decreased the production levels. This might have been due to limitation of oxygen in the central biomass of the pellets, and exhaustion of nutrients other than energy sources. Martins et al. [2008] and Steiner et al. [1993] also demonstrated that carboxymethycellulose or cereal straw (1%, w/w) would be the best carbon source compared to sawdust for CMCase and β-glucosidase production using *Chaetomium globosum* as the producer organism. In contrast, 3% malt extract or water hyacinth was found optimum for CMCase, FPase and β-glucosidase as observed with lactose as an additional carbon sources [Mukhopadhyey and Nandi, 1999]. However, the saccharification of alkali-treated bagasse at higher substrate levels of 4% w/v was also reported [Singh et al., 1990]. Interestingly, higher concentrations (2.5–6.2% w/v) of carbon source were observed to be suitable for maximum saccharification when cellobiose was supplemented into the medium containing delignified rice straw, news print or other paper wastes as substrates [Wu and Ju, 1998; Ju and Afolabi, 1999].

Effect of Nitrogen Sources

The effects of nitrogen sources on cellulase production were variable with respect to the fungi and compounds tested [Kachlishvili et al., 2006]. The enzyme production was affected significantly under different concentrations of nitrogen sources [Panagiotou et al., 2003]. With different nitrogen sources, results showed that the enzyme activities were higher with organic nitrogen as shown in Table 3. Maximum cellulase activity was obtained with yeast extract [Gao et al., 2008], though other researchers found that inorganic nitrogen sources were the optimal [Kalogeris et al., 2003a].

The effect of different inorganic nitrogen sources such as ammonium sulfate, ammonium nitrate, ammonium ferrous sulfate, ammonium chloride and sodium nitrate have been studied. Among these, ammonium sulfate (0.5 g /L) led to maximum production of cellulases [Singh et al., 1991]. In contrast, Menon et al. [1994] observed a significant reduction in enzymatic levels in the presence of ammonium salts as the nitrogen source. However, an increase in the level of β-glucosidase was reported when corn steep liquor (0.8% v/v) was added into the production medium. Corn steep liquor also resulted in 3-5 fold induction of endo- and exoglucanase levels with synthetic cellulose, wheat straw and wheat bran as the substrates. Enzyme production was sensitive to corn steep liquor (0.88 g/L), and production increased significantly when mixed nitrogen sources (corn steep liquor and ammonium nitrate) were supplied [Steiner et al., 1993]. However, additional incorporation of nitrogen sources into medium scale up the cost of the process.

Phosphorus Sources

Phosphorus is an essential requirement for fungal growth and metabolism. It is an important constituent of phospholipids involved in the formation of cell membranes. Besides its role in linkage between the nucleotides forming the nucleic acid strands, it is involved in the formation of numerous intermediates, enzymes and coenzymes essential in carbohydrate metabolism, other oxidative reactions and intracellular processes [Singh et al., 1991]. Different phosphate sources such as potassium dihydrogen phosphate, tetra-sodium pyrophosphate, sodium β-glycerophosphate and dipotassium hydrogen phosphate have been evaluated for their effect on cellulases production [Garg and Neelkantan, 1982]. It has been widely accepted that potassium dihydrogen phosphate is the most favorable phosphorus source for cellulase production.

Table 3. Effect of carbon and nitrogen sources on cellulases production of *A. terreus* M11 [Source: Gao et al., 2008].

Source	Enzyme yield (U/g dry carbon source)		
	CMCase	FPase	β-Glucanase
Carbon source			
Sugar cane	267	48	8
Rape straw	122	12	11
Bulrush straw	255	98	16
Wheat straw	417	166	87
Wheat bran	315	94	79
Corn stover	440	198	91
Nitrogen source			
Beef paste	368	183	58
Yeast extract	467	201	93
Peptone	356	177	89
Urea	318	171	65
(NH4)2SO4	186	82	41
NH4NO3	142	61	25
NaNO3	113	49	30
Yeast extract (%, w/v)			
0.5	432	175	86
1.0	469	207	102
1.5	443	184	96

Physical Factors

pH

Different physical parameters influence the cellulose bioconversion, and pH is an important factor affecting cellulase production [Pardo and Forchiassin, 1999]. The effect of pH on cellulase production was analysed using *Aspergillus niger*, and found that pH 5.5 was optimal for maximum cellulase production. On other side, the pH range of 5.5–6.5 was optimal for β-glucosidase production from *Penicillium rubrum* [Menon et al., 1994]. Eberhart et al. [1977] had reported that production and release of cellulase from *Neurospora crassa* depends on pH of the medium and maximum release occurs at pH 7, whereas the enzyme remained accumulated in the cell at pH 7.5. Similarly, pH 7 was suitable for extracellular production of cellulase from the *Humicola fuscoatra* [Rajendran et al., 1994]. Further, the adsorption behavior of cellulases was also found to be affected by pH of the medium. Kim et al. [1988] had reported maximum adsorption of cellulase from *Aspergillus phoenicus* at pH

4.8–5.5. The pH range 4.6–5.0 was found suitable for CMCase, filterpaperase (FPase) and β-glucosidase production with *Aspergillus ornatus* and *Trichoderma reesei* AYCC-26921 [Mukhopadhyey and Nandi, 1999].

Temperature

Temperature has a profound effect on lignocellulosic bioconversion. The temperature for assaying cellulase activities are generally within 50–65 °C for a variety of microbial strains [Menon et al., 1994; Steiner et al., 1993], whereas growth temperature of these microbial strains was found to be 25–30 °C [Macris et al., 1989]. Similarly *Penicillium purpurogenum*, *Pleurotus florida* and *Pleurotus cornucopiae* showed higher growth at 28 °C but maximum cellulase activities at 50 °C [Steiner et al., 1993] and about 98, 59 and 76% of the CMCase, FPase and β-glucosidase activities, respectively, retained after 48 h at 40 °C. Researchers have shown that temperature influences the cellulose-cellulase adsorption behaviour. A positive relationship between adsorption and saccharification of cellulosic substrate was observed at temperature below 60 °C. The adsorption activities beyond 60 °C decreased possibly because of the loss of enzyme configuration leading to denaturation of the enzyme activity [Van-Wyk, 1997]. Bronnenmeier and Staudenbauer [1988] reported that extracellular as well as cell bound β-glulcosidase from *Clostridium stercorarium* required an identical temperature of 65 °C for their activity. Further increase in temperature led to a sharp decrease in the enzyme activity. Some of the thermophilic fungi having maximum growth at or above 45–50 °C had produced cellulase with wide temperature optima (50–78 °C) [Wojtczak et al., 1987].

LIMITATION OF CELLULASE ACTION ON CELLULOSIC BIOMASS

Various serious obstacles in the biotechnological application of lignocellulosic biomass have been explored and one of the important constraints is structural complexity of cellulose itself. X-ray diffraction analysis revealed that cellulose exists in several crystalline forms [Blackwell, 1992] which are highly resistant to microbial and enzymatic degradation. In contrast, the amorphous regions of cellulose are hydrolyzed much faster. The rate of enzymatic hydrolysis of cellulose is greatly affected by its degree of crystallinity [Cohen et al., 2005]. Dunlap et al. [1976] had analysed the relationship between the cellulose crystallinity and its digestibility by cellulases. Cellulases degrade readily the accessible amorphous regions of regenerated cellulose but are unable to attack the less accessible crystalline region. Caulfied and Moore [1974] measured the degree of crystallinity of the ball milled cellulose before and after partial hydrolysis and observed that mechanical action (ball milling) increased the susceptibility of both the amorphous and crystalline components of cellulose. Therefore, crystallinity of natural lignocellulosic biomass is the major hinderance to its utilization to produce fermentable sugar economically.

A wide spectrum of pretreatment protocols has been investigated for hydrolysis and few have been developed to technology levels [Bisaria and Ghose, 1981; Kim et al., 1988]. The suitability of pretreatment procedures varies depending on the raw material used. Different chemical pretreatments which are generally practiced include sodium hydroxide, perchloric acid, peracetic acid, acid hydrolysis using sulfuric and formic acids and ammonia freeze explosion. In addition, a number of organic solvent like *n*-propylamine, ethylene diamine, *n*-butylamine etc. are also used for same purpose [Martinez et al., 2005]. Besides these chemical treatments, steam or acid/alkali-steam pretreatment has also been found suitable. However, use of chemicals in the pretreatment procedures is a major drawback since it affects the total economy of the bioconversion of the lignocellulosic biomass.

Hypercellulolytic Enzyme Production

There are various methods/procedures available to enhance cellulase activity as well as production which are discussed below:

Table 4. Specific activity of commercial cellulase preparations [Source: Nieves et al., 1988; Howard et al., 2003].

Preparation	Microbial source	FPase	β-Glucosidase	CMCase	Cellobiase
Biocellulase TRI	T. reesei	0.24	0.72	5.5	0.059
Biocellulase A	A. niger	0.01	1.4	3.6	ND
Cellulast 1.5L	T. reesei	0.37	0.16	5.1	0.018
Cellulase TAP10	T. viride	0.13	5.2	14	ND
Cellulase AP30K	A. niger	0.03	10	21	ND
Cellulase TRL	T. reesei	0.57	1.0	13	0.016
Econase CE	T. reesei	0.42	0.48	8.5	0.038
Multifect CL	T. reesei	0.42	0.20	7.1	0.015
Multifect GC	T. reesei	0.43	0.39	13	0.025
Spezyme #1	T. reesei	0.54	0.35	15	0.026
Spezyme #2	T. reesei	0.57	0.42	15	0.029
Spezyme #3	T. reesei	0.57	0.46	25	0.031
Ultra-low Microbial	T. reesei	0.48	0.96	ND	ND

ND = not determined

Mutagenesis

The production of cellulases by the microbial cell is regulated by genetic and biochemical controls involving induction and catabolite repression, or end product inhibition. These controls are operative under cellulase production conditions, thereby resulting in limited yields of the enzymatic constituents. The first catabolite repressed *Bacillus pumilus* with cellulase yielded four times higher than the wild type strain that was created through mutagenesis [Kotchoni et al., 2003]. Mutagenic treatments of *Trichoderma reesei* Qm 6a, a wild type strain isolated at US Army Natick Research and Development Command, Natick, USA led to the development of mutants with higher cellulolytic activity [Bisaria and Ghose,

1981]. A hypercellulolytic mutant NTG-19 from *Fusarium oxysporum* was developed [Kuhad et al., 1994] by ultraviolet treatment followed by chemical mutagenesis using NTG (100 μg ml^{-1}). The resultant mutant strain had substantially higher (80%) cellulolytic activity than its parent strain. NTG treatment of *Cellulomonas flavigena* also produced four mutants (M4, M9, M11 and M12) with improved xylanolytic activities [Reyes and Noyola, 1998]. A mutant creAd30 with the end product inhibition resistance and improved levels of D-glucose metabolism was constructed from *Aspergillus nidulans* [Veen et al., 1995]. Specific activities (U/mg) of various commercial preparations of cellulases are given in Table 4 [modified from Nieves et al., 1988]. However, these efforts did not result in robust strains capable of consistently producing ethanol at high yields under a broad range of conditions and in the hands of different investigators [Lynd et al., 2002].

Genetic Manipulations Techniques

The cellulase coding genes are located on chromosomes in bacteria and fungi, both. In fungi, cellulase genes are usually randomly distributed over the genome, with each gene having its own transcription regulatory elements [Tomme et al., 1995]. Only in exceptional cases, such as for *P. chrysosporium*, are the three cellobiohydrolase-like genes clustered [Covert et al., 1992]. A comparison of the promoter regions of *cbh1, cbh2, eg1*, and *eg2* of *T. reesei* revealed the presence of CRE1-binding sites through which catabolite repression is exerted [Kubicek et al., 1998]. ACEI and ACEII activate transcription by binding to at least the *cbh1* promoter region [Saloheimo et al., 2000]. In bacteria, the cellulase genes are either randomly distributed or clustered on the genome. The cellulase gene cluster of *C. cellulovorans* is approximately 22 Kbp in length and contains nine cellulosomal genes with a putative transposase gene in the 3' flanking region. Similar arrangements have also been found in the chromosome of *C. cellulolyticum* and *C. acetobutylicum*, suggesting the presence of a common bacterial ancestor to these mesophilic clostridia or the occurrence of transposon-mediated horizontal gene transfer events. Transcriptional terminators could be identified within these large gene clusters; however, promoter sequences have not yet been found [Tamaru et al., 2000]. Both cellulolytic bacteria and fungi (aerobic and anaerobic) primarily contain multidomain cellulases, with single-domain cellulases being the exception (e.g., EGIII of *T. reesei* and EG 28 of *P. chrysosporium* [Henriksson, 1999]). The most common modular arrangements involve catalytic domains attached to CBMs through flexible linker-rich regions. The CBM module can be either at the N or C terminus; the position is of little relevance when considering the tertiary structure of the protein. This arrangement is found predominantly in noncomplexed cellulase systems. The enzymes of complexed systems (anaerobic bacteria and fungi) are more diverse. Cellulosomal enzymes contain at least one catalytic domain linked to a dockerin. However, some enzymes contain multiple CBMs, an immunoglobulin- like domain (e.g., for CelE of *C. cellulolyticum*) [Gaudin et al., 2000] and a fibronectin type III domain (CbhA of *C. thermocellum*) [Zverlov et al., 1998]. The most complex enzymes are those of the extremely thermophilic bacteria [Bergquist, 1999]. The megazymes of the anaerobic hyperthermophile *Caldicellulosiruptor* isolate Tok7B1 often have two catalytic domains; usually a cellulase and a hemicellulase domain linked through several CBM domains [Gibbs et al., 2000].

The large number of homologous cellulase genes observed within cellulolytic organisms, between related organisms, or between distant organisms within a niche environment, such as the rumen, suggests that chromosomal rearrangements and horizontal gene transfer has contributed to the presently existing rich repertoire of cellulase systems. The presence of CBH1-like gene clusters in *P. chrysosporium* [Covert et al., 1992] and the highly homologous CelK and CbhA exoglucanases in *C. thermocellum* [Zverlov et al., 1998] suggests more recent gene duplication events. The formation of cellulases from the same family within a species but with different cellulase activity, such as EGI (Cel6B) and CBHII (Cel6A) of *T. reesei*, could represent more distant gene duplications, followed by substrate specificity divergence. The development of polyspecific families, such as the cellulases and hemicellulases in family 5, may represent common ancestor genes that underwent gene duplication followed by substantial divergence with regard to substrate specificity. Examples are the CelE (endoglucanase) and CelO (cellobiohydrolase) of *C. thermocellum* [Shoham et al., 1999] as well as EGIII (endoglucanase) [Saloheimo et al., 1988] and MANI (mannanase) in *T. reesei* [Stalbrand et al., 1995]. The different arrangement of catalytic domains and CBMs in the megazymes of the hyperthermophilic bacteria in all likelihood originated from intergenic domain shuffling through homologous or unequal crossover recombination events [Bergquist et al., 1999]. The role of horizontal gene transfer in the evolution of cellulase systems has been expected, but only recently has evidences of such events started to accumulate. The possibility that the cellulosomal gene cluster of *C. cellulovorans* could have been acquired through a transposase-mediated transfer event was discussed by Tamaru et al. [2000]. The absence of introns in the glycoside hydrolase genes of the anaerobic fungi (in contrast to aerobic fungi, which contain introns in their glycoside hydrolase genes) raised suspicion that the anaerobic fungi acquired their genes from bacteria. The high microbial density in the rumen (10^{10} to 10^{11} cfu/ml ruminal fluid) and the consequent close proximity between ruminal bacteria and fungi, provide ideal conditions for horizontal gene transfer events to occur. Horizontal gene transfer has been demonstrated in the rumen [Netherwood et al., 1999], suggesting genome plasticity in this niche that could also allow the anaerobic fungi to acquired new genes [Martin, 1999].

Genetic engineering of cellulolytic microorganisms for cellulose production will benefit from the observations obtained over the past two decades pursuant to engineering of an end product metabolism in noncellulolytic anaerobes. Examples of these results include enhancement of ethanol production in *E. coli* and *K. oxytoca* [Ingram et al., 1999], solvent production in *C. acetobutylicum* [Mitchell, 1998], and lactic acid production in yeasts [Porro et al., 1999]. In these and other cases, metabolic flux is altered by blocking undesirable pathways, typically via homologous recombination-mediated "gene knockout" [Kubo et al., 2000] and/or overexpression of genes associated with desirable pathways [Dequin et al., 1999; Harris et al., 2000]. Various microbial strains have been metabolically engineered to produce lactic acid, succinic acid, ethanol and butanol [Ishida et al., 2006; Lee et al., 2006; Romero et al., 2007]. *Corynebacterium glutamicum* was metabolically engineered to broaden its lignocellulosic substrate utilization for the production of fermentable sugar. While significant progress has been made using physical and chemical mutagens to increase the production of lignocellulolytic enzymes, RDT and protein engineering are also being used as a powerful modern approach for efficient lignocellulosic bioconversion. RDT offers significant potential for improving various aspects of lignocellulolytic enzymes such as production, specific activity, pH and temperature stability as well as creating "synthetic" designer enzymes for specific applications [Katahira et al., 2006, Hong et al., 2007]. It may also prove possible to fuse different lignocellulolytic genes or sections of genes from different organisms to produce novel chimeric proteins/enzymes with altered properties. For example, a heterologously expressed *Neocallimastrix patriciarum* CelD encoding a multi-domain, multi-functional enzyme possessing endoglucanase, cellobiohydrolase and xylanase activity exhibited higher specific activities on Avicel than cellobiohydrolase and endoglucanase of *T. reesei* [Aylward et al., 1999]. A number of designer enzymes, also called glycosynthases, including cellulases and hemicellulases, have been engineered by replacing nucleophilic residues resulting in higher yields of different oligosaccharides [Fairweather et al., 2002]. RDT can improve our understanding of the molecular mechanisms of lignocellulose degradation and the development of the bioprocessing potential of lignocellulolytic microorganisms. It is expected that for industrial applications, cellulases must have high adsorption capacities and catalytic efficiencies, high thermal stabilities and lower end product inhibition. It is therefore essential to put efforts to clone cellulose genes with desirable molecular properties. A large number of fungal and bacterial genes have been cloned in *E. coli*, recently [Wulff et al., 2006; Feng et al., 2007]. In addition, cellulase genes have also been expressed efficiently in other microbial systems such as *Penicillium crysogenum*, *Trichoderma reesei*, *Pseudomonas xuorescens* and yeast [Ouyang et al., 2006; Li et al., 2006; Hou et al., 2007; Hong et al., 2007]. The cloning and sequencing of various cellulolytic genes will help in characterizing the potential systems for economizing the process of biotechnological applications of lignocellulosic biomass in future.

Conclusion and Future Prospects

Cellulosic bioconversion, a multi-step process, requires a multi-enzyme complex for its efficient bioconversion into fermentable sugars. However, there is no known organism capable of producing all the necessary enzymes in sufficient quantities. With escalating energy demands and shrinking energy resources, the utilization of lignocellulosic biomass for biofuel production offers a renewable alternative. Apart from fermentable sugars and biofuels, other value-added products such as organic acids, solvents, drink softeners etc. may also be produced from lignocellulosic biomass using appropriate technologies. With theoretically possible looking system, there exist a number of technological lacunas. Morphological complexity and crystallinity of the lignocellulosic biomass is one of the major hurdles in the bioconversion processes. Apart from that, physical and chemical conditions required for efficient enzymatic adsorption and hydrolysis of lignocellulosic biomass are somewhat different (i.e. higher temperature) than the optimum for enzyme biosynthesis. Most of the lignocellulose degrading organisms have end product inhibition which reduces the rate of enzyme synthesis resulting in incomplete utilization of lignocellulosic biomass. Various biotechnological approaches are being used for efficient biomass conversion with limited success. Therefore, to combat the problem, various mutant strains are being developed and used at the laboratory scale. Metabolic engineering including blocking of undesirable pathways and induction of gene expression associated with desirable pathways to enhance the production of biofuels and organic acids using lignocellulosic biomass is under progress. However, no single cost effective and efficient technology is currently available to meet the challenges of large-scale utilization of lignocellulosic biomass. Further, strain improvement for enhanced cellulases biosynthesis using mutagenesis, metabolic engineering and genomics approaches, should be used for the lignocellulosic bioconversion processes into a powerful technology to produce the value added and industrially significant products in future. In nut shell, the major goals for future cellulase research would be: (1) Reduction in the cost of cellulase production and (2) Improving the performance/activity of cellulases to reduce the enzyme input. More strategic research is needed to make designer cellulase enzymes (synthetic cellulases) suited for specific applications.

REFERENCES

Adsul, M.G., Bastawde, K.B., Varma, A.J. & Gokhale, D.V. (2007). *Strain improvement of Penicillium janthinellum NCIM 1171 for increased cellulase production*. Bioresour Technol, 98, 1467–1473.

Akhtar, M. (1994), *Biochemical pulping of aspen wood chips with three strains of Ceriporiopsis subvermispora*. Holzforschung, 48, 199-202.

Amon, T., Amon, B., Kryvoruchko, V., Machmuller, A., Hopfner-Sixt, K., Bodiroza, V., Hrbek, R., Friedel, J., Potsch, E., Wagentristl, H., Schreiner, M. & Zollitsch, W. (2007). *Methane production through anaerobic digestion of various energy crops grown in sustainable crop rotations*. Bioresour Technol, 98, 3204–3212.

Antoni, D., Zverlov, V.V. & Schwarz, W.H. (2007). *Biofuels from microbes*. Appl Microbiol Biotechnol, 77, 23–35.

Aylward, J.H., Gobius, K.S., Xue, G.P., Simpson, G.D. & Dalrymple, B.P. (1999). *The Neocallimastrix patriciarum cellulase, CelD, contains three almost identical catalytic domains with high specific activity on Avicel.* Enzyme Microb Technol, 24, 609–614.

Bayer, E.A., Belaich, J.P., Shoham, Y. & Lamed, R. (2004). *The cellulosomes: multienzyme machines for degradation of plant cell wall polysaccharides*. Annu Rev Microbiol, 58, 521–554.

Beauchemin, K.A., Colombatto, D., Morgavi, D.P. & Yang, W.Z. (2003). *Use of exogenous fibrolytic enzymes to improve animal feed utilisation by ruminants*. J Animal Sci, 81(E. Suppl.2), E37-E47.

Beauchemin, K.A., Morgavi, D.P., Mcallister, T.A., Yang, W.Z. & Rode, L.M. (2001). *The use of enzymes in ruminant diets.* In Wiseman J. & Garnsworthy P.C. (Eds.), Recent Advances in Animal Nutrition (pp. 296-322) Nottingham, Nottingham University Press.

Beauchemin, K.A., Rode, L.M. & Sewalt, V.J.H. (1995). *Fibrolytic enzymes increase fibre digestibility and growth rate of steers fed dry forages*. Can J Animal Sci, 75, 641-644.

Bedford, M.R., Morgan, A.J., Fowler, T., Clarkson, K.A., Ward, M.A., Collier, K.D. & Larenas, E.A. (2003). *Enzyme feed additive and animal feed including it.* US Pat 6562340 (to Genencore International Inc. USA).

Bergquist, P.L., Gibbs, M.D., Morris, D.D., Teo, V.S.J., Saul, D.J. & Morgan, H.W. (1999). *Molecular diversity of thermophilic cellulolytic andn hemicellulolytic bacteria*. FEMS Microbiol Ecol, 28, 99–110.

Bhat, M.K. (2000). *Cellulases and related enzymes in biotechnology*. Biotechnol Adv, 18, 355-383.

Bisaria, V.S. & Ghose, T.K. (1981). *Biodegradation of cellulosic material: substrate, microorganism, enzyme and products*. Enzyme Microbiol Technol, 3, 91–104.

Blackwell, J. (1992). *Cellulose and other natural polymer system: biogenesis, structure and degradation*. In Martin A.M. (Eds.), Bioconversion of wastes material to industrial products (SEI 8 HN, pp. 197–246). London, Blackie Acad Prof.

Bronnenmeier, K. & Staudenbauer, W.L. (1988). *Purification and properties of an extracellular β-glucosidase from the cellulolytic thermophilic Clostridium stercorarium*. Appl Microbiol Biotechnol, 28, 380–386.

Campbell, C.J. & Laherrere, J.H. (1998). *The end of cheap oil*. Sci Am, 3, 78-83.

Caulfied, D.F. & Moore, W.E. (1974). *Effect of varying crystallinity of cellulose on enzymatic hydrolysis*. Wood Sci, 6, 375–379.

Chaabouni, S.E., Belguith, H., Hassairi, I., M'Rad, K. & Ellouz, R. (1995). *Optimization of cellulase production by Penicillium occitanis*. Appl Microbiol Biotechnol, 43, 267-269.

Cohen, R., Suzuki, M.R. & Hammel, K.E. (2005). *Processive endoglucanase active in crystalline cellulose hydrolysis by the brown rot basidiomycete Gloeophyllum trabeum*. Appl Environ Microbiol, 71, 2412–2417.

Cortez, J.M., Ellis, J. & Bishop, D.P. (2001). *Cellulase finishing of woven, cotton fabrics in jet and winch machines*. J Biotechnol, 89, 239-245.

Covert, S.F., Vanden, A.W. & Cullen, D. (1992). *Structure, organization, and transcription of a cellobiohydrolase gene cluster from Phanerochaete chrysosporium*. Appl Environ Microbiol, 58, 2168–2175.

Das, M., Royer, T.V. & Leff, L.G. (2007). *Diversity of fungi, bacteria, and actinomycetes on leaves decomposing in a stream*. Appl Environ Microbiol, 73, 756–767.

De Ruyck, J., Allard, G. & Maniatis, K. (1996). *An externally fired evaporative gas turbine cycle for small scale biomass CHP production*. In Chartier P. et al. (Eds.), Proceedings of the 9th European Bioenergy conference. Pergamon, Oxford.

Demain, A.L., Newcomb, M. & Wu, J.H.D. (2005). *Cellulase, Clostridia, and ethanol microbiology*. Mol Biol Rev, 69, 124–154.

Dequin, S., Baptista, E. & Barre, P. (1999). *Acidification of grape musts by Saccharomyces cerevisiae wine yeast strains genetically engineered to produce lactic acid*. Am J Enol Vitic, 50, 45–50.

Doppelbauer, P., Esterbauer, H., Steiner, W., LaVerty, R.M. & Steinmuller, H. (1987). *The use of lignocellulosic wastes for production of cellulose by Trichoderma reesei*. Appl Microbiol Biotechnol, 26, 485–494.

Dunlap, C.E., Thomson, J. & Chiang, L.C. (1976). *Treatment processes to increase microbial digestibility*. AICHE Symp Ser, 72, 58–63.

Eberhart, B.M., Beek, R.S. & Goolsby, K.M. (1977). *Cellulase of Neurospora crassa*. J Microbiol, 130, 181–186.

Fairweather, J.K., Faijes, M., Driguez, H. & Planas, A. (2002). *Specific studies of Bacillus 1,3–1–4- β-glucanase and application to glycosynthase-catalyzed transglycosylation*. Chem Biochem, 3, 866–873.

FAOSTAT (2006). *FAO statistical databases*. http://faostat.fao. org/

Farrell, A.E., Plevin, R.J., Turner, B.T., Jones, A.D., O'Hare, M. & Kammen, D.M. (2006). *Ethanol can contribute to energy and environmental goals*. Science, 113, 506–508.

Feng, Y., Duan, C.J., Pang, H., Mo, X.C., Wu, C.F., Yu, Y., Hu, Y.L., Wei, J., Tang, J.L. & Feng, J.X. (2007). *Cloning and identification of novel cellulase genes from uncultured microorganisms in rabbit cecum and characterization of the expressed cellulases.* Appl Microbiol Biotechnol, 75, 319–328.

Fennington, G., Lupo, D. & Stutzenberger, F. (1982). *Enhanced cellulase production in mutants of Thermonospora curvata.* Biotechnol Bioeng, 24, 2487-2497.

Fowler, T., Carkson, K.A., Michael, W., Collier, K.D. & Edmond, L. (1999).*Cellulase enzymes and systems for their expressions.* US Pat. 5861271 (to Genencor International, Inc. USA).

Galante, Y.M. & Formantici, C. (2003). *Enzyme applications in detergency and in manufacturing industries.* Curr Org Chem, 7, 1399-1422.

Galante, Y.M., De Conti, A & Monteverdi, R. (1998). *Application of Trichoderma enzymes in food and feed industries.* In Harman G.F. & Kubicek C.P. (Eds.), Trichoderma & Gliocladium - Enzymes, Biological Control and Commercial Applications (vol 2, pp. 327-342). London, Taylor & Francis.

Gao, J., Weng, H., Zhu, D., Yuan, M., Guan, F. & Xi, Y. (2008). *Production and characterization of cellulolytic enzymes from the thermoacidophilic fungal Aspergillus terreus M11 under solid-state cultivation of corn stover.* Bioresour Technol 99, 7623–7629.

Garg, S.K. & Neelkantan, S. (1982). *Effect of nutritional factors on cellulose enzyme and microbial protein production by Aspergillus terrus and its evaluation.* Biotechnol Bioeng, 24, 109–125.

Gaudin, C., Belaich, A., Champ, S. & Belaich, J.P. (2000). *CelE, a multidomain cellulase from Clostridium cellulolyticum: a key enzyme in the cellulosome?* J Bacteriol, 182, 1910–1915.

Gibbs, M.D., Reeves, R.A., Farrington, G.K., Anderson, P., Williams, D.P. & Bergquist, P.L. (2000). *Multidomain and multifunctional glycosyl hydrolases from the extreme thermophile Caldicellulosiruptor isolate Tok7B.1.* Curr Microbiol, 40, 333–340.

Gomes, I., Gomes, J., Gomes, D.J. & Steiner, W. (2000). *Simultaneous production of high activities of thermostable endoglucanase and betaglucosidase by the wild thermophilic fungus Thermoascus aurantiacus.* Appl Microbiol Biotechnol, 53, 461–468.

Graham, H. & Balnave, D. (1995). *Dietary enzymes for increasing energy availability.* In: Wallace R.J. & Chesson A. (Eds.) Biotechnology in Animal Feeds and Animal Feedings (pp. 296-309). Weinheim, VHC.

Grigorevski de-Limaa, A.L., do-Nascimento, R.P., da-Silva, B.E.P. & Coelho, R.R. (2005). *Streptomyces drozdowiczii cellulase production using agro-industrial by-products and its potential use in the detergent and textile industries.* Enz Microb Technol, 37, 272-277.

Gunju, R.K., Vithayuthil, P.J. & Murthy, S.K. (1990). *Factors influencing production of cellulases by Chaetomium thermophile var. coprophile.* Indian J Exp Biol, 28, 259–264.

Harris, L.M., Desai, R.P., Welker, N.E. & Papoutsakis, E.T. (2000). *Characterization of recombinant strains of the Clostridium acetobutylicum butyrate kinase inactivation mutant: need for new phenomenological models for solventogenesis and butanol inhibition.* Biotechnol Bioeng, 67, 1–11.

Heck, J.K., Hertz, P.F. & Ayub, M.A.Z. (2002). *Cellulase and xylanase production by isolated Amazon Bacillus strains using soybean industrial residue based solid-state cultivation.* Braz J Microbiol, 33, 213-218.

Henriksson, G., Nutt, A., Henriksson, H., Pettersson, B., Stahlberg, J., Johansson, G. & Pettersson, G. (1999). *Endoglucanase 28 (Cel12A), a new Phanerochaete chrysosporium cellulase.* Eur J Biochem, 259, 88–95.

Hong, J., Tamaki, H., Akiba, S., Yamamoto, K. & Kumagai, H. (2001). *Cloning of a gene encoding a highly stable endo-β-1, 4-glucanase from Aspergillus niger and its expression in yeast.* J Biosci Bioeng, 92, 434–441.

Hong, J., Wang, Y., Kumagai, H. & Tamaki, H. (2007). *Construction of thermotolerant yeast expressing thermostable cellulase genes.* J Biotechnol, 130, 114–123.

Hou, Y., Wang, T., Long, H. & Zhu, H. (2007). *Cloning, sequencing and expression analysis of the first cellulase gene encoding cellobiohydrolase-1 from a cold-adaptive Penicillium chrysogenum FS010.* Acta Biochim Biophys Sin, 39, 101–107.

Howard, R.L., Abotsi, E., Jansen van Rensburg, E.L. & Howard, S. (2003). *Lignocellulose biotechnology: issues of bioconversion and enzyme production.* African J Biotechnol, 2, 602–619.

Hreggvidsson, G.O., Kaiste, E., Holst, O., Oggertsson, G., Palsdottir, A. & Kristjansson, A.J. (1996). *An extremely thermostable cellulase from the thermophilic eubacterium Rhodothermus marinus.* Appl Environ Microbiol, 62, 3047-3049.

Hsu, J.C. & Lakhani, N.N. (2002). *Method of making absorbent tissue from recycled waste paper.* US Pat. 6413363 (to Kimberly-Clark Worldwide, Inc, Wisconsin, USA).

Hutnan, M., Drtil, M. & Mrafkova, L. (2000). *Anaerobic biodegradation of sugar beet pulp.* Biodegradation, 11, 203–211.

Ingram, L.O., Aldrich, H.C., Borges, A.C.C., Causey, T.B., Martinez, A., Morales, F., Saleh, A., Unverwood, S.A., Yomano, L.P., York, S.W., Zaldivar, J. & Zhou, S.D. (1999). *Enteric bacterial catalysts for fuel ethanol production.* Biotechnol Prog, 15, 855–866.

Ishida, N., Saitoh, S., Ohnishi, T., Tokuhiro, K., Nagamori, E., Kitamoto, K. & Takahashi, H. (2006). *Metabolic engineering of Saccharomyces cerevisiae for efficient production of pure L-(+)-lactic acid.* Appl Biochem Biotechnol, 129–132, 795–807.

Jorgensen, H., Eriksson, T., Borjesson, J., Tjerneld, F. & Oisson, L. (2003). *Purification and characterization of five cellulases and one xylanase from Penicillium brasilianum IBT* 20888. Enz Microb Technol, 32, 851-861.

Ju, L.K. & Afolabi, O.A. (1999). *Waste papers hydrolysate as soluble inducing substrate for cellulase production in continuous culture of Trichoderma reesei.* Biotechnol Prog, 15, 91–97.

Kachlishvili, E., Penninckx, M.J., Tsiklauri, N. & Elisashvili, V. (2006). *Effect of nitrogen source on lignocellulolytic enzyme production by white-rot basidiomycetes under solid-state cultivation.* World J Microbiol Biotechnol, 22, 391–397.

Kalogeris, E., Christakopoulos, P., Katapodis, P., Alexiou, A., Vlachou, S., Kekos, D. & Macris, B.J. (2003a). *Production and characterization of cellulolytic enzymes from the thermophilic fungus Thermoascus aurantiacus under solid state cultivation of agricultural wastes.* Process Biochem, 38, 1099–1104.

Katahira, S., Mizuike, A., Fukuda, H. & Kondo, A. (2006). *Ethanol fermentation from lignocellulosic hydrolysate by a recombinant xylose- and cellooligosaccharide-assimilating yeast strain.* Appl Microbiol Biotechnol, 72, 1136–1143.

Kato, S., Haruta, S., Cui, Z.J., Ishii, M. & Igarashi, Y. (2005). *Stable coexistence of five bacterial strains as a cellulose-degrading community.* Appl Environ Microbiol, 71, 7099–7106.

Kato, S., Haruta, S., Cui, Z.J., Ishii, M., Yokota, A. & Igarashi, Y. (2004). *Clostridium straminisolvens sp. nov., a moderately thermophilic, aerotolerant and cellulolytic bacterium isolated from a cellulose degrading bacterial community.* Int J Syst Evol Microbiol, 54, 2043–2047.

Kaur, G., Kumar, S. & Satyanarayana, T. (2004). *Production, characterization and application of a thermostable polygalacturonase of a thermophilic mould Sporotrichum thermophile Apinis.* Bioresour Technol, 94, 239–243.

Kaur, J., Chadha, B.S., Kumar, B.A. & Saini, H.S. (2007). *Purification and characterization of two endoglucanases from Melanocarpus sp.* MTCC 3922. Bioresour Technol, 98, 74–81.

Kerr, R.A. & Service, R.F. (2005). *What can replace cheap oil—and when?* Science, 309, 101.

Khandelwal, S.K. (2004). *Degradation of organic wastes under anaerobic conditions using chemical inducers and microorganisms for biogas production.* In Ph.D. thesis sudmitted to Department of Biochemistry, Rajasthan College of Agriculture, Maharana Pratap University of Agriculture & Technology, Udaipur, India.

Kim, D.W., Yang, J.H. & Jeong, Y.K. (1988). *Adsorption of cellulose from Trichoderma viride on microcrystalline cellulose.* Appl Microbiol Biotechnol, 28, 148–154.

Kim, J. & Yun, S. (2006). *Discovery of cellulose as a smart material.* Macromolecules, 39, 4202–4206.

Kotchoni, O.S., Shonukan, O.O. & Gachomo, W.E. (2003). *Bacillus pumilus BpCRI 6, a promising candidate for cellulase production under conditions of catabolite repression.* Afr J Biotechnol, 2, 140–146.

Kubicek, C.P. & Penttila, M.E. (1998). *Regulation of production of plant polysaccharide degrading enzymes by Trichoderma.* In Harman G.E. & Kubicek C.P. (Eds.), Trichoderma and Gliocladium (Vol. 2, pp. 49-72). London, Taylor & Francis Ltd.

Kubo, Y., Takagi, H. & Nakamori, S. (2000). *Effect of gene disruption of succinate dehydrogenase on succinate production in a sake yeast strain.* J Biosci Bioeng, 90, 619–624.

Kuhad, R.C., Kumar, M. & Singh, A. (1994). *A hypercellololytic mutant of Fusarium oxysporum.* Lett Appl Microbiol, 19, 397–400.

Kumar, R., Singh, S. & Singh, O.V. (2008). *Bioconversion of lignocellulosic biomass: biochemical and molecular perspectives.* J Ind Microbiol Biotechnol, 35, 377-391.

Kundu, S., Ghose, T.K. & Mukhopadhyey S.N. (1983). *Bioconversion of cellulose into ethanol by Clostridium thermocellum – product inhibition.* Biotechnol Bioeng, 25, 1109-1126.

Kwon, K.S., Kang, H.G. & Hah, Y.C. (1992). *Purification and characterization of two extracellular β-glucosidases from Aspergillus nidulans.* FEMS Microbiol Lett., 97, 149-154.

Lamed, R. & Bayer, E.A. (1988). *The cellulosome of Clostridium thermocellum.* Adv Appl Microbiol, 33, 1–46.

Lee, S.J., Song, H. & Lee, S.Y. (2006). *Genome-based metabolic engineering of Mannheimia succiniciproducens for succinic acid production.* Appl Environ Microbiol, 72, 1939–1948.

Levin, D.B., Zhu, H., Beland, M., Cicek, N. & Holbein, B.E. (2007). *Potential for hydrogen and methane production from biomass residues in Canada.* Bioresour Technol, 98, 654–660.

Li, Y.H., Ding, M., Wang, J., Xu, G.J. & Zhao, F. (2006). *A novel thermoacidophilic endoglucanase, Ba-EGA, from a new cellulose degrading bacterium, Bacillus sp.* AC-1. Appl Microbiol Biotechnol, 70, 430–436.

Lopez-Contreras, A.M., Gabor, K., Martens, A.A., Renckens, B.A.M., Claassen, P.A.M., van der Oost, J & de Vos, W.M. (2004). *Substrate-induced production and secretion of cellulases by Clostridium acetobutylicum.* Appl Environ Microbiol, 70, 5238-5243.

Lynd, L.R., Weimer, P.J., van Zyl, W.H. & Pretorius, I.S. (2002). *Microbial cellulose utilization: fundamentals and biotechnology.* Microbiol Mol Biol Rev, 66, 506–577.

Macris, B.J., Kekos, D. & Evangelidou, E. (1989). *A simple and inexpensive method for cellulose and β-glucosidase production by Neurospora crassa.* Appl Microbiol Biotechnol, 31, 150–151.

Maheshwari, P., Mishra, B.K., Kumar, S. & Doshi, A. (2009). *Cellulolytic enzyme profile of different mushrooms species on maize residue.* Environ. Ecol., 27, 689-691.

Martin, W. (1999). *Mosaic bacterial chromosomes: a challenge on route to a tree of genomes.* Bioessays, 21, 99–104.

Martinez, A.T., Speranza, M., Ruiz-Duenas, F.J., Ferreira, P., Camarero, S., Guillen, F., Martinez, M.J., Gutierrez, A. & del Rio, J.C. (2005). *Biodegradation of lignocellulosics: microbial, chemical, and enzymatic aspects of the fungal attack of lignin.* Int Microbiol, 8, 195–204.

Martins, L.F., Kolling, D., Camassola, M., Dillon, A.J., Ramos, L.P. (2008). *Comparison of Penicillium echinulatum and Trichoderma reesei cellulases in relation to their activity against various cellulosic substrates.* Bioresour Technol, 99, 1417–1424.

Mawadza, C., Hatti-Kaul, R., Zvauya, R. & Mattiasson, B. (2000). *Purification and characterization of cellulases production by two Bacillus strains.* J Biotechnol, 83, 177-187.

Menon, K., Rao, K.K. & Pushalkar, S. (1994). *Production of β-glucosidase by Penicillium rubrum O stall.* Indian J Exp Biol, 32, 706–709.

Mishra, B.K. & Nain, L. (2010). *Rice straw as a substrate for lignocellulolytic enzymes production form Phanerochaete chrysosporium and cellulolytic bacteria.* J Mycol Pl Pathol, 40, 110-114.

Mishra, B.K., Joshi, A. & Maloo, S.R. (2009). *Cellulolytic enzyme activity of mushrooms grown on carboxymethylcellulose as sole carbon source.* Adv Experimental Agric, 1, 139-143.

Mishra, B.K., Pandey, A. & Lata (2007). *Ligocellulolytic enzyme production from submerged fermentation of paddy straw.* Indian J Microbiol, 47, 176-179.

Mitchell, W.J. (1998). *Physiology of carbohydrate to solvent conversion by Clostridia.* Adv Microbiol Physiol, 39, 31–130.

Mo, H., Zhang, X. & Li, Z. (2004). *Control of gas phase for enhanced cellulase production by Penicillium decumbens in solid-state culture.* Process Biochem, 39, 1293-1297.

Mukhopadhyey, S. & Nandi, B. (1999). *Optimization of cellulose production by Trichoderma reesei ATTCC 26921 using a simplified medium on water hyacinth biomass.* J Sci Ind Res, 58, 107–111.

Nakamura, K. & Kitamura, K. (1983). *Purification and some properties of a cellulase active on crystalline cellulose from Cellulomonas uda.* J Ferment Technol, 61, 379-382.

Netherwood, T., Bowden, R., Harrison, P., O'Donnell, A.G., Parker, D.S. & Gilbert, H.J. (1999). *Gene transfer in the gastrointestinal tract.* Appl Environ Microbiol, 65, 5139–5141.

Nieves, R.A., Ehrman, C.I., Adney W.S., Elander, R.T. & Himmel, M.E. (1998). *Technical communication: survey and commercial cellulase preparations suitable for biomass conversion to ethanol.* World J Microbiol Biotechnol, 14, 301-304.

Nochure, S.V., Roberts, M.F. & Demain, A. (1993). *True cellulase production by Clostridium thermocellum grown on different carbon sources.* Biotech Lett, 15, 641-646.

Oinonen, A.M., Londesborough, J., Joutsjoki, V., Lantto, R. & Vehmaanpera, J. (2004). *Three cellulases from Melanocarpus albomyces for textile treatment at neutral pH.* Enz Microb Technol, 34, 332-341.

Okeke, B.C. & Paterson, A. (1992). *Simultaneous production and induction of cellulolytic and xylanolytic enzymes in the Streptomyces sp.* World J Microbiol Biotechnol, 8, 483-487.

Ong, L.G., Abd-Aziz, S., Naraini, S., Karim, M.I. & Hassan M.A. (2004*). Enzyme production and profile by Aspergillus niger during solid substrate fermentation using palm kernal cake as substrate.* Appl Biochem Biotechnol, 118, 73-79.

Ortega, J. (1990). *Production of extracellular cellulolytic enzymes by Fusarium oxysporum lycopersici.* Tex J Sci, 42, 405-410.

Ouyang, J., Yan, M., Kong, D. & Xu, L. (2006). *A complete protein pattern of cellulase and hemicellulase genes in the filamentous fungus Trichoderma reesei.* Biotechnol J, 1, 1266–1274.

Panagiotou, G., Kekos, D., Macris, B.J., Christakopoulos, P. (2003). *Production of cellulolytic and xylanolytic enzymes by Fusarium oxysporum grown on corn stover in solid state fermentation.* Ind Crop Prod, 18, 37–45.

Pardo, A.G. & Forchiassin, F. (1999). *Influence of temperature and pH on cellulase activity and stability in Nectria catalinensis.* Rev Argent Microbiol, 31, 31–35.

Paul, J. & Verma, A. (1990). *Influence of sugars on endoglucanase and β-xylanase of a bacillus strain.* Biotech Lett, 22, 61–64.

Porro, D., Bianchi, M.M., Brambilla, L., Menghini, R., Bolzani, D., Carrera, V., Lievense, J., Liu, C.L., Ranzi, B.M., Frontali, L. & Alberghina, L. (1999). *Replacement of a metabolic pathway for large-scale production of lactic acid from engineered yeasts.* Appl Environ Microbiol, 65, 4211–4215.

Qin, C., Zhou, B., Zeng, L., Zhang, Z., Liu, Y., Du, Y. & Xiao, L. (2004). *The physicochemical properties and antitumor activity of cellulase-treated chitosan,* Food Chem, 84, 107-115.

Rajaram, S. & Verma, A. (1990). *Production and characterization of xylanase from Bacillus thermoalkalophilus growth on agricultural wastes.* Appl Microbiol Biotechnol, 34, 141–144.

Rajendran, A., Gunasekaran, P. & Lakshmanan, M. (1994). *Cellulase activity of Humicola fuscoatra.* Indian J Microbiol, 34, 289–295.

Rajoka, M.I. & Malik K.A. (1997). *Cellulase production by Cellulomonas biazotea cultured in media containing different cellulosic substrates.* Bioresour Technol, 59, 21-27.

Reyes, L.M. & Noyola, T.P. (1998). *Isolation of a hyperxylanolytic Cellulomonas flavigena mutant growing on continuous culture on sugarcane bagasse*. Biotechnol Lett, 20, 443–446.

Roberto, I.C., Mussatto, S.I. & Rodrigues, R.C.L.B. (2003). *Dilute-acid hydrolysis for optimization of xylose recovery from rice straw in a semi-pilot reactor*. Ind Crops Prod, 17, 171–176.

Rohit, D., Jie, H., Pin-Ching, M., Ali, M., Stefan, C. & Esteban, C. (2007). *Hydrogen production from the fermentation of corn stover biomass pretreated with a steam-explosion process*. Int J Hydrogen Energy, 32, 932–939.

Romero, S., Merino, E., Bolivar, F., Gosset, G. & Martinez, A. (2007). *Metabolic engineering of Bacillus subtilis for ethanol production: lactate dehydrogenase plays a key role in fermentative metabolism*. Appl Environ Microbiol, 73, 5190–5198.

Saloheimo, A., Aro, N., Ilmen, M. & Penttila, M. (2000). *Isolation of the ace1 gene encoding a Cys2-His2 transcription factor involved in regulation of activity of the cellulase promoter cbh1 of Trichoderma reesei*. J Biol Chem, 275, 5817–5825.

Saloheimo, M., Lehtovaara, P., Penttila, M., Teeri, T.T., Stahlberg, J., Johansson, G., Pettersson, G., Claeyssens, M., Tomme, P. & Knowles, J.K.C. (1988). *EGIII, a new endoglucanase from Trichoderma reesei the characterization of both gene and enzyme*. Gene, 63, 11–21.

Schulein, M. (1988). *Cellulases of Trichoderma reesei*. In Wood W.A. & Abelson J.N. (Eds.), Methods in Enzymology (Vol 160, pp. 234-242). New York, Academic Press.

Schulein, M. (1997). *Enzymatic properties of cellulases from Humicola insolens*. J Biotechnol, 57, 71-81.

Schwarz, W.H. (2001). *The cellulosome and cellulose degradation by anaerobic bacteria*. Appl Microbiol Biotechnol, 56, 634–649.

Sharma, A., Khare, S.K. & Gupta, M.N. (2001). *Hydrolysis of rice hull by crosslinked Aspergillus niger cellulase*. Bioresour Technol, 78, 281–284.

Sharma, S.K., Kalra, K.L. & Kocher, S. (2004). *Fermentation of enzymatic hydrolysate of sunflower hulls for ethanol production and its scale up*. Biomass Bioenergy, 27, 399–402.

Shen, H., Meinke, A., Tomme, P., Damude, H.G. & Kwan, E. (1996). *Cellulomonas fimi cellobiohydrolases*. In Saddler J.N. & Penner M.H. (Eds.), Enymatic Degradation of Insoluble Carbohydrates (pp. 174-196), London, Oxford University Press.

Shiang, M., Linden, J.C., Mohagheghi, A., Grohmam, K. & Himmel, M.E. (1991). *Characterization of eng F, a gene for a non-cellulosomal Clostridium cellulovoras endoglucanase*. Gene, 182, 163–167.

Shoham, Y., Lamed, R. & Bayer, E.A. (1999). The cellulosome concept as an efficient microbial strategy for the degradation of insoluble polysaccharides. Trends Microbiol, 7, 275–281.

Singh, A., Abidi, A.B., Darmwal, N.S. & Agrawal, A.K. (1989). *Production of protein and cellulase by Aspergillus niger AS101 in solid state culture*. MIRCEN J, 5, 451–456.

Singh, A., Abidi, A.B., Darmwal, N.S. & Agrawal, A.K. (1990). *Saccharification of cellulosic substrates by Aspergillus niger cellulase*. World J Microbiol Biotechnol, 6, 333–336.

Singh, A., Abidi, A.B., Darmwal, N.S. & Agrawal, A.K. (1991). *Influency of nutritional factors on cellulose production from natural cellulosic residues by Aspergillus niger* AS101. Agri Biol Res, 7, 19–27.

Smith, J.E., Anderson, J.G. & Senior, E.K. (1987). *Bioprocessing of lignocelluloses.* Phil Trans R Soc Lond Ser A, 321, 507-521.

Stalbrand, H., Saloheimo, A., Vehmaanpera, J., Henrissat, B. & Penttila, M. (1995). *Cloning and expression in Saccharomyces cerevisiae of a Trichoderma reesei β-mannanase gene containing a cellulose binding domain.* Appl Environ Microbiol, 61, 1090–1097.

Steiner, J., Saccha, C. & Enzyaguirre, J. (1993). *Culture condition for enhanced cellulose production by a native strain of Penicillium purpurogenum.* World J Microbiol Biotechnol, 10, 280–284.

Sukumaran, R.K., Singhania, R.R. & Pandey, A. (2005). *Microbial cellulases: Production, application and challenges.* J Sci Ind Res, 64, 832-844.

Sun, Y. & Cheng, J. (2002). *Hydrolysis of lignocellulosic material from ethanol production: A review.* Bioresour Technol, 83, 1-11.

Szijarto, N., Szengyel, Z., Liden, G. & Reczey, K. (2004). *Dynamics of cellulase production by glucose grown cultures of Trichoderma reesei Rut-C30 as a response to addition of cellulose.* Appl Biochem Biotechnol, 113–116, 115–124.

Takashima, S., Likura, H., Nakamura, A., Hidaka, M., Masaki, H. & Uozumi T. (1998). *Overproduction of recombinant Trichoderma reesei celluloses by Aspergillus oryzae and their enzymatic properties.* J Biotechnol, 65, 163-171.

Takashima, S., Nakamura, A., Masaki, H. & Uozomi, T. (1996). *Purification and characterization of cellulases from Humicola grisea.* Biosci Biochem, 60, 77-82.

Tamaru, Y., Karita, S., Ibrahim, A., Chan, H. &. Doi, R.H. (2000). *A large gene cluster for the Clostridium cellulovorans cellulosome.* J Bacteriol, 182, 5906–5910.

Taylor, L.E., Henrissat, B., Coutinho, P.M., Ekborg, N.A., Hutcheson, S.W. & Weiner, R.M. (2006). *Complete cellulase system in the marine bacterium Saccharophagus degradans strain 2-40T.* J Bacteriol, 188, 3849–3861.

Theberge, M., Lacaze, P., Shareck, F., Morosoli, R. & Kluepfel, D. (1992). *Purification and characterization of an endoglucanase from Streptomyces lividans 66 and DNA sequence of the gene.* Appl Environ Microbiol, 58, 815-820.

Tomme, P., Heriban, V. & Claeyssens, M. (1990). *Adsorption of two cellobiohydrolases from Trichoderma reesei to Avicel: evidence for "exo-exo"synergism and possible "loose complex" formation.* Biotechnol Lett, 12, 525–530.

Tsai, G.J., Wu, Z.Y. & Su, W.H. (2000). *Antibacterial activity of a chitooligosaccharide mixture prepared by cellulase digestion of shrimp chitosan and its application to milk preservation.* J Food Prot, 63, 747-752.

Tucker, M.P., Mohagheghi, M., Grohmann, K. & Himmel, M.E. (1989). *Ultra-thermostable cellulases from Acidothermus celluloyticus: Comparison of temperature optima with previously reported cellulases.* Biol Technol, 7, 817-820.

Uhlig, H. (1998). *Industrial Enzymes and their Applications* (Eds.). (pp. 435). New York, John Wiley & Sons Inc.

Valaskova, V. & Baldrian, P. (2006). *Degradation of cellulose and hemicelluloses by the brown rot fungus Piptoporus betulinus production of extracellular enzymes and characterization of the major cellulases.* Microbiology, 152, 3613–3622.

Van-Wyk, J.P.H. (1997). *Cellulose adsorption–desorption and cellulose saccharification during enzymatic hydrolysis of cellulose material.* Biotech Lett, 19, 775–778.

Veen, P.W.D., Ruijter, G.J.G. & Visser, J. (1995). *An extreme cre A mutation in Aspergillus nidulans has severe effects on D-glucose utilization.* Microbiol, 141, 2301–2306.

Wilson, D.B. (1988). *Cellulases of Thermomonospora fusca.* In Abelson J.A. et al. (Eds.), Methods in Enzymology: Biomass, Part A: Cellulose and Hemicellulose (Biomass Vol.160 pp. 314-323), San Diego, CA, Academic Press.

Wojtczak, G., Breuil, C., Yamuda, J. & Saddler, J.N. (1987). *A comparision of the thermostability of cellulose from various thermophilic fungi.* Appl Miocrobiol Biotechnol, 27, 82–87.

Wood, T.M. & McCrae, S.I. (1977). *Cellulase from Fusarium solani: Purification and properties of C1 componen*t. Carbohydr Res, 57, 117-133.

Wu, J. & Ju, L.K. (1998). *Enhancing enzymatic saccharification of waste news print by surfactant addition.* Biotechnol Prog, 14, 649–652.

Wulff, N.A., Carrer, H. & Pascholati, S.F. (2006). *Expression and purification of cellulase Xf818 from Xylella fastidiosa in Escherichia coli.* Curr Microbiol, 53, 198–203.

Yamane, K., Suzuki, H. & Nisizawa, K. (1970). *Purification and properties of extracellular and cellbound cellulase components of Pseudomonas fluorescens var. cellulasa.* J Biochem, 67, 19-35.

Yeoh, H.H., Tan, T.K. & Koh, S.K. (1986*). Kinetic properties of β-glucosidase from Aspergillus ornatus.* Appl Microbiol Biotechnol, 25, 25–28.

Yu, H., Zeng, G., Huang, H., Xi, X., Wang, R., Huang, D., Huang, G. & Li, J. (2007). *Microbial community succession and lignocellulose degradation during agricultural waste composting*. Biodegradation, 18, 793–802.

Zhou, S. & Ingram, L.O. (2000). *Synergistic hydrolysis of carboxymethyl cellulose and acid-swollen cellulose by two endoglucanases (CelZ and CelY) from Erwinia chrysanthemi.* J Bacteriol, 182, 5676–5682.

Zverlov, V.V., Velikodvorskaya, G.A., Schwarz, W.H., Bronnenmeier, K., Kellermann, J. & Staudenbauer, W.L. (1998). *Multidomain structure and cellulosomal localization of the Clostridium thermocellum cellobiohydrolase CbhA.* J Bacteriol, 180, 3091–3099.

In: Cellulase: Types and Action, Mechanism, and Uses
Editor: Adam E. Golan ISBN: 978-1-61761-983-0

Chapter 3

CELLULASES: FROM PRODUCTION TO BIOTECHNOLOGICAL APPLICATIONS

Rodrigo Pires do Nascimento and Rosalie Reed Rodrigues Coelho
Dep. Microbiologia Geral, Instituto de Microbiologia Prof. Paulo de Goes,
Universidade Federal do Rio de Janeiro, Rio de Janeiro, Brazil

ABSTRACT

Cellulases are well established in different industrial areas, and are currently the third largest industrial enzyme worldwide, by dollar volume, mainly because of their use in cotton processing and paper recycling, as detergent industry enzymes, and in juice extraction and animal feeding additives as well. Nowadays the cellulases are the most important enzyme group for studies aiming at the so called second generation ethanol production and others chemicals products. The cellulase group involves three different enzymes: β-1,4-endoglucanase (EC 3.2.1.4), β–1,4-exoglucanase (EC 3.2.1.91) and cellobiase (EC 3.2.1.21), that are produced by an array of microorganisms, including bacteria and fungi. For cellulase production economically viable the raw material needs to be cheap. There are many types of low cost carbon sources that could be used for cellulases production, such as sugar cane bagasse, sugar cane straw, wheat straw, wheat bran, corn cobs, etc, reducing the costs effects and being friendly environmentally. In this chapter, the importance of using agro-industrial by-products as raw material for cellulase production will be addressed, as well as its biotechnological application in industry.

LIGNOCELLULOSIC BIODEGRADATION: USE OF PLANT BIOMASS FOR CELLULASES PRODUCTION

The use of biomass as fuel by humans is a very ancient activity but the use of steam technology was widespread only in the nineteenth century, when the burning of wood was used to move trains and boats. Latter the advance of technology allowed the use of other types of energy, such as vegetable oils and, more recently, the production of ethanol from vegetables, specially sugar cane, corn and beets. However, activities related to the use of biomass, such as processing of agricultural products as in the paper industry, constantly generate a lot of waste. These by-products, with high energy and hard to degrade, end up causing a variety of environmental problems. Considering these aspects, and also the low value of these wastes, studies concerning the use of this biomass to produce enzymes, ethanol and other biofuels has been stimulated in recent years. Indeed the concept of producing products from agricultural commodities (i.e., biomass) is not new. However, using biomass as an input to produce multiple products using complex processing methods, an approach similar to a petroleum refinery where fossil fuels are used as input, is relatively new (Fernando et al., 2006).

The anthropogenic activities generate tons of residues annually from plant biomass. The biomass of the world is synthesized by the photosynthetic process pathway, which converts atmospheric carbon dioxide to sugar. The amount of solar energy received at the earth's surface is 56,212 joules /year, more than 12,000 times the present human requirement of 55,991 joules/year, and approximately 4,000 times the energy humans are projected to use in 2050 (Kumar et al., 2008, Demain et al. 2005). Plants use glucose to synthesize complex materials as biomass. These consist of carbohydrates (cellulose, hemicelluloses, pectin, starch, etc.), lignin, proteins, fats, and to a lesser extent, various other chemicals, such as vitamins, dyes, and flavors. The goal of a biorefinery is to transform such plentiful biological materials into useful products using a combination of technologies and processes. Figure 1 describes the elements of a biorefinery in which biomass feedstocks are used to produce various useful products such as fuel, power, and chemicals using biological and chemical conversion processes (Fernando et al., 2006). Biomass in the biorefinery could include grains such as corn, wheat and barley, oils, agricultural residues (sugar cane bagasse and straw, sisal waste, wheat bran) and waste wood (Kumar et al., 2008).

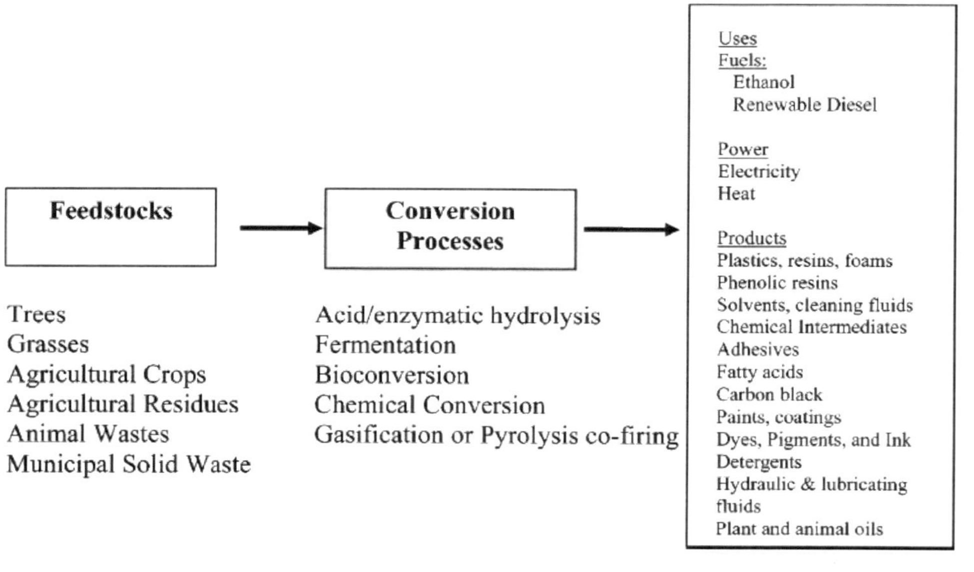

Figure 1. Simple three-step biomass-process-products procedure (Sokhansanj et al., 2005, Fernando et al. 2006).

The use of grains and oils for energy reduces their availability for use as food or feed. Corn stover and sugar cane bagasse are the leading candidate as a biomass source to support a lignocellulosic biorefinery because of the large quantities available. It has been estimated that in the USA there is a potential supply of corn stover between 60 to 100 million tons per year (Kadam & McMillan, 2003).

Brazil has various agro-products such as coffee, soybean, cassava, corn, fruits, sugarcane, and several others. However, sugarcane has been one of the main products for several decades. In 1970, 50 million tons of sugarcane was produced, yielding approximately 5 million tons of sugar. At that time the Brazilian government developed a huge program aiming at ethanol production, called Pró-alcool. Sugarcane was chosen as the substrate for due to a number of reasons, including its great adaptation to the Brazilian soil and weather conditions. In this phase, the anhydrous alcohol was mixed to gasoline using up to 20% (Soccol et al., 2010).

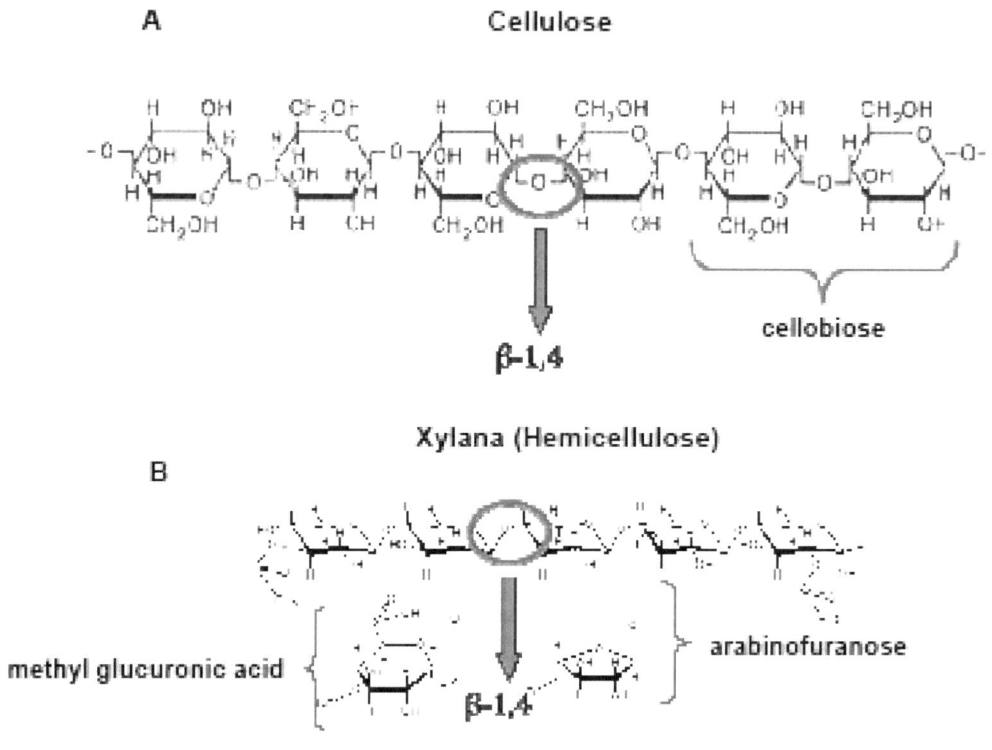

Figure 2. Schematic structure of cellulose (A) and hemicellulose (B) polymers.

All forms of biomass are formed by three main polymeric constituents: cellulose, hemicellulose, and lignin. Cellulose is the largest fraction (40 to 50%), hemicellulose is the next (20 to 30%) and lignin is usually 15 to 20% of biomass. The structures of these substances are shown in Figure 2. The plant primary cell wall (PCW) is a highly organized network of lignocellulose components, composed of cellulose microfibrils (9-25%) and an interpenetrating matrix of hemicelluloses (25-50%), pectins (10-35%) and proteins (10%).

Keegstra et al (1973) and Albersheim (1975) describe the PCW composition as cellulose fibers bound together by molecules made of sugar units. Approximately 90% of the PCW consists of carbohydrates (mostly pentose and hexose units) and the remaining 10% is protein. Cellulose forms the framework of the PCW while hemicelluloses cross-link non-cellulosic and cellulosic polymers. Pectins provide cross-links and structural support to the PCW whereas proteins can function either structurally (extensin) or enzymatically (Bidlack et al., 1992). Secondary CW of plants contains cellulose (40-80%), hemicellulose (10-40%) and lignin (5-25%). The arrangement of these components allows cellulose microfibrils to be embedded in lignin, much as steel rods are embedded in concrete to form pre-stressed concrete (Fig. 3). As a definition, secondary walls are derived from primary walls by thickening and inclusion of lignin into the CW matrix and occur inside the primary wall (Talmadge et al., 1973, Preston, 1979, Bidlack et al., 1992). As the main component of the structural support, the PCW is built to resist microbial degradation (Aristidou & Penttila, 2000, Goyal et al., 2008). The distribution of lignocellulosic components of plant cell wall depends on the species of plant and stage of growth and development of the same (Prade, 1995). The monomeric composition of the lignocellulosic material can vary widely depending on the source of biomass (Table 1).

Table 1. Composition of different plant biomass in relation to the presence (%) of carbohydrates (pentoses and hexoses, C5 or C6) and other components*.

Content (%)	Corn stover	Wheat straw	Rice straw	Rice hulls	Sugarcane Bagasse	Cotton gin trash
Carbohydrate						
Glucose (C6)	39.0	36.6	41.0	36.1	38.1	20.0
Mannose (C6)	0.3	0.8	1.8	3.0	nd	2.1
Galactose (C6)	0.8	2.4	0.4	0.1	1.1	0.1
Xylose (C5)	14.8	19.2	14.8	14.0	23.3	4.6
Arabinose (C5)	3.2	2.4	4.5	2.6	2.5	2.3
Total C6	40.1	39.8	43.2	39.2	39.2	81.0
Total C5	18.0	21.6	19.3	16.6	25.8	5.1
Non-carbohydrate						
Lignin	15.1	14.5	9.9	18.4	18.4	17.6
Ash	4.3	9.6	2.4	2.8	2.8	14.8
Protein	4.0	3.0	nd	3.0	3.0	3.0

*Source: Aristidou and Pentilla, 2000.

Cellulose and hemicelluloses are the major polysaccharides components in the cell wall, with hemicelluloses representing up to 20–35% of the total lignocellulosic plant biomass (de Vries and Visser, 2001, Pastor et al., 2007). Hemicellulose is a complex of polymeric carbohydrates including xylan, xyloglucan (heteropolymer of D-xylose and D-glucose), glucomannan (heteropolymer of D-glucose and D-mannose), galactoglucomannan (heteropolymer of D-galactose, D-glucose and D-mannose) and arabinogalactan (heteropolymer of D-galactose and arabinose). Plant biomass is an alternative natural source for chemical feedstocks with a replacement cycle short enough to meet the demand in the world fuel market.

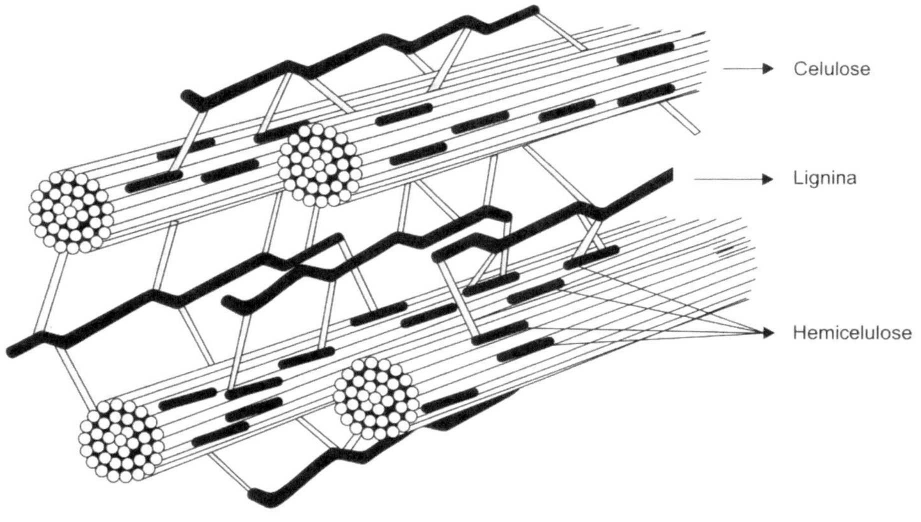

Figure 3. Secondary cell-wall structure. Components are arranged so that the cellulose microfibrils and hemicellulosic chains are embedded in lignin (Shleser, 1994).

Lignocellulose biodegradation is a central step for carbon recycling in land ecosystems. Moreover, fungal decay of wood in service results in billion-euro losses. Basidiomycetes are the main wood rotters due to their ability to degrade or modify lignin, an enzymatic process that originated in the Upper Devonian period in parallel with the evolution of vascular plants (Eriksson et al., 1990). Wood-rotting basidiomycetes are classified as white-rot and brown-rot fungi based mainly on macroscopic aspects. Basidiomycetes can overcome difficulties in wood decay, including the low nitrogen content of wood and the presence of toxic and antibiotic compounds. White-rot basidiomycetes, the most frequent wood-rotting organisms, are characterized by their ability to degrade lignin, hemicelluloses, and cellulose, often giving rise to a cellulose-enriched white material. Due to the ability of white-rot basidiomycetes to degrade lignin selectively or simultaneously with cellulose, two white-rot patterns have been described in different types of wood, namely selective delignification, also called sequential decay, and simultaneous rot (Otjen and Blanchette, 1986, Schwarze et al., 2000, Martínez et al., 2005). Many others microorganisms are capable to degraded lignocellulosic biomass, as bacteria and actinomycetes and use as carbon source for your metabolic pathways. A ton of lignocellulosic biomass as corn cob, wheat bran, wheat straw, wheat germ, rice straw, sugar cane bagasse, sugar cane straw, sisal bagasse, are produced by year. Due to structural complexity of lignocellulosic biomass, several agro-industrial by-products could be used as feedstock for microbial enzyme production with biotechnological potential.

The utilization of lignocellulose biomass for the production of enzymes, fuels and chemicals has the potential to change the world economically, socially, and environmentally. Today roughly 87% of the energy used in the world is derived from non-renewable sources such as oil, natural gas, and coal, with total energy consumption increasing at approximately 4% per annum. The long-term cost of continued use of these finite fuel sources can already be seen in increased conflict over their control and distribution, climate change linked to increased greenhouse gas emissions, and increasing prices, all of which negatively impact people around the world every day.

The cost of various fermentation products (sugars, organic acids, glues, solvents or drink softeners, etc.) largely depends on the cost of the carbohydrate raw material, and lignocellulosic residues from forests and agriculture still comprise the prominent carbohydrate source. Technologies need to be developed that are capable of handling a billion tons of biomass per year for the production of biofuels. According to the DOE-USDA Billion-Ton Study, corn stover and perennial crops such as switchgrass and hybrid poplar could provide about 1.3 billion tons of biomass by the mid-twenty-first century for utilization in bioenergy generation (Kumar et al., 2008, Perlack et al., 2005).

Brazil is the largest producer of sugarcane with 495 billion tons (Unica, 2009). The centre-south region of Brazil accounts for almost 80% of feedstock production (Zarrilli, 2006). The Brazilian bioethanol industry was poised for a major jump during 2006–2008 as a part of new national plan to increase the sugarcane production by 40% by 2009 (Renewable Energy Policy Network, 2006). Sugarcane bagasse (or, ''bagasse'' as it is generally called, Fig. 4), is a porus residue of cane stalks left over after the crushing and extraction of the juice from the sugarcane (Pandey et al., 2000). It presents a great morphological heterogeneity and consists of fiber bundles and other structural elements such as vessels, parenchyma, and epithelial cells (Sanjuan et al., 2001). It is composed by 19–24% of lignin, 27–32% of hemicellulose, 32–44% of cellulose and 4.5–9.0% of ashes (Jacobsen and Wyman, 2002). Sugar mills generate approximately 270–280 kg of bagasse (50% moisture) per metric ton of sugarcane (Rodrigues et al., 2003). The Brazilian annual production of sugarcane bagasse is currently estimated at 186 million tons (Soccol et al., 2010). Part of the sugarcane bagasse is burned to generate energy, while another part is used as animal feed and organic fertilizer for agricultural practices. However, a large part of energy stored as glucose in cellulose is thrown away without proper use of energy purposes.

Figure 4. Sugar cane bagasse mountain from Ethanol Refinery in Ribeirão, PE – Brazil.

Another type of substrate which could be used for enzyme production is cassava. Cassava (*Manihot esculenta* Cranz) is considered an important source of food and dietary calories for a large population in tropical countries in Asia, Africa and Latin America (Soccol, 1996, Laukevics et al., 1985). About 60% of the cassava produced all over the world is used for human consumption. Another large consumer of cassava is the animal food industry, using about 33% of the world production. With the advent of biotechnological approaches, focus has shifted to widening the application of cassava and its starch for newer applications with the aim of value addition (Pandey et al., 2000).

Cassava bagasse is a fibrous residue, which contains about 50% starch on a dry weight basis (Carta et al., 1999). Table 2 shows the composition of cassava bagasse as determined by various authors. These analyses (Table 2) were conducted on the bagasse samples obtained from different processing units at different times in the State of Parana, Brazil. The composition shows variation probably due to the fact that most of the processing is done under poorly controlled technological conditions. In addition, the composition may also differ due to the use of different crop varieties. Starch is the main component determined as carbohydrates. Cassava bagasse does not show any cyanide content (Pandey et al., 2000). Due to its physico-chemical characteristics, the cassava bagasse could be used as potential raw material for production of microbial enzymes with biotechnological potential, such as amylases and cellulases.

Table 2. Physico-chemical composition of cassava bagasse (g/100 g dry weight)*.

Composition	Cereda (1994)	Vandenberghe (1998)	Sterz (1997)
Moisture	9.52	11.20	10.70
Protein	0.32	1.61	1.60
Lipids	0.83	0.54	0.53
Fibers	14.88	21.10	22.20
Ash	0.66	1.44	1.50
Carbohydrates	63.85	63.00	63.40

*Source: Pandey et al., 2000.

Cellulase production is the most important step in the economical production of ethanol, single cell protein and other chemicals from renewable cellulosic materials. To date, the production of cellulase has been widely studied in submerged culture processes, but the relatively high cost of enzyme production has hindered the industrial application of cellulose bioconversion. It has been reported that solid state fermentation is an attractive process to produce cellulase economically due to its lower capital investment and lower operating expenses. Another approach to reduce the cost of cellulase production is the use of lignocellulosic materials as substrates rather than expensive pure cellulose. In prior publications, abundant agricultural residue such as corn stover, wheat straw, rice straw, bagasse, etc. were used in cellulase production (Xia and Cen, 1999).

Plant cellulose exists in a highly crystalline form. In addition, it is associated with hemicelluloses and surrounded by lignin, which may also be covalently bound to hemicellulose. Several microorganisms are capable of degrading cellulose. However, the complete degradation of cellulose requires cooperation of different microbial populations, especially between fungi and bacteria.

Strategies for Cellulases Detection and Production

Many microorganisms that produce various cellulolytic enzymes have been studied for several decades. The genus *Trichoderma* has been especially famous for producing cellulolytic enzymes with relatively high enzymatic activity. However, it is also well-known that the *Trichoderma* enzymes do not effectively hydrolyze cellulose biomass alone because of their enzyme composition. The microbial (bacterial or fungi) cellulases production depends on the type of inducer substrate (agro-industrial by products) and microbial cultivation conditions. There are several types of agro-industrial waste that can be used as raw material for the production of microbial enzymes. The production of enzymes such as cellulases, amylases and xylanases at low cost is extremely important for processes that require their use in large scale. Studies have been performed with substrates of low cost inductors in order to minimize the impact of the cost of raw material and processing in the production of enzymes, both in bench scale and pilot scale using bioreactors of different scales (Grigorevski-Lima et al., 2005, Nascimento et al., 2009). Currently various industrial processes using enzymes are large-scale, particularly paper industries, food and textiles.

The world market for enzymes rose by nearly 1.45 billion dollars in 1995 to almost 3.7 million dollars in 2004 with growth forecast global demand of 6.5% per year until 2010. Only for the chemical processing enzyme grains (which currently has approximately 28-30% of the total sale of enzymes), is expected to increase by volume in the market value of 715 million dollars in 2001 to 1 billion in 2010 (Godfrey et al., 2003). Currently, the technical industries, such as detergent, starch, textile, fuel alcohol, account for the majority of the total enzyme market, alongside those of food and feed industry, totaling only about 35% of the market. However, sales in some key technical industries has stagnated (fall 3% in 2001), while sales in the food and feed are increasing with an annual growth rate expected for approximately 4-5% (Godfrey, 2003). The hydrolases constitute 75% of the market for industrial enzymes such as glucosidases, constituting the second largest group after the peptidases (Bhat, 2000). Xylanases comprise the largest proportion of commercial hemicellulases, but represents only a small percentage of total sales of enzymes. However, we expect increased sales, since these enzymes have attracted increasing attention because of its potential for use in various applications. Thus, studying the production of lignocellulases aiming at minimizing the process costs and targeting their marketing is of great value to the world market for enzymes.

Cellulases are a complex mixture of enzymes necessary for complete solubilization of cellulose into sugars that serve in nature as a carbon source for microbial metabolism. Cellulases are produced, among other microorganisms, by different genera of fungi and actinomycetes, and the highest producers identified so far in the genus *Trichoderma*, *Penicillium*, *Fusarium*, *Acremonium* and *Aspergillus* (Martins et al., 2008, Fang et al., 2009, Ahamed and Vermette, 2009) in the case of fungi and *Streptomyces* species in the case of actinomycetes (Grigorevski-Lima et al., 2005, Nascimento et al., 2009). In general, levels of this enzyme complex secreted by microorganisms meet, in nature, the needs of decomposition of the lignocellulosic material and availability of fermentable sugars. However, the industrial use of cellulases requires reaching enzyme preparations with high activity levels and stability, and in some cases the use of resources of molecular biology to improve promising natural strains. Work in this direction has been developed in different national and international laboratories, universities and companies.

To obtaining preparations with high cellulolytic activity, it is necessary to perform the optimization of the fermentation process, especially regarding the composition of culture medium (to avoid sources of carbon and nitrogen repressing), to choose how to drive the process (single or batch fed batch) and the use of inducers of the genes coding for cellulolytic complex. Considering that the physical-chemical environment to which the organism is exposed is of fundamental importance for the production of enzymes, studies on the composition of culture medium represent a key role in product yield. Several studies have been conducted with cellulases and xylanases aiming at industrial application, particularly in pulp and paper industry, textiles, detergents and biofuels. Some studies (Nascimento et al., 2009, Grigorevski et al., 2005, Nascimento et al., 2003, Chen et al., 2007) on production of xylanases and cellulases on different substrates inductors have been performed with strains of actinomycetes isolated from Brazilian soils. Ikeda et al (2007) observed a high titer (15.5 U/mL, 114.2 U/mL, respectively) of FPase and Carboxymethylcellulase (CMCase) production using Solka Floc by *Acremonium cellulolyticus*, at 8 fermentation-days. These cellulase activities are very high in comparison to others fungal species. Liming and Xueliang (2004) observed a maximum of CMCase production (5.48 U/mL) using corn cob as carbon source by *Trichoderma reesei* ZU-02 at 4 fermentation-days. Nascimento et al. (2009), observed a good CMCase titer (0.71 U/mL) for *Streptomyces malaysiensis* when using brewer's spent grain and corn steep liquor as inducer substrates, after 4 fermentation-days.

Many natural and anthropogenic environments could be used as microbial source with biotechnological potential. Bioprospecting of microorganisms depends on biotechnology target (cellulase, xylanase, amylase, laccase, protease), and hence the appropriate choice of selective techniques. There are some important techniques that are used to select cellulolytic microorganisms, among which we highlight cultivation in solid medium containing mineral salts supplemented with 1% carboxymethylcellulose (CMC) and the cellulose-azure test. In the first case the microorganism is inoculated as spots in the surface of the medium in a Petri dish, and the system is incubated for 2-10 days. After this period, a solution of Congo Red it is added and expected to react with the cellulose (-1,4 linkages). After washing with 1.0 M NaCl solution hydrolysis zones circumscribing the colonies can be observed, indicating the production of carboxymethylcellulase (endoglucanase), that acts randomly on the amorphous region of the cellulose fiber (Sazci et al.,1986). In the second case, the microorganism is inoculated in test tubes containing a mineral basal medium covered with the same medium added of cellulose-azure. After incubation for 2-10 days the cellulase production is observed by the liberation of the azure dye in the lower part of the medium, due to the releases of the dye into the medium, which turns it blue. Non-cellulolytic microorganisms do not degrade the cellulose-azure and therefore the dye does not migrate to the medium, remaining transparent (Plant et al., 1988) (Fig. 5).

The use of microbial fermentations for the production of a wide range of enzymes, such as cellulases, xylanases, laccases, amylases and proteinases, for use in several industries as the pulp and paper, food, textile and detergent, has turn this technology an important role in contemporary manufacturing. Specially considering the cellulases system in the textile, detergent and bioenergetics industries, its importance has promoted the beginning of research on cellulases and accessories enzymes (xylanases, β-glucosidases, β-xylosidases) production by microorganisms. Filamentous fungi and actinomycetes are particularly interesting producers of cellulases, since they are extra-cellularly excreted and their enzyme levels are much higher than those of yeast and other bacteria. The cellulase production by actinomycetes and fungi is a very important way to obtain enzymes with special biochemical characteristics. An efficient production of cellulolytic enzymes involves the choice of an appropriate inducing substrate, an optimum medium composition and culture conditions, such as pH, temperature, air supply and agitation (Fig. 6).

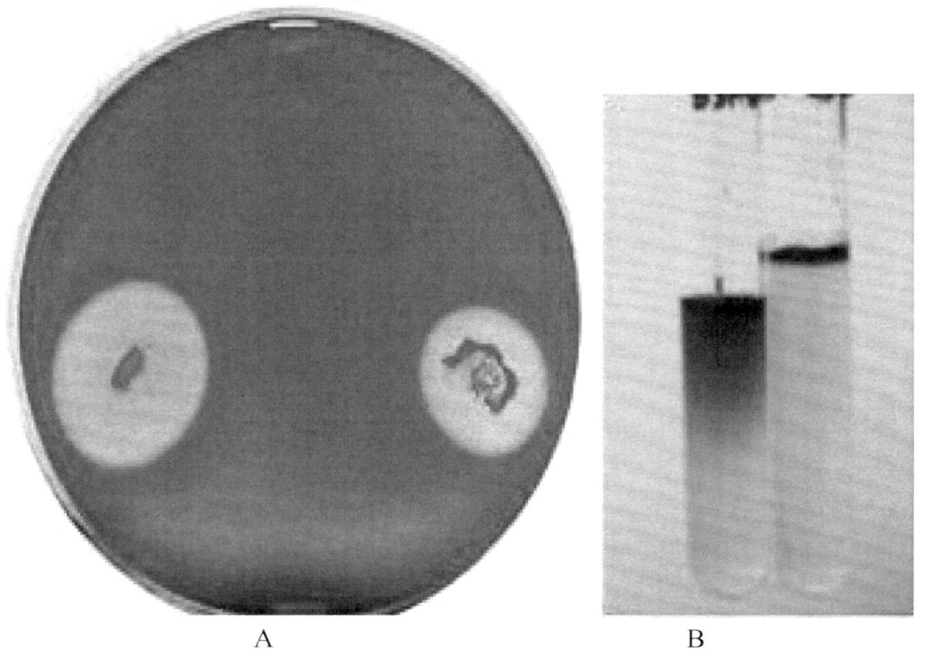

Figure 5. Selection of cellulolytic microorganism. (A) salt mineral medium using carboxymethylcellulose as sole carbon source; after a incubation period a Congo Red solution was used to reveal the hydrolytic zones. (B) salt mineral medium using cellulose-azure in the upper part of the medium, after a incubation period the positive result is indicated by dye migration in the transparent medium.

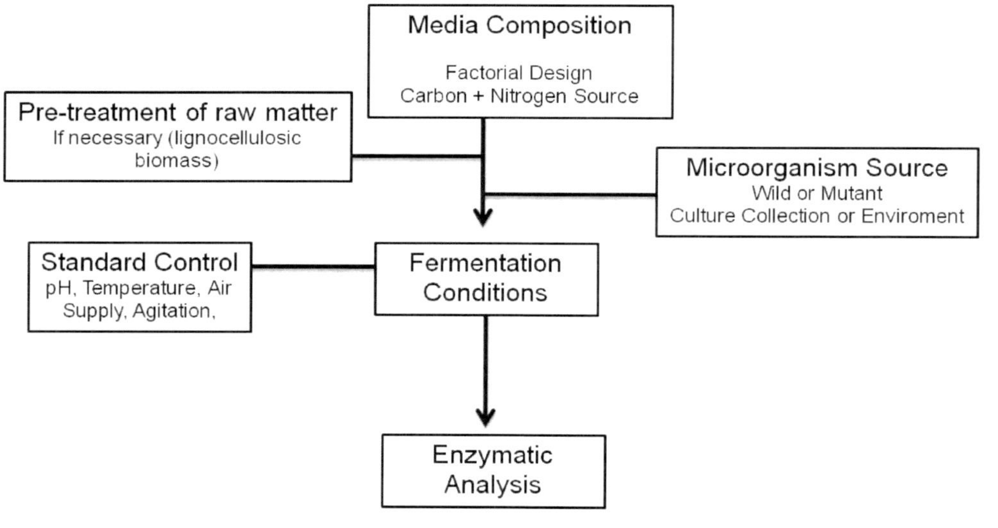

Figure 6. Schematic representation for cellulases production using different factors and standard controls.

Cellulose hydrolysis results in glucose, cellobiose and cellooligosaccharides, which are readily determined by HPLC or by chemical and enzymatic assays. But these results depend on the type of cellulase (endo- exoglucanase, β-glucosidase) produced, by the microorganism chosen, and by the inducing substrate used in the process. To obtain good results on cellulases production it's necessary to observe, firstly, what substrate and what kind of fermentation medium you will be use. Different ions present in the composition of the medium could be causing any kind of effect on cellulase activity and production. What concentration of phosphate or trace elements (S, Na^+, K^+, Mg^{2+}, Cu^{2+}, Fe^{3+}, Zn^{2+}, Ca^{2+}, Ba^{2+}), for example, is indicated to obtain a high cellulase production? What kind of carbon and nitrogen source, and concentration of these substrates are indicated to obtain the best result? For those questions, and many others such as best temperature, pH, inoculum size and agitation, a factorial design can be use as a statistical tool to help in the production of high titers of cellulase.

Considering the nutrient medium for cellulase production, conventional methods based on the "change-one-factor-at-a-time", in which one independent variable is studied while fixing all others at a specific level, may lead to unreliable results and inaccurate conclusion. This experimental procedure is also expensive and time consuming for large number of variables. The mathematical experimental design finds wide application in nutrient media optimization for microbial enzyme production (Dobrev et al., 2007). If you adopt an experimental factorial design tool to study the cellulase production, you can chose as independent variables carbon, nitrogen, mineral salt concentrations, temperature, pH, agitation, and air supply and inoculums size. For instance, if you have obtained a promising cellulolytic fungi strain, and you are going to study its cellulase production in optimizing conditions, you can choose as independent variables carbon and nitrogen concentration, temperature, pH and inoculums size. How can the mathematical matrix be developed to study these conditions? Firstly it's necessary to determine the range of each independent variable (Tables 3 and 4), and which kind of experimental design you will use. In this example we will use a fractional experimental design (2^{5-1}).

Table 3. Real values of independent variables in correspondence to codified value.

Independent Variables	Levels		
	-1	0	+1
Carbon Concentration ($g.L^{-1}$)	10.0	20.0	30.0
Nitrogen Concentration ($g.L^{-1}$)	5.0	10.0	15.0
Temperature (°C)	25	30	35
pH	5.0	6.0	7.0
Inoculum size ($spores.mL^{-1}$)	10^6	10^7	10^8

Once the variables have been chosen, it's necessary to select the type of fermentation to be used, Submerged (SmF) or Solid-State Fermentation (SSF). The choice depends on the aim of the study. Both processes have advantages and disadvantages. Filamentous fungi and actinomycetes are great microorganisms for SSF system. SSF holds tremendous potential for the production of enzymes. It can be of special interest in those processes where the crude fermented product may be used directly as the enzyme source. This system offers numerous advantages over submerged fermentation (SmF) system, including high volumetric productivity, relatively higher concentration of the products, less effluent generation, requirement for simple fermentation equipments, etc. However, the main disadvantages are the great difficulties to control parameters such as pH, temperature and aeration, which may be well controlled in submerged fermentation. Agro-industrial residues are generally considered the best substrates for the SSF and SmF processes, and use of SSF for the production of enzymes is no exception to that. A number of such substrates (sugar cane bagasse, brewer's spent grain, wheat bran, rice straw, corn cob, wheat germ, coconut oil cake, mustard oil cake, cassava flour,) have been employed for the cultivation of microorganisms to produce several enzymes using both systems. Wheat bran however holds the key, and has most commonly been used in various processes (Pandey et al., 2000).

Table 4. Matrix of fractional experimental design (2^{5-1}) using codified values.

Run	[Carbon]	[Nitrogen]	Temp. (°C)	pH	Inoculum
1	-1	-1	-1	-1	+1
2	+1	-1	-1	-1	-1
3	-1	+1	-1	-1	-1
4	+1	+1	-1	-1	+1
5	-1	-1	+1	-1	-1
6	+1	-1	+1	-1	+1
7	-1	+1	+1	-1	+1
8	+1	+1	+1	-1	-1
9	-1	-1	-1	+1	-1
10	+1	-1	-1	+1	+1
11	-1	+1	-1	+1	+1
12	+1	+1	-1	+1	-1
13	-1	-1	+1	+1	+1
14	+1	-1	+1	+1	-1
15	-1	+1	+1	+1	-1
16	+1	+1	+1	+1	+1
17	0	0	0	0	0
18	0	0	0	0	0
19	0	0	0	0	0
20	0	0	0	0	0

The selection of a substrate for enzyme production in a SSF or SmF process depends upon several factors, mainly related with cost and availability of the substrate, and thus may involve screening of several agro-industrial residues. In a SSF process, the solid substrate not only supplies the nutrients to the microbial culture growing in it but also serves as an anchorage for the cells (Pandey et al., 2000). By the other hand, in SmF process, the insoluble substrate is submerged in the culture medium containing nitrogen and trace elements to improve enzyme production. The substrate that provides all the needed nutrients to the microorganisms growing in it should be considered as the ideal substrate. However, in SSF process, some of the nutrients may be available in sub-optimal concentrations, or even absent in the substrates. In such cases, it would become necessary to supplement them externally with these. It has also been a practice to pre-treat (chemically or mechanically) some of the substrates before using in SSF processes (e.g. ligno-cellulose), thereby making them more easily accessible for microbial growth (Pandey et al., 2000).

SSF has been considered superior in several aspects to SmF due to various advantages it renders. It is cost effective due to the use of simple growth and production media comprising agro-industrial residues, uses little amount of water, which consequently releases negligible or considerably less quantity of effluent, thus reducing pollution concerns. SSF processes are simple, use low volume equipment (lower cost), and are yet effective by providing high product titers (concentrated products). Further, aeration process (availability of atmospheric oxygen to the substrate) is easier since oxygen limitation does not occur as there is an increased diffusion rate of oxygen into moistened solid substrate, supporting the growth of aerial mycelium. These could be effectively used at smaller levels also, which makes them suitable for rural areas also. SSF systems resemble the natural habitats of microbes and, therefore, may prove efficient in producing certain enzymes and metabolites (Pandey et al., 2000).

The SmF system, on the other hand, may represent a strategy of a more viable production when you consider the availability of extracellular enzyme in liquid medium. While in SSF system is necessary the addition of a buffer to extract the enzyme, which almost always becomes trapped in the substrate due to chemical affinity resulting in losses, in SmF system the enzyme is already dissolved in the supernatant liquid, representing a simple phase extraction of the enzyme by separating the solid phase (biomass) of the liquid (supernatant enzyme).

Once you have determined what type of fermentation which the process will be conducted, you should select variables, as previously described, to assess the production of cellulases under optimal conditions. The use of the SmF can assess the effect of agitation, aeration and pH on enzyme production, while the SSF system allows us to evaluate the effect of moisture, some salts in the culture medium used to correct the moisture and the inoculum size. To study the effect of different variables in cellulase production, the experimental design could be use, e.g. a factorial design with 5 independent variables (temperature, inoculums size, carbon concentration, nitrogen concentration, aeration, pH, etc.). After that, it is necessary to observe the mainly effects on cellulase production and selected the most important independent variables and use another type of experimental design using 2 or 3 independent variables and analyzing the results.

Biotechnological Applications of Microbial Cellulases

Lignocellulosic biomass, in the form of plant materials such as grasses, woods, and crop residues, offers a renewable, geographically distributed, greenhouse-gas neutral source of sugars that can be converted to ethanol or other liquid fuels via microbial fermentation (Merino and Cherry, 2007). There are so many enzymes involved in lignocellulosic transformation of biomass in sugars (cellulases, xylanases, amylases, pectinases, laccases, lignin peroxidase, etc) that it would be impossible to name them all. In fact, scientists have yet to discover many enzymes, or fully understand their structure and properties. On the other hand, many other enzymes have been successfully studied and applied to industrial and commercial uses, especially cellulases. The cellulases could be applied in processes that break down cellulose, which is the basic raw material used to make products such as paper, cotton, and other textiles.

Active research on cellulases and related polysaccharidases began in the early 1950's, owing to their enormous potential to convert lignocellulose, the most abundant and renewable source of energy on Earth, to glucose and soluble sugars (Coughlan, 1985, Mandels, 1985, Reese and Mandels, 1984, Reese, 1976). Extensive basic and applied research during the 1970's and 1980's have shown that the enzyme-induced bio-conversion of lignocellulose to soluble sugars was rather difficult and uneconomical (Bhat, 2000, Coughlan, 1985, Mandels, 1985, Ladisch et al., 1983, Ryu and Mandels, 1980). However, new technologies have been developed aiming to separate the lignin from the cellulose fiber, thus allowing greater accessibility of enzymes to polysaccharides.

Biotechnology of cellulases and hemicellulases began in early 1980's, first in animal feed followed by food applications. Subsequently, these enzymes were used in the textile, laundry as well as in the pulp and paper industries (Bhat, 2000). During the last three decades, the use of cellulases and hemicellulases has increased considerably, especially in textile, food, brewing and wine as well as in pulp and paper industries and bioethanol production (Godfrey and West, 1996, Saddler, 1993, Uhlig, 1998). Today, these enzymes account for approximately 20% of the world enzyme market (Mantyla et al., 1998), mostly from *Trichoderma* and *Aspergillus* (Godfrey and West, 1996, Uhlig, 1998). At the present time several commercial enzyme products are available, which are marketing tailor-made enzyme preparations, suitable for biotechnology. The updated details of those, can be found in their respective company web pages.

Cellulases are currently the third largest industrial enzyme worldwide, by dollar volume, because of their use in juice extraction and wine, in cotton processing, paper recycling, as detergent enzymes, as animal feed additives and more recently for bioethanol productions. However, cellulases will become the largest volume industrial enzyme, if ethanol, butanol, or some other fermentation product of sugars, produced from biomass by enzymes, becomes a major transportation fuel (Wilson, 2009).

Cellulases, as well as hemicellulases and pectinases, have a wide range of potential applications in food biotechnology. The production of fruit and vegetable juices is important both from the human health and from commercial standpoints. The production of fruit and vegetable juices requires methods for extraction, clarification and stabilization (Bhat, 2000). Cellulases and pectinases from Trichoderma reesei has been used to liquefied mashed fruit and vegetables resulting from the extraction of juices (Bhat, 2000). The use of two enzymes is also suitable for extraction of juice with more consistent structures and when the nutritional constituents are retained in the fraction of the pulp.

The commercial enzyme preparation, Olivex (a pectinase preparation with low levels of cellulase and hemicellulase from *Aspergillus aculeatus*) was the first enzyme mixture used to improve the extraction of olive oil (Fantozzi et al., 1977). Systematic studies carried out in the 1980's revealed that indeed no single enzyme was adequate for the efficient maceration and extraction of oil from olives. In fact cellulases and hemicellulases, besides pectinases, were found to be really essential for this purpose (Galante et al., 1998). Also, a combination of enzymes, consisting of pectinases (from *Aspergillus*), cellulases and hemicellulases (from *Trichoderma*), performed better than the enzymes from a single micro-organism (Bhat, 2000, Galante et al., 1993).

Beer brewing and wine making are old technologies and have an ancient history. In simple terms, beer brewing involves malting the barley in a malt house followed by the preparation and fermentation of the wort in the brewery, while wine making requires the extraction of juice from grapes and subsequent fermentation of the juice by yeast. The production of wine involves the maintenance of the pulp of grapes before fermentation at 50-60°C to improve the extraction of skin color. The viscosity can be decreased by the addition of pectinase to the folder, which facilitates the clarification and filtration of wine (Galante et al., 1998). Enzyme technology plays a central role in both these processes (Table 5). The use of lignocellulolytic enzymes is very important to facilitate maceration by increasing the pressing juice yield and speed the process. The addition of exogenous glucanases and related polysaccharidases are known to improve not only the beer and wine qualities, but also their overall production efficiency (Bhat, 2000, Galante et al., 1998). This technology is based on the action of enzymes activated during malting and fermentation. Malting of barley depends on seed germination, which initiates the biosynthesis and activation of α- and β-amylases, carboxypeptidase and β-glucanase that hydrolyze the seed reserve. All these enzymes should act in synergy under optimal conditions to produce high quality malt. Nevertheless, many breweries end up using un-malted or poor quality barley, due to seasonal variations, different cultivars or poor harvest, which contains low levels of endogenous β-glucanase activity. Microbial β-glucanases, which hydrolyse β-glucan and reduce the viscosity of the wort are added either during mashing or primary fermentation. The commonly used β-glucanases are from Penicillium emersonii, Aspergillus niger, Bacillus subtilis and Trichoderma reesei (Bhat, 2000, Galante et al., 1998).

Table 5. Cellulases and hemicellulases in brewing, wine and animal feed biotechnology*.

Enzyme	Function	Application	Reference
Macerating enzymes (cellulases, hemicellulases and pectinases)	Hydrolysis of plant cell wall polysaccharides	Improvement in skin maceration and colour extraction of grapes; quality, stability, filtration and clarification of wines	Galante et al., 1998 Grassin and Fauquembergue, 1996 Uhlig, 1998
β-Glucosidase	Modification of aromatic residues	Improvement in the aroma of wines	Caldini et al., 1994 Gunata et al., 1990
Cellulases and hemicellulases	Partial hydrolysis of lignocellulosic materials; dehulling of cereal grains; hydrolysis of β-glucans; decrease in intestinal viscosity; better emulsifica-tion and flexibility of feed materials	Improvement in the nutritional quality of animal feed and thus the performance of ruminants and monogastrics	Beauchemin et al., 1995; Cowan, 1996; Galante et al., 1998; Lewis et al., 1996
Cellulases, hemicellulases and pectinases	Partial hydrolysis of plant cell wall during silage and fodder preservation; expression of preferred genes in ruminant and monogastric animals for high feed conversion efficiency.	Production and preservation of high quality fodder for ruminants; improving the quality of grass silage; production of transgenic animals	Ali et al., 1995; Hall et al., 1993;

* Source: Bhat, 2000.

Cellulases and hemicellulases are widely used for supplementing diets rich in nonstarch polysaccharides (NSP) feed to monogastric animals (Table 5). The use of hydrolases is either to (1) eliminate anti-nutritional factors (ANF) present in grains or vegetables; (2) degrade certain cereal components in order to improve the nutritional value of feed; or (3) to supplement animals own digestive enzymes (e.g. proteases, amylases and glucanases), whenever these enzymes are inadequate during post-weaning period, as it is often the case with broilers and piglets (Bhat, 2000, Galante et al., 1998). In addition, the products of cellulase and hemicellulase activity are more prone to fermentation by the microbial organisms that colonize the last compartments of the gastrointestinal tract, and more energy is consequently absorbed from the hydrolysis of NSP (Ponte et al., 2004). Dietary application of cellulase complex in monogastric and ruminant animals has been studied by methods of feeding, balance, metabolism, production, simulated rumen, sensation assessment, electronic microscope, atomic spectrum analysis, rumen fistula, economy benefit, etc. Cellulases and hemicellulases could contribute to a significant depolymerization of plant cell wall polysaccharides resulting in a considerable release of energy, otherwise not available to the animal (Ponte et al., 2004). Enzyme preparations containing high levels of cellulase, hemicellulase and pectinase have been used to improve the nutritive quality of forages (Graham and Balnave, 1995; Kung et al., 199; Lewis et al., 1996).

Cellulases have achieved their worldwide success in textile and laundry because of their ability to modify cellulosic fibres in a controlled and desired manner, so as to improve the quality of fabrics. Cellulases are used in textile processing of cellulosic fibers with the goal of eliminating microfibril surface, creating a smoother surface, increasing the brightness and avoid the formation of pellets (pilling) and fade pieces made paints (mainly denim) and get a point used (stone-wash effect). Although cellulases were introduced in textile and laundry only a decade ago, they have now become the third largest group of enzymes used in these applications. Bio-stoning and bio-polishing are the best-known current textile applications of cellulases (Table 6). Cellulases are also increasingly used in household washing powders, since they enhance the detergent performance and allow the removal of small, fuzzy fibrils from fabric surfaces and improve the appearance and colour brightness (Bhat, 2000).

Table 6. Cellulases in textile and laundry biotechnology*.

Enzyme	Function	Application	Reference
Cellulase, preferably neutral	Removal of excess dye from denim fabrics; soften the cotton fabrics without damaging the fibre	Bio-stoning of denim fabrics; production of high quality and environmentally friendly washing powders	Galante et al., 1998; Godfrey, 1996; Uhlig, 1998
Cellulase, preferably acid	Removal of excess microfibrils from the surface of cotton and non-denim fabrics	Bio-polishing of cotton and non-denim fabrics	Galante et al., 1998; Godfrey, 1996; Kumar et al., 1996
Cellulase	Restoration of softness and colour brightness of cotton fabrics	Production of high quality fabrics	Galante et al., 1998; Godfrey, 1996; Kumar et al., 1994

* Source: Bhat, 2000.

In the mid 1980's, biotechnology provided a perfect alternative for stone-washing using microbial cellulases, later known as "biostoning". During the bio-stoning process, cellulases act on the cotton fabric and break off the small fibre ends on the yarn surface, thereby loosening the indigo, which is easily removed by mechanical abrasion in the wash cycle. The advantages in the replacement of pumice stones by a cellulase based treatment include: (i) reduced wear and tear of washing machines and short treatment times; (ii) increased productivity of the machines because of high loading; (iii) substantial decrease of second quality garments; (iv) less work-intensive and safer working conditions; (v) safe environment, since pumice powder is not produced; (vi) flexibility to create and consistently reproduce new finished products; and (vii) the possibility to automate the process with computer-controlled dosing devices when using liquid cellulase preparations (Bhat, 2000, Galante et al., 1998, Cavaco-Paulo, 1998).

The main objective of the "biopolishing" is the removal of the yarn hairiness and thus reduce the tendency of "pilling" or the formation of pellets. The biopolishing is usually performed after the processing of fabrics and textile pieces made during the textile wet processing stage and includes desizing, scouring, bleaching, dyeing and finishing. Visually, the fabrics have a bio-polished surface noticeably cleaner and texture of the fabric becomes more apparent. During this process, the cellulases act on small fiber ends that protrude from the fabric surface, where the mechanical action removes these fibers and polishes the fabrics. The fabric is softer and the water absorption is not hindered, as in many softeners. The effect is permanent, since the tips are fibrous anchors for the accession of other fibers and the development of balls. Even after repeated washings, the tissue remains almost free of pellets (Cavaco-Paulo, 1998). Serious problems occur in mixtures of synthetic fibers (polyester and polyamide) with cotton, because synthetic fibers serve as an anchor for the short fibers of cotton. The main advantages of using cellulases are: (i) removal of short fibres and surface fuzziness; (ii) smooth and glossy appearance; (iii) improved colour brightness and uniformity; (iv) high hydrophilicity and moisture absorbance; (v) new and improved finishing and fashionable effects; and (vi) environmentally friendly process. In fact, bio-polishing is currently a key step in the textile industry for producing high quality garments (Bhat, 2000). Since 1989, the cellulases were introduced to replace partially or completely the pumice stones and avoid some problems observed in this process. Besides giving a faded look to the articles, they decrease the flexural strength and make use of the more enjoyable. The hydrolytic action of cellulases occurs directly in the structure of cellulose, reduces the degree of polymerization of cellulose and causes a loss of mass and tensile strength as mentioned before. Cellulases break the fibrils protruding from the surface of the wires. This action, together with the mechanical effect of the machine, causes the shedding of the superficial layer of indigo. The abrasion suffered by the yarn dyed with indigo dye releases for the bath. The amount of indigo is released by the level of abrasion, as well as enzyme concentration, length of wash cycle, mechanical action, article type denim, etc (Bhat, 2000, Cavaco-Paulo, 1998).

The cellulase preparations capable of modifying the structure of cellulose fibrils are added to laundry detergents to improve the colour brightness, hand feel and dirt removal from cotton and cotton blend garments. Most cotton or cotton blend garments, during repeated washings, tend to become fluffy and dull. This is mainly due to the presence of partially detached microfibrils on the surface of garments that can be removed by cellulases in order to restore a smooth surface and original colour to the garment. Also, the degradation of microfibrils by cellulase, softens the garment and removes dirt particles trapped in the microfibril network. This is currently accomplished by adding a commercial cellulase preparation from *H. insolens*, active under mild alkaline conditions (pH 8.5–9.0), and at temperatures over 508C in washing powders (Uhlig, 1998). Although, the amount of cellulase added represents approximately 0.4% of the total detergent cost, it is considered rather expensive and hence, alternative cellulase preparations are required to attract the worldwide laundry market (Bhat, 2000).

Cellulases and hemicellulases have been used in the pulp and paper industry for different purposes. Commercial enzyme preparations contain various enzyme activities, where some may be vital, while others may be detrimental for a specific application. Therefore, enzyme mixtures or purified enzymes should be well characterized with respect to their substrate specificity and mode of action before using for a particular application in pulp and paper industry (Table 7).

Table 7. Cellulases and hemicellulases in pulp and paper biotechnology*.

Enzyme	Function	Application	Reference
Cellulases and hemicellulases	Modification of coarse mechanical pulp and handsheet strength properties; partial hydrolysis of carbohydrate molecules and the release of ink from fiber surfaces; hydrolysis of colloidal materials in paper mill drainage	Bio-mechanical pulping; modification of fiber properties; de-inking of recycled fibers; improving draining and runnability of paper mills	Akhtar, 1994; Buchert et al., 1998; Pere et al., 1996; Rahkamo et al., 1996; Saddler, 1993;
Purified cellulase and hemicellulose components	Partial or complete hydrolysis of pulp fibers	Bio-characterization of pulp fibers	Buchert et al., 1997; Oksanen et al., 1997; Suurnakki et al.,1996;

* Source: Bhat, 2000.

Cellulase and hemicellulase mixtures have been used for the modification of fiber properties with the aim of improving drainage, beatability and runnability of the paper mills. In these applications, the enzymatic treatment was performed either before or after beating of the pulps. The aim of cellulase and hemicellulase treatment prior to the refining process is either to improve the beatability response or to modify the fiber properties. A commercial cellulase/hemicellulase preparation, named Pergalase –A40, from *Trichoderma* has been used by many paper mills around the world for the production of release papers and wood-containing printing papers (Bhat, 2000).

The biggest challenge for biotechnological applications of cellulases on a large scale remains to their use to obtain glucose from the great mass of renewable cellulosic waste available throughout the year. The fermentation of glucose into solvents and fuels, particularly ethanol and buthanol, might provide a partial replacement for fossil fuels (Bon et al., 2008). The problem of bioethanol has been object of great attention in Europe. There, the fuel use goes through two important factors. Firstly, through the Kyoto Protocol, the EU agreed to limit the emission of gases that cause global warming, while increasing demand for fuel use. Secondly, the dependence on oil from the Middle East, a politically unstable region, and generating concern about the fluctuation of prices charged and a possible disruption in supply. Thus, the use of alternative fuels like ethanol has been widely discussed, it would be an opportunity to reduce greenhouse gas emissions and secure energy supply. At the same time, the development of biofuels could create new jobs, mainly in rural areas already in decline. Currently, the production of biofuels by the European countries is still not economically viable in comparison with fossil fuels (Ryan et al., 2006). The production of fuel ethanol from lignocellulosic biomass includes biomass pre-treatment, cellulose hydrolysis, fermentation of hexoses, separation, effluent treatment, and, depending upon the feedstock, gathering, which may have an additional cost (Ojeda and Kafarov, 2009). Intensive efforts have been made in recent years to develop efficient technologies for the pre-treatment of bagasse, developments enzymes for enhanced cellulose/hemicelluloses saccharification and suitable technologies for the fermentation of both C6 and C5 sugars (Soccol et al., 2010).

Although the pre-treatment is required to make the biomass accessible to the enzymes action, it is desirable to use mild conditions that minimize the degradation of the sugars and lignin into inhibitory by-products (Almeida et al., 2007). Therefore, to improve the enzymatic hydrolysis process and offsetting the low severity applied during the pre-treatment, the trend is the use enzyme mixtures containing xylanase and other accessory enzymes such as feruloyl esterase (Meyer et al., 2009). The use of these enzymes, naturally secreted by cellulolytic fungi, in the deconstruction of biomass has been considered an interesting approach. The enzymatic hydrolysis can be carried out separately from the alcoholic fermentation, a process known as Separate Hydrolysis and Fermentation (SHF) or both processes can run together as Simultaneous Saccharification and Fermentation (SSF). In the SHF process, hydrolysis can be done at temperatures as high as 50 °C, taking advantage of enzymes stability at this temperature to increase rates and minimize bacterial contamination. It also allows easy separation of the sugar syrups from the hydrophobic lignin that can be used as solid fuel. Nevertheless, SHF leads to the accumulation of the glucose derived from the hydrolysis of cellulose that can inhibit the endo-and exo-glucanases and β-glucosidase, affecting the reaction rates and yields. As the subsequent fermentation step is run separately from the hydrolysis step the yeast cells can be recycled or used as animal feed, a usual and well regarded practice in the Brazilian ethanol industry. In the SSF process the producing ethanol is faster, as the glucose formed is simultaneously fermented to ethanol. Besides, the risk of contamination is lower due to the presence of ethanol, the anaerobic conditions and the continuous withdrawal of glucose (Soccol et al., 2010). The process also presents a lower cost as only one reactor is necessary. In this context, it is interesting to note that the ethanol that accumulates in the medium does not significantly affect the enzymes activity. The difficulty of this process relates to the different optimum temperature for enzymatic hydrolysis (45–50 °C) and alcoholic fermentation (28–35 °C).

In the future, cellulases may be applied in the production of glucose syrups from cellulose materials that will compete with starch and sucrose in the production of alternative sweeteners for use in beverage and food industries and biofuels. Municipal solid waste and waste from wood processing and from forest thinning operations are additional sources of biomass for use in producing fuel, power and products in biorefineries, an ecological alternative for clean energy and enzyme production. Hydrolyzed cellulose can also be used as nutrients and fuels in fermentations for the production of various chemicals, including enzymes for food processing and food ingredients such as citric and acetic acids, and amino acids, ethanol, buthanol, methane. The abundant amounts of cellulolytic materials available (> 10^{12} tons annually on a global basis) offer an inexpensive alternative and renewable source of biomass.

REFERENCES

[1] Ahamed, A. and Vermette, P. *Effect of culture medium composition on Trichoderma reesei's morphology and cellulase production*. Biores. Technol. 100 (2009), 5979–5987.

[2] Akhtar, M. *Biochemical pulping of aspen wood chips with three strains of Ceriporiopsis subvermispora*. Holzforschung 48 (1994), 199–202.

[3] Albersheim, P. *The Walls of Growing Plants*. Sci. Am. 232 (1975), 80–95.

[4] Ali, S., Hall, J., Soole, K.L., Fontes, C.M.C.A., Hazlewood, G.P., Hirst, B.H. and Gilbert, H.J. *Targeted expression of microbial cellulases in transgenic animals. In: Petersen, S.B., Svensson, B., Pedersen, S., editors*. In: Carbohydrate Bioengineering. Progress in Biotechnology, Vol. 10. Amsterdam: Elsevier (1995), 279–293.

[5] Almeida, J.R., Modig, T., Petersson, A., Hahn-Hägerdal, B., Liden, G. and Gorwa-Grauslund, M.-F., *Increased tolerance and conversion of inhibitors in lignocellulosic hydrolysates by Saccharomyces cerevisiae*. J. Chem. Technol. Biotechnol. 82 (2007), 340–349.

[6] An, W. *Studies on dietary application of cellulases complex in monogastric and ruminant animals*. J. North. Agric. Univ. 3 (1998),

[7] Aristidou, A. and Penttilä, M. *Metabolic engineering applications to renewable resource utilization*. Curr. Opin. Biotechnol. 11 (2000), 187-198.

[8] Beauchemin, K.A., Rode, L.M. and Sewalt, V.J.H. *Fibrolytic enzymes increase fiber digestibility and growth rate of steers fed dry forages*. Can. J. Anim. Sci. 75 (1995), 641–644.

[9] Bhat, M.K. *Cellulases and related enzymes in biotechnology*. Biotechnol. Adv. 18 (2000), 355–383.

[10] Bidlack, J., Malone, M. and Benson, R. *Molecular structure and component integration of secondary cell walls in plants*. Proc. Okla. Acad. Sci. 72 (1992), 51–56.

[11] Bon, E.P.S., Gírio, F. and Pereira Junior, N.. *Enzimas na produção de etanol*. In: Enzimas em Biotecnologia: Produção, Aplicações e Mercado. Bon, E. P. S.; Corvo, M. L., Vermelho, A. B., Paiva, C. L. A, Ferrara, M. A. and Coelho, R. R. R., editors, Interciência Brasil. (2008), 241-271.

[12] Buchert, J., Bergnor, E., Lindbland, G., Viikari, L. and Ek, M. *Significance of xylan and glucomannan in the brightness reversion of kraft pulps*. Tappi J. 80 (1997), 165–175.

[13] Buchert, J., Oksanen, T., Pere, J., Siika-aho, M., Suurnakki, A. and Viikari, L. *Applications of Trichoderma reesei enzymes in the pulp and paper industry*. In: Harman

GF, Kubicek CP, editors. Trichoderma & Gliocladium–Enzymes, biological control and commercial applications, Vol. 2. (1998), 343–363.

[14] Caldini, C., Bonomi, F., Pifferi, P.G., Lanzarini, G. and Galante, Y.M. *Kinetic and immobilization studies on fungal glycosidases for aroma enhancement in wine*. Enz. Microb. Technol. 16 (1994), 286–291.

[15] Carta, F.S., Soccol, C.R., Ramos, L.P. and Fontana, J.D. *Production of fumaric acid by fermentation of enzymatic hydrolysates derived from cassava bagasse*. Bioresource Technol. 68 (1999), 23–28.

[16] Cavaco-Paulo A. *Processing Textile Fibres with Enzymes: An overview* In: K. E. Eriksson & A. Cavaco-Paulo (Eds.); ACS Symp. Ser. 687 (1998), 180-189.

[17] Chen, M.; Xia, L. & Xue, P. 2007. *Enzymatic hydrolysis of corncob and ethanol production from cellulosic hydrolysate.* Bioterior. Biodegrad. 59 (2007), 85-89.

[18] Coughlan M.P. *Cellulases: production, properties and applications*. Biochem. Soc. Trans. 13 (1985), 405–406.

[19] Cowan, W.D. *Animal feed*. In: Godfrey T, West S, editors. Industrial Enzymology. 2nd ed. London; Macmillan Press, (1996) 360–371.

[20] De Vries, R.P., Kester, H.C.M., Poulsen, C.H., Benen, J.A.E. and Visser, J. *Synergy between enzymes from Aspergillus involved in the degradation of plant cell wall polysaccharides.* Carbohydr. Res. 327 (2000), 401–410.

[21] Demain, A.L., Newcomb, M. and David Wu. J.H. *Cellulase, clostridia, and ethanol microbiol.* Mol. Biol. Rev. 69 (2005), 124–154.

[22] Eriksson K-EL, Blanchette R.A. and Ander P. *Microbial and enzymatic degradation of wood components.* Springer-Verlag (1990), Berlin.

[23] Fantozzi, P., Petruccioli, G. and Montedoro, G. *Trattamenti con additivi enzimatici alle paste di oliva sottoposte ad estrazione per pressione unica: Influenze delle cultivars, dell'epoca di raccolta e della conservazione*. Grasse 54 (1977), 381–388.

[24] Fernando, S., Adhikari, S., Chandrapal, C. and Murali, N. *Biorefineries: Current Status, Challenges, and Future Direction.* Energy & Fuels 20 (2006), 1727-1737.

[25] Galante, Y.M., De Conti, A. and Monteverdi, R. *Application of Trichoderma enzymes in textile industry.* In: Harman, G.F., Kubicek, C.P., editors. Trichoderma & Gliocladium—Enzymes, biological control and commercial applications. Vol. 2. London: Taylor & Francis, (1998), 311–342.

[26] Godfrey, T. and West, S. *Industrial Enzymology*, 2nd ed. London: Macmillan Press, (1996).

[27] Godfrey, T. *The enzymes market for grain processing* In: Recent Advances in Enzymes in Grain Processing. Courtin, C.M., Veraverbeke, W.S. and Delcour, J.A. (eds.), (2003), 401–406. Kat. Univ. Leuven, Leuven.

[28] Grassin, C. and Fauquembergue, P. *Fruit juices*. In: Godfrey, T., West, S., editors. Industrial enzymology, 2nd ed. UK: Macmillan (1996), 226–383.

[29] Grigorevski-Lima, A.L., Nascimento, R.P., Bon, E.P.S. and Coelho, R.R.R. *Cellulase activity produced by Streptomyces drozdowiczii using low cost agri-industrial by-products and tests for biotechnological application.* Enz. Microbial. Technol. 37 (2005), 272-277.

[30] Gunata, Y.Z., Bayonove, C.L., Cordonnier, R.E., Arnaud, A. and Galzy, P. *Hydrolysis of grape monoterpenyl glycosides by Candida molischiana and Candida wickerhamii β-glucosidases*. J. Sci. Food. Agric. 50 (1990), 499–506.

[31] Hall, J., Simi, A., Surani, M.A., Hazlewood, G.P., Clark, A.J., Simons, J.P., Hirst, B.H. and Gilbert, H.J. *Manipulation of the repertoire of digestive enzymes secreted into the gastrointestional tract of transgenic mice*. Bio. Technol. 11 (1993), 376–379.

[32] Harman, G.E. and Kubicek, C.P. *Trichoderma and Gliocladium: Enzymes, biological control and commercial applications*,Vol. 2. London: Taylor & Francis Ltd, (1998), 393.

[33] Ikeda, Y., Hayashi, H., Okuda, N. and Park, E.Y. *Efficient cellulase production by the filamentous fungus Acremonium cellulolyticus*. Biotechnol. Prog. 23 (2007), 333-338.

[34] Jacobsen, S.E., Wyman, C.E. *Xylose monomer and oligomer yields for uncatalyzed hydrolysis of sugarcane bagasse hemicellulose at varying solids concentration*. Ind. Eng. Chem. Res. 41 (2002), 1454–1461.

[35] Kadam, K.L. and McMillan, J.D. *Availability of corn stover as a sustainable feedstock for bioethanol production*. Biores. Technol. 88 (2003), 17-25.

[36] Keegstra, K., Talmadge, K.W., Bauer, W.D., and Albersheim, P. *The Structure of Plant Cell Walls. III. A Model of the Walls of Suspension-Cultured Sycamore Cells Based on the Interconnections of the Macromolecular Components*. Plant Physiol. 51 (1973), 188–196.

[37] Kumar, A., Lepola, M. and Purtell, C. *Enzyme finishing of man-made cellulosic fabrics*. Textile Chem. Colourist. 26 (1994), 25–28.

[38] Kumar, A., Purtell, C. and Yoon, M.Y. *Performance characterisation of endo-riched cellulase enzymes in the treatment of 100% lyocell and lyocell-blended fabrics*. In: Proceedings of the Textile Institute's 77th World Conference, May 21–24, Tampere, Finland. (1996), 177–89.

[39] Kumar, R., Singh, S. and Singh, O.V. *Bioconversion of lignocellulosic biomass: biochemical and molecular perspectives*. J. Ind. Microbiol. Biotechnol. 35 (2008), 377–391.

[40] Ladisch, M.R., Lin, K.W., Voloch, M. and Tsao, G.T. *Process considerations in the enzymatic hydrolysis of biomass*. Enz. Microb. Technol. 5 (1983), 82–100.

[41] Laukevics, J.J., Aptise, A.F., Viesturs, *U.S. and Tengerdy, R.P. Biotechnol*. Bioengng. 27 (1985), 1687–1691.

[42] Lewis, G.E., Hunt, C.W., Sanchez, W.K., Treacher, R., Pritchard, G.T. and Feng, P. *Effect of direct-fed fibrolytic enzymes on the digestive characteristics of a forage-based diet fed to beef steers*. J. Animal Sci. 74 (1996), 3020–3028.

[43] Liming, X. and Xueliang, X. *High-yield cellulase production by Trichoderma reesei ZU-02 on corn cob residue*. Biores. Technol. 91 (2004), 259–262.

[44] Mantyla, A., Paloheimo, M. and Suominen, P. *Industrial mutants and recombinant strains of Trichoderma reesei*. In: Harman, G.F., Kubicek, C.P., editors. Trichoderma & Gliocladium – Enzymes, biological control and commercial applications, Vol. 2. London: Taylor & Francis, (1998), 291–309.

[45] Martínez, A.T., Speranza, M., Ruiz-Dueñas, F.J., Ferreira, P., Camarero, S., Guillén, F., Martínez, M.J., Gutiérrez, A. and Del Río, J.C. *Biodegradation of lignocellulosics:*

microbial, chemical, and enzymatic aspects of the fungal attack of lignin. Int. Microbiol. 8 (2005),195–204.

[46] Martins, L.F., Kolling, D., Camassola, M., Dillon, A.J.P. and Ramos, L.P. *Comparison of Penicillium echinulatum and Trichoderma reesei cellulases in relation to their activity against various cellulosic substrates.* Biores. Technol. 99 (2008), 1417–1424.

[47] Merino, S.T. and Cherry, J. *Progress and Challenges in Enzyme Development for Biomass Utilization.* Adv. Biochem. Engin. Biotechnol. 108 (2007), 95–120.

[48] Meyer, A.S., Rosgaard, L., Sørensen, H.R. *The minimal enzyme cocktail concept for biomass processing.* J. Cereal Sci. 50 (2009), 337–344.

[49] Nascimento, R. P., Marques, S., Alves, L., Girio, F.M., Amaral-Collaço, M.T., Sacramento, D.R., Bon, E. P. S. and Coelho, R. R. R. *A novel strain of Streptomyces malaysiensis isolated from Brazilian soil produces high endo-β-1,4-xylanase titres.* W. J. Microb. Biotechnol. 19 (2003), 879-881.

[50] Nascimento, R. P., Junior, N. A., Pereira-Jr., N., Bon, E. P. S. and Coelho, R. R. R. *Brewer's spent grain and corn steep liquor as substrates for cellulolytic enzymes production by Streptomyces malaysiensis.* Lett. App. Microbiol. 48 (2009), 529-535.

[51] Ojeda, K. and Kafarov, V. *Energy analysis of enzymatic hydrolysis reactors for transformation of lignocellulosic biomass to bioethanol.* Chem. Eng. J. (2009) doi:10.1016/j.cej.2009.05.032.

[52] Oksanen, T., Pere, J., Buchert, J., Viikari, L. *The effect of T. reesei cellulases and hemicellulases on the paper technical properties of never-dried bleached kraft pulp.* Cellulose 4 (1997), 329–339.

[53] Otjen L., Blanchette R.A. *A discussion of microstructural changes in wood during decomposition by white rot basidiomycetes.* Can. J. Bot. 64 (1986), 905–911.

[54] Pandey, A., Soccol, C.R., Nigam, P., Soccol, V.T. *Biotechnological potential of agro-industrial residues. Part I. Sugarcane bagasse.* Bioresour. Technol. 74 (2000), 69–80.

[55] Pandey, A., Soccol, C.R., Nigam, P., Soccol, V.T., Vandenberghe, L.P.S. and Mohan R. *Biotechnological potential of agro-industrial residues. II: cassava Bagasse.* Bioresource Technology 74 (2000), 81–87.

[56] Pastor, F.I.J., Gallardo, O., Sanz-Aparicio, J. and Días, P. Xylanases: *Molecular properties and applications. In: Industrial Enzymes: structure, function and applications.* Polaina, J. and MacCabe, A. (editors). Springer, The Netherlands, (2007), 65–82.

[57] Pere, J., Siika-Aho, M., Buchert, J. and Viikari, L. *Effects of purified T. reesei cellulases on the fibre properties of kraft pulp.* Tappi J. 78 (1995), 71–78.

[58] Perlack, R.D., Wright, L.L., Turhollow, A.F., Graham, R.L., Stokes, B.J., Erbach, D.C. *Biomass as feedstock for a bioenergy and bioproducts industry: the technical feasibility of a billion-ton annual supply* (Oak Ridge Natl. Lab., Oak Ridge, TN), (2005) ORNL Publ. No.TM-2005_66.

[59] Plant JE, Attwell RW, Smith CA (1988) *A semi-micro quantitative assay for cellulolytic activity in microorganisms.* J Microbiol Methods 7:259-263

[60] Ponte, P.I.P., Ferreira, L.M.A., Soares, M.A.C., Aguiar, M.A.N.M., Lemos, J.P.C., Mendes, I. and Fontes, C. M. G. A. *Use of Cellulases and Xylanases to Supplement Diets Containing Alfalfa for Broiler Chicks: Effects on Bird Performance and Skin Color.* J. Appl. Poult. Res. 13 (2004), 412–420.

[61] Preston, R.D. *Polysaccharide Conformation and Cell Wall Function*. Ann. Rev. Plant Physiol. 30 (1979), 55–78.

[62] Rahkamo, L., Siika-Aho, M., Vehvilainen, M., Dolk, M., Viikari, L., Nou-Siainen, P. and Buchert, J. *Modification of hardwood dissolving pulp with T. reesei cellulases*. Cellulose 3 (1996), 153–163.

[63] Reese, E.T. and Mandels, M. *Rolling with the time: production and applications of Trichoderma reesei cellulase*. Ann. Rep. Ferment. Proc. 7 (1984), 1–20.

[64] Reese, E.T. *History of the cellulase program at the US Army Natick development centre.* Biotechnol. Bioeng. Symp. 6 (1976), 9–20.

[65] Renewable Energy Policy Network (REN21). *Renewables-2006: global status report.* REN21 and Worldwatch Institute, Paris and Washington, DC (2006).

[66] Rodrigues, R.C.L.B., Felipe, M.G.A., Sil, J.B.A., Vitolo, M. *Response surface methodology for xylitol production from sugarcane bagasse hemicellulosic hydrolyzate using controlled vacuum evaporation process variables.* Process Biochem. 38 (2003), 1231–1237.

[67] Ryan, L., Convery, F. and Ferreira, S. *Stimulating the use of biofuels in the European Union: implications for the climate change policy. Energy Policy*. 34 (2006), 3184-3194.

[68] Ryu, D.D. and Mandels, M. *Cellulases: biosynthesis and applications*. Enz. Microb. Technol. 2 (1980), 91–101.

[69] Sachaal, B.A. and Olsen, K.M., Cassava's origin. *Proceedings of National Academy of Sciences, USA,* 11 May 1999. Cited from Quadnet, 4 June 1999, Cassava6-4WU (www.quad.net.com/archive/CASSAVA6-4wu).

[70] Saddler, J.N. *Bioconversion of forest and agricultural plant residues Biotechnol.* Agriculture, no. 9. UK: C.A.B. International, Wallingford, Oxon, (1993), 349.

[71] Sanjuan, R., Anzaldo, J., Vargas, J., Turrado, J., Patt, R. *Morphological and chemical composition of pith and fibres from Mexican sugarcane bagasse.* Holz als Roh-und Werkstoff 59 (2001), 447–450.

[72] Sazci A, Radford A, Erenler K (1986) *Detection of cellulolytic fungi by using Congo red as an indicator: a comparative study with the dinitrosalicyclic acid reagent method.* J Appl Bacteriol 61:559-562

[73] Schwarze F.W.M.R., Engels J., Mattheck C. *Fungal strategies of decay in trees*. (2000) Springer, Berlin.

[74] Shleser, R. *Ethanol production in Hawaii: processes, feedstocks and current economic feasibility of fuel grade ethanol production in Hawaii*. Hawaii State Department of Business, Economic Development & Tourism. (1994) http://hawaii.gov/dbedt/ info/ energy/transportation/ethanol/ studies/ethanol94.pdf (acesso em 13/06/2010).

[75] Soccol, C.R. *Biotechnological products from cassava roots by solid state fermentation.* J. Sci. Indust. Res. 55 (1996), 358–364.

[76] Soccol, C.R., Vandenberghe, L.P.S., Medeiros, A.B.P., Karp, S.G., Buckeridge, M., Ramos, L.P., Pitarelo, A.P., Ferreira-Leitão, V., Gottschalk, L.M.F., Ferrara, M.A., Bon. E.P.S., Moraes, L.M.P., Araújo, J.A. and Torres, F.A.G. *Bioethanol from lignocelluloses: Status and perspectives in Brazil*. Biores. Technol. 101 (2010), 4820–4825.

[77] Sokhansanj, S.; Cushman, J.; Wright, L. *Collection and delivery of biomass for fuel and power production.* Available from http://www.tennesseebiomass.com/storage.php (June 20, 2006).

[78] Suurnakki, A., Heijnesson, A., Buchert, J., Tenkanen, M., Viikari, L. and Westermark, U. *Effect of pulp surfaces on enzyme-aided bleaching of kraft pulps*. J Pulp Paper Sci 22 (1996), 91–96.

[79] Talmadge, K.W., Keegstra, K., Bauer, W.D., and Albersheim, P. *The Structure of Plant Cell Walls. I. The Macromolecular Components of the Walls of Suspension-Cultured Sycamore Cells with a Detailed Analysis of the Pectic Polysaccharides.* Plant Physiol. 51 (1973), 158–173.

[80] Uhlig H. *Industrial enzymes and their applications,* New York: John Wiley & Sons, Inc., (1998), 435.

[81] Unica, (2009) *União da Indústria da Cana de Açúcar.* (http://www.unica.com.br/downloads/estatisticas/processcanabrasil.xls).

[82] Wilson, D.B. *Cellulases and biofuels*. Curr. Opin. Biotechnol. 20 (2009), 295–299.

[83] Xia, L. and Cen, P. *Cellulase production by solid state fermentation on lignocellulosic waste from the xylose industry*. Process Biochem. 34 (1999), 909–912.

[84] Zarrilli, S. *The emerging biofuels market: regulatory, trade and development implications.* In: UNCTAD Intergovernmental Expert Meeting on BioFuels, Geneva, November 30, (2006).

In: Cellulase: Types and Action, Mechanism, and Uses
Editor: Adam E. Golan
ISBN: 978-1-61761-983-0

Chapter 4

SOLID-STATE FERMENTATION FOR PRODUCTION OF MICROBIAL CELLULASE: AN OVERVIEW

Ramesh C. Ray*
Microbiology Laboratory, Central Tuber Crops Research Institute (Regional Centre), Bhubaneswar 751 019, Orissa, India

ABSTRACT

Cellulose present in renewable lignocellulosic material is considered to be the most abundant organic substrate on earth for the production of hexoses and pentoses, for fuel and other chemical feed stock. Research on cellulase has progressed very rapidly in the past few decades, emphasis being on enzymatic hydrolysis of cellulose to hexose sugars. The enzymatic hydrolysis of cellulose requires the use of cellulase [1,4-(1,3:1,4)-β-D-glucan glucanohydrolase, EC 3.2.1.4], a multiple enzyme system consisting of endo-1,4,-β-D-glucanases [1,4-β-D-glucanases (CMCase, EC 3.2.1.4)], exo-1,4,-β-D-glucanases [1,4-β-D glucan cellobiohydrolase, FPA, EC 3.2.1.91] and β – glucosidase (cellobiase) (β-D-glucoside glucanohydrolase, EC 3.2.1.21). Major impediments to exploiting the commercial potential of cellulases are the yield, stability, specificity, and the cost of production. In the past few decades focus has been on submerged fermentation (SmF) and very little attention has been given to solid-state fermentation (SSF). SSF refers to the process whereby microbial growth and product fermentation occurs on the surface of the solid materials. This process occurs in the absence of "free" water, where the moisture is absorbed to the solid matrix. The direct applicability of the product, the high product concentration, lower production cost, easiest product recovery and reducing energy requirement make SSF a promising technology for cellulase production. This review highlights the research carried out on the production of cellulase in SSF using various lignocellulosic substrates, microorganisms, cultural conditions, process parameters (i.e., moisture content and water activity, mass transfer processes: aeration and nutrient diffusion, substrate particle size, temperature, pH, surfactant, etc), bioreactor design, and the strategies to improve enzyme yield. Also, the biotechnological

* Fax/Tel: 91-674-2470528; E-mail: rc_rayctcri@rediffmail.com

potentials of microbial cellulases produced in SSF for bioconversion of agricultural wastes –providing a means to a “greener” technology, have been discussed.

1. Introduction

Cellulose present in renewable lignocellulosic materials, represent about 1.5 x10^{12} tons of total annual biomass production through photosynthesis especially in the tropics and is considered to be an almost inexhaustible source of raw material for the production of glucose, bio-fuels (ethanol, methanol and hydrogen), and other chemical feed stocks (Krishna, 1999; Sukumaran *et al.*, 2005; Zang and Lynd, 2005). It is the most abundant and renewable biopolymers on earth and the dominating waste materials from agriculture.

Microbial degradation of lignocellulosic waste and the down-streaming products resulting from it is accompanied by a concerted action of several enzymes, the most prominent of which are the cellulases, which are produced by a number of microorganisms and comprise several different enzyme classifications. Cullulases hydrolyze cellulose (β-1, 4-D-glucan linkages) and produce as primary products: glucose, cellobiose and cello-oligosaccharides (Himmel, 1994). There are three major groups of cellulase enzymes:

1) Exo-glucanase (Cellobiohydrolase) (CBH) (1, 4-β-D-glucan cellobiohydrolase, EC 3.2.1.91),
2) Endo-β-1,4-glucanase (EG) or endo-1,4-β-D-glucan 4-glucanohydrolase (endo-glucanase, EC 3.2.1.4), and
3) β-glucosidase or β-D-glucoside glucanohydrolase [(BG), EC 3.2.1.21).

Enzymes within these classifications can be separated into individual components, such as microbial cellulase compositions may consist of one or more CBH components, one or more EG components and possible β- glucosidases. These enzymes act in a synergistic fashion to carry out the complete hydrolysis of cellulose, as follows (Wither, 2001; Soni *et al.*, 2010).

- Endo-glucanase acts internally on the chain of cellulose clearing 1, 4-β-linked bonds and liberating oligosaccharides of varying degrees of polymerization. It does not attack cellobiose but hydrolyzes cellodextrins, phosphoric acid- swollen cellulose and substituted celluloses like CMC (carboxymethyl cellulose) and HEC (hydroxyethyl cellulose). It is also claimed that some endo-glucanases act on crystalline cellulose.
- Exo-glucanases (cellobiohydrolase) act progressively from reducing and non-reducing ends reducing cellobiose in a sequential manner.
- Finally, β-glucosidase completes the saccharification by splitting cellobiose and small cello-oligosaccharides into glucose molecules (Figure 1).

Cellulases find extensive use in extraction of green-tea components, modification of food tissues, removal of soybean seed coat, improving cattle feed quality, recovering juice as well as other products from plant tissues and as component of digestive aid (Lonsane *et al.*, 1985). Cellulases can be produced by submerged or solid state fermentations. The later technique is generally preferred as it offers many advantages such as two-three times higher enzyme production as well as protein rate, higher concentration of the product in the medium, direct use of air-dried fermented solids as source of enzyme, which lead to elimination of expenses on downstream processing, employment of natural cellulosic wastes as substrate in contrast to the necessity of using pure cellulose in submerged fermentation (SmF), and the possibility of carrying out fermentation in non-aseptic conditions (Bhat and Bhat, 1997). The biosynthesis of cellulases in SmF process is strongly affected by catabolic and end-product repressions (Gallo *el al.,* 1978; Ryu and Mandels, 1980) and the recent reports on the overcoming of these repressions to significant extent in solid state fermentation (SSF) system (Bhat and Bhat, 1997; Ray *et al.*, 2008), therefore, are of economic importance. The amenability of SSF technique to use up to 20-30% substrate, in contrast to the maximum of 5% in SmF process, has been documented (Pamment *et al.,* 1978; Vandevoorde and Verstraete, 1987).

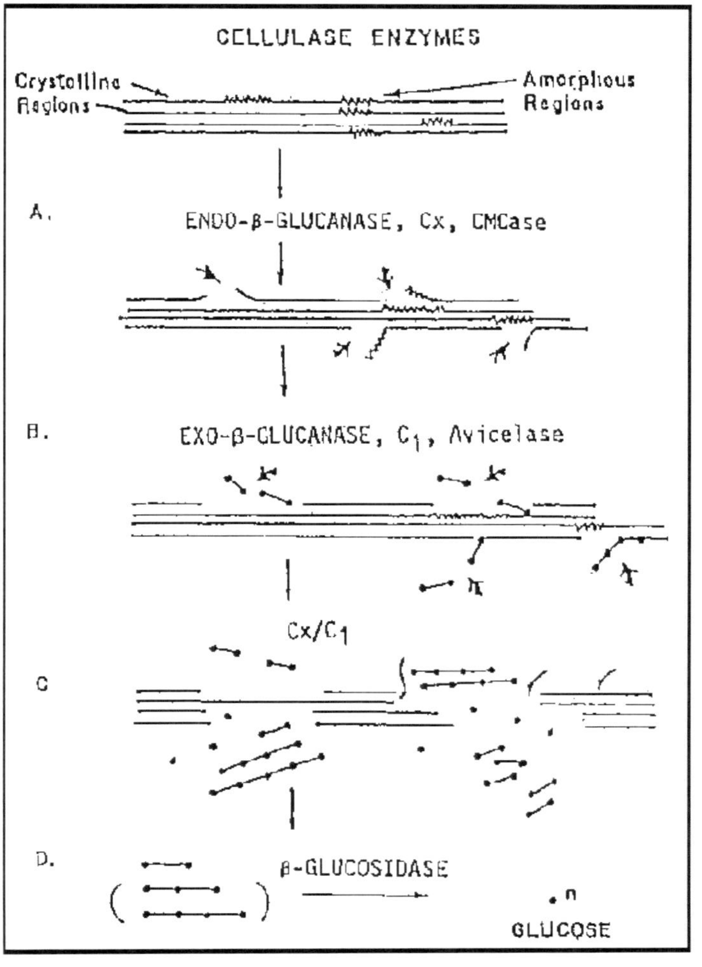

Figure 1. Schematic representation of sequential stages in cellulolysis.

2. Cellulase Producing Microorganisms

Cellulolytic microbes are primarily carbohydrate degraders and are generally unable to use proteins or lipids as energy source for growth. Cellulolytic microorganisms notably the bacteria *Cellulomonas* and *Cytophaga* and most fungi can utilize a variety of other carbohydrates in addition to cellulose, while the anaerobic cellulolytic species have a restricted carbohydrate range limited to cellulose and/or its hydrolytic products. The ability to secrete large amounts of extra-cellular protein is characteristic of certain fungi and bacteria, and such strains are most suited for production of higher levels of extra-cellular cellulases. One of the most extensively studied fungi is *Trichoderma reesei*, which converts native as well as derived cellulose to glucose. Commonly studied cellulolytic organisms of industrial interest include (Table 1):

Table 1. Major microorganisms employed in cellulase production.

Major groups	Microorganisms	
	Genus	Species
Fungi	*Aspergillus*	*A. niger*
		A. nidulans
		A, oryzae
		A. fumigatus
		A. phoenicis
	Fusarium	*F. solani*
		F. oxysporum
	Humicola	*H. insolens*
		H. grisea
	Melanocarpus	*M.albomyces*
	Paecilomyces	*P. themophila*
	Penicillium	*P. brasilianum*
		P. decumbans
		P. occitanis
	Phanerochaete	*P. chrysosporium*
	Trichoderma	*T. reesei*
		T. longibrachiatum
		T. harzianum
		T. viride
Bacteria	*Acidothermus*	*cellulolyticus*
	Bacillus	*B. subtilis*
		pumilus
	Clostridium	*C. acetobutylicum*
		C. thremocellum
	Cellulomonas	*C. fimi*
		C. bioazotea
		C. uda
Actinomycetes		
	Streptomyces	*S. drozdowiczii*
		S. lividans
	Thermonospora	*T. fusca*
		T. curvata

- Fungi- *Trichoderma* (*T. harzianum*, *T.koningii*, *T. longibrachiatum* and *T. reesei*), *Humicola* (*H. grisea*, *H. insolens*), *Penicillium* (*P. funiculosum*, *P. iriensis*, *P. verruculosum*), *Aspergillus* (*A. niger*, *A. terreus*, *A. awamori*, *A. phoenicis*), *Fusarium* (*F. solani* and *F. oxysporum*), *Phanerochaete* (*P. chrysosporium*), *Myrothecium* (*M. verrucaria*).
- Bacteria- Aerobic bacteria: *Bacillus* (*B. subtilis*, *B. pumilus*), *Cellulomonas* *C. bioazotea*, *C. uda*), *Cellovibrio*, *Cytophaga*, *Pseudomonas* (*P. fulvus*); Anaerobic bacteria: *Clostridium* (*C. thermocellum*), *Ruminococcus* (*R. flavefaciens*).
- Actinomycetes- *Actinomucor*, *Streptomyces*, *Thermomonospora* (*T. fusca*).

While several fungi can metabolize cellulose as an energy source, only few strains are capable of secreting a complex of cellulase enzymes, which could have practical applications in the enzymatic hydrolysis of cellulose. Besides *Trichoderma reesei*, other fungi like *Humicola*, *Aspergillus*, and *Penicillium* have the ability to yield high levels of extra-cellular cellulases. Aerobic bacteria such as *Cellulomonas*, *Cellovibrio* and *Cytophaga* are capable of cellulose degradation in pure cultures. However, the microorganisms commercially exploited for cellulase production are mostly limited to *T. reesei*, *Humicola insolens*, *Aspergillus niger*, *Thermomonospora fusca* and few other organisms (Bhat and Bhat, 1997).

3. Solid- State Fermentation (SSF)

Solid- state fermentation [also called as solid state bioprocessing (SSB)] refers to the process where microbial growth and product formation occurs on the surface of solid materials. This process occurs in the absence of "free" water, where the moisture is absorbed to the solid matrix (Zheng and Shetty, 1999; Suryanarayan, 2003). Solid state fermentation has a series of advantages over submerged fermentation (SmF) including lower cost, improved product characteristics, higher product yield, easiest product recovery and reduced energy requirement (Raimbault, 1998; Pandey *et al.*, 2000; Krishna, 2005; Ray *et al.*, 2008).

3.1 Lignocellulosic Residues/Wastes as Solid Substrate

The agro-industrial lignocellulosic residues/wastes form a most important renewable reservoir of carbon for a variety of vitally important chemical feedstock and fuel in the overall economy of any country. Their unlimited availability and environmental pollution potential if not disposed of properly, dictate renewed efforts for their efficient and economic utilization.

A number of such substrates (Table 2) have been employed for the cultivation of microorganisms to produce cellulase. Some of the substrates that have been used include sugar cane bagasse, cassava bagasse, rice bran, wheat bran, maize bran, wheat straw, rice straw, rice husk, soy hull, grapevine trimming dust, saw dust, corncobs, coir pith, banana waste, etc. Wheat bran however, holds the key and has most commonly been used in production of cellulase (Pandey *et al.*, 1999; Ray *et al.*, 2008).

Table 2. Spectrum of microbial cultures employed for production of cellulases in solid state fermentation systems.

Types of Wastes	Microorganisms	Enzyme
Agro wastes (saw dust, waste paper, rice husk, rice bran, coconut waste, etc)	*Aspergillus niger, Spiecellum roseum,Trichoderma reesei Trichoderma* sp., *Pestalotiopsis versicolor*	Cellulases, βellulases, o
Banana fruit stalk waste	*Bacillus subtilis* CBTK 106	Cellulase
Cassava waste	*Trichoderma harzianum*	Cellulases, xylanase
Cellulose, starch	*Trichoderma viride, Aspergillus niger*	Cellulase, amylase
Coconut coir pith	*Aspergillus niger*	Cellulases + βellulases +
Corn cob	*Trichoderma reesei*	Cellulase
Grape vine trimming dust Grape vine cutting waste	*Cerrena unicolor, Trichoderma koningii*	Cellulases, xylanase, lignanse
Palm cornel meal	*Aspergillus wentii, A. niger, A. oryzae, Penicillium* sp. and *Trichoderma reesei*	Cellulase, xylanase, mannanase
Palm empty fruit branch	*Trichoderma harzianum*	Cellulase
Palm oil mill waste	*Cerrena unicolor*	Cellulases
Pumpkin oil cake	*Penicillium roqeforti*	Cellulase
Rice husk	*Penicillium citrinum,* Mesophilic fungi (10 species)	Cellulases, CMCase, xylanase, laccase
Rice straw, spent wheat bran	*Botrytis* sp., *Aspergillus ustus, Sporotrichum pulverulentum*	Cellulase
Sago hampas	*Pleurotus sajor-caju*	Cellulases
Saw dust + Wheat bran	*Trichoderma reesei*	Cellulases, xylanase, laccase
Soyhull	*Phanerochaete chrysosporium*	Cellulases
Sweet sorghum silage, wheat straw	*Gliocladium* sp., *Trichoderma* sp., *Penicillium* sp.	Cellulase, xylanase
Sweet sorghum pulp, wheat straw	Ligninolytic fungal cultures	CMCase
Sweet sorghum silage	*T. reesei, A. niger*	Cellulases, xylanase
Steam pre-treated willow	*T. reesei*	Cellulases
Sweet sorghum	*Gliocladium* sp.	Cellulases
Wheat straw + Wheat bran	*Trichoderma harzianium*	Cellulase
Wheat bran, wheat bran + rice straw	*Trichoderma* sp., *Botrytis* sp., *Aspergillus ustus, Sporotrichum pulverulentum*	Cellulase, ligninase
	T. reesei, S. pulverulentum	

Source: Pandey *et al.,* 1999; Krishna, 1999; Latifian *et al.*, 2007; Yang *et al.*, 2006; Asha Poorna and Prema, 2007; Pericin *et al.*, 2008; Alam *et al.,* 2009.

The selection of a substrate for cellulase production in a SSF process depends upon several factors, mainly related with cost and availability of the substrate and thus may involve screening of several agro-industrial residues. In a SSF process, the solid substrate not only supplies the nutrients to the microorganism(s) growing on it but also serves as anchorage for the cells. The substrate that provides all the needed nutrients to the microorganism growing on it should be considered as the ideal substrate. However, some of the nutrients may be available in sub-optimal concentrations, or even absent in the substrates. In such circumstances, it would become necessary to supplement them eventually with these nutrients. Furthermore, lignin is one of the major deterrents to wide spread utilization of lignocellulosic residues for microbial conversion. It has been a practice to pre-treat (physically, mechanically and/or chemically) some of the substrates (i.e., lignocelluloses) before using in SSF processes, thereby making them easily accessible for microbial growth. However, in large scale production of cellulase, the pre-treatment of agricultural residues is practically not possible because of the enormous expenditure that may incur, which would escalate the cost of enzyme production by 100 to 150% (Cen and Xia, 1999; Liu and Yang, 2007).

3.2 Pretreatment of Agricultural Residues

A. Physical Methods

1) Steam Explosion. It is initiated at a temperature of 160-260^0 C with a corresponding pressure 0.69 to 4.83 MPa for several seconds to several minutes before the material is exposed to atmosphere for cooling (Heerah *et al.*, 2008).
2) Grinding/Milling. Grinding or milling the agricultural residues to a small particle size markedly enhances its susceptibility to microbial influence.
3) Irradiation. Gamma rays and high velocity electric irradiations substantially improves the digestibility of wood, straw or husk bran by microorganisms (Detroy and Julian, 1983). Han *et al.* (1981) combined chemical pretreatment with low dosage of irradiation to solubilize cellulose in sugarcane bagasse, news paper, cotton linter and saw dust.
4) Thermal. Dry heat modified cellulose structure for modest benefits. About 200^0 C is the optimum temperature to produce a maximum rate of acid hydrolysis. However, a 32 h pretreatment is necessary to effect maximum hydrolysis of 35% with a yield of 27% sugar (Datroy and Julian, 1983).

B. Chemical Method

Cellulose and lignocelluloses have been transformed with alkali, acid, ethylamine and ammonia. There is a wide range of differences in the manner in which alkali, acid or ammonia affect the cellulose in wood chips, rice straw or bran and wheat straw or husk, due primarily to the extent of lignifications in the plant materials treated (Detroy and Julian, 1983).

NaOH (4%) pre-treatment of wheat straw maximized FPAse (Filter Paper Activity) of *T. reesei* mutants in SSF. The FPA is a relative measure of the overall cellulose hydrolyzing capacity of microbial cellulase preparations, thus reliable and comparable data may be obtained only under standardized conditions (Urbànszki *et al.* (2000).The cellulase system produced was capable of hydrolyzing over 80% de-lignified wheat straw. Combinations of NaOH pre-treatment with stream explosion did not entrance the activity of the cellulase systems of two mutants of *T. reesei* (Awafo *et al.*, 2000). Similar results were obtained with *Cellulomonas biazotea*, when kaller grass (*Leptochloa fusca*) was used as solid substrate for cellulase production (Rajoka and Malik, 1997).

In contrast, Aiello *et al* (1996) reported no difference in cellulase yield from *T. reesei* QM 9414 between NaOH - treated and un- treated sugarcane bagasse.

3.3 Selections of Microorganisms in SSF

The ability of the microorganisms to grow on solid substrate is a function of their water activity requirements, their capacity to adhere to and penetrate into the substrate and their ability to assimilate mixtures of different polysaccharides from complex heterogeneous substrates (Raimbault, 1998; Pêrez- Guerra *et al.*, 2001).

The filamentous cellulolytic fungi such as *Trichoderma reesei*, *Aspergillus*, etc, are the best adapted microorganisms for SSF owing to their physiological, enzymological and biochemical properties. Growth pattern of these cellulolytic fungi in SSF have been studied in detail in many cases (Xia and Cen, 1999; Awafo *et al.*, 2000) and these can be summarized in two phases: (a) germination, germ tube elongation and mycelial branching to loosely cover most of the substrate, and (b) increase in mycelial density with aerial and penetrative hyphal development (Aiello *et al.*, 1996; Alam *et al.,* 2009). These features also give them a major advantage over bacteria for their colonization of the substrate and the utilization of the available nutrients. In addition, their ability to grow at lower water activity (a_w) and under high osmotic pressure conditions (high nutrient conditions) makes fungi efficient and competitive in the natural microbial ecosystem for bioconversion of solid substrates in to cellulases (Liu and Yang, 2007).

3.4 Substrates and Nutrient Source in SSF

In SSF, two types of substrates can be distinguished depending on the nature of the solid phase:

1) SSF processes that use natural solid substrates from agriculture or by- products from the agro-food industry, which serve as the source of carbon and nutrients for microbial growth (Tengerdy and Szakacs, 2003). Their basic macromolecular structures (i.e. cellulose, lignin, pectin, starch and fiber) confer the properties of a solid to the substrate. For these complex characteristics, the agro- based substrates should be pre-treated to convert the raw substrate into a suitable form to increase nutrient availability and its utilization by the microorganisms. This includes (as discussed under section 3.2):

- Size reduction by cutting, milling, grinding, or rasping.
- Damage to outer substrate layers by cracking, grinding or pearling.
- Chemical (acid/alkali treatment) or enzymatic hydrolysis of lignocelluloses.
- Cooking (gelatinization) or hydrolyzing starch and other polysaccharides for easy growth of microorganisms.
- Supplementation with nutrients (nitrogen, phosphorus, sulphur and other minerals) and adjusting the pH and moisture content. For example, the addition of urea at 2% (w/w) could enhance the cellulase yields by *Phanerochaete chrysosporium* to 74.8 IU (International Units)/gds (gram dry substrate) (CMCase) and 29.1 IU/gds (FPAse) from the initial values of 27.5 IU/gds [CMCase (carboxymethylcellulase)] and 12.2 IU/gds (FPAse) in a SSF system using soy hull as the substrate (Jha *et al.*, 1995). An increase of almost 2-5-fold in activity was achieved. A similar effect has also been observed by Elshafei *et al.* (1990). Krishna (1999) reported that it was essential to supplement banana fruit stalk waste with $(NH_4)SO_4$ or $NaNO_3$ or glucose at 1% (w/v) to enhance cellulase yield by *Bacillus subtilis* CBTK 106 under SSF conditions. In a recent study, Wen *et al.* (2005a) reported that elimination of $CaCl_2$, Mg SO_4, nitrogen sources (NH^+ or urea) and trace elements (Fe^{2+}, Zn^{2+}, CO^{2+} and Mn^{2+}) had no negative influence on the cellulase production, while phosphate elimination did reduce cellulase production.

2) SSF processes that use an inert supports (sugar cane bagasse, jute butts, hemp, inert fibres, resins, polyurethane foam and vermiculite) impregnated with a liquid medium, which contains all the nutrients [sugars (carbon source), nitrogen, phosphorus, etc] required for the growth of the microorganism. This strategy is less used, but it has some advantages (Ray *et al.*, 2008). The use of a defined liquid medium and an inert support with a homogeneous physical structure improves controlling and monitoring of the process and the reproducibility of fermentations. In any case, the use of inert supports has economical disadvantages (Ooijkaas *et al.*, 2000; Gervais and Molin, 2003).

In both cases, the success of the SSF process for cellulase production is directly related to the physical characteristics of the support, which favours both gases and nutrients diffusion and the anchorage of the microorganisms. From a practical point of view, the physical characteristics of the solid matrix must be taken into account because of their influence on the development of SSF, namely particle size and shape, porosity and consistency of the material (Mitchell *et al.*, 2003).

3. 5 Measurement of Cellulase Activity in SSF

The FPA (filter paper activity) is a relative measure of the overall cellulose hydrolyzing capacity of microbial cellulase preparations, thus reliable and comparable data may be obtained only under standardized conditions. The conditions of the FPA assay were standardized for SSF (Urbànszki *et al.*, 2000). The standardization developed for submerged fermentation (SmF) cannot be translated directly to SSF. In SSF, the FPA is strongly dependent on the extraction volume and on the dilution of the enzyme in the assay. The optimal extraction volume was substrate dependent in SSF of corn fiber, spent brewing grains and wheat straw for cellulase production by *Trichoderma reesei* Rut C30. Other cellulolytic enzyme assays (endoglucanase (CMCase), β-glucosidase and xylanase) were much less sensitive to the extraction volume.

4. Cellulase Production in SSF System

Currently, industrial demand for cellulases is being met by production methods mostly using submerged fermentation (SmF) processes, employing generally genetically modified strains of *Trichoderma*. The cost of production in SmF system is however high. Therefore, it necessitates in reduction of production cost by deploying alternative methods, i.e., the SSF system. Tengerdy (1996) compared cellulase production in SmF and SSF systems. While the production in the crude fermentation by SmF was about $20/kg, by SSF it was only $ 0.2/kg. The enzyme in SSF crude product was concentrated; thus it could be used directly in such agro-biotechnological applications as silage or feed additive lingocellulosic hydrolysis, and natural fibre (i.e., jute) processing. Similarly, Vintila *et al.* (2009) found the cost of production of cellulase from *Trichoderma* was cheaper in solid state culture than submerged culture.

Xia and Cen (1998) reported cellulase production through SSF using corn cob residue, a lingocellulosic waste from the xylose industry, as the substrate for *Trichoderma reesei* Zu-02. The cellulase *koji* produced in the process could be used directly to hydrolyze corn cob residue effectively when the cellulase dosage was above 20 IU FPAse/gds; the saccharification yield could be over 84%. Rocky-Salimi and Esfahani (2009) reported 11.65, 99.76 and 94.21 IU/gds of FPAse, Avicelase (exo-glucanase) and CMCase (carboxymethylcellulase) activity, respectively by *T. reesei* QM 9414 grown on rice bran in SSF. Likewise, Liu and Yang (2007) reported FPAse activity of 6.90 IU/gds and CMCase activity of 23.76 IU/gds by *Trichoderma koningii* AS 3.4262 obtained after 84 h of fermentation with media containing vinegar waste. A fungal strain, *Trichoderma harzinum* T2008 was used to evaluate the solid- state bioconversion of palm empty fruit branches for cellulase production. The study was conducted in two systems: an Erlenmeyer flask (500 ml) and a horizontal rotary drum bioreactor (50l). The highest cellulase activity on the 4th day of fermentation in the Erlenmeyer flask was 8.2 FPAse/gds, while its activity from the rotary drum bioreactor was 10.1 FPAse/gds on the 2nd day of fermentation (Alam *et al.,* 2009). Sugarcane bagasse was used as substrate for SSF for cellulase production by *T. reesei* RUT C30. Maximum cellulase production (25.6 IU/gds) was obtained with incubation temperature and time were 33^0 C and 67 h, respectively (Mekala *et al.*, 2008). Apple pomace was also used for cellulase production by *Trichoderma* (Sun *et al.*, 2010).

Apart from *Trichoderma*, several other fungi were employed for cellulase production in SSF. In a study on the ligninolytic system of *Cerrena unicolor* 062- a higher basidiomycetes - upon supplementation of the medium with carbon sources and phenolic compounds in SSF system, it was observed that the growth of *C. unicolor* 062 could be regulated by the exogenous addition of these compounds (Elisashvilli *et al.*, 2002). Production of cellulases and hemicellulases from *Aspergillus niger* KK2 in SSF was studied by using different ratios of rice straw and wheat bran. When *A. niger* KK2 was grown on rice straw alone as a solid support in SSF, the maximum FPAse activity was 19.5 IU/g in 4 days. Also, CMCase (124 IU/g), β- glucosidase (100 IU/g), xylanase (5070 IU/g) and β- xylosidase (193 IU/g) activities were concurrently obtained after 5-6 days of fermentation (Kang *et al.*, 2009). Pothiraj *et al* (2006) using cassava bagasse as solid substrate, reported *Rhizopus stolonifier* as more efficient in bio-converting the bagasse into fungal protein (9%) than *Aspergillus niger* and *A. terreus*. Wheat bran served as the best carbon source for CMCase activity by *Penicillium roquefortii* as it gave the highest enzyme activity (53.06 IU/gds) as compared to different oil cakes. Further, reasonably good quantities of cellulase by *P. roquefortii* was produced on pumpkin oil cake and pumpkin oil cake + wheat bran which were 37.07 and 48.49 IU/gds, respectively (Pericin *et al.,* 2008). Palm kernel meal has been used as substrate in SSF for production of cellulase, xylanase and mannanase from *Aspergillus wentii* TISTR 3075, *A. niger, A. oryzae, Trichoderma reesei* and *Penicillium* sp. during palm kernel meal fermentation; all the fungal strains produced these enzymes but mannanase activity was high (Lee, 2007). In a recent study, Soni *et al.* (2010) reported the optimization of cellulase production by a versatile *Aspergillus fumigatus* fresenius strain capable of efficient de-inking and enzymatic hydrolysis of Solka floc and bagasse. The culture produced maximum levels of cellulase on basal salt medium containing rice straw as carbon and beef extract as nitrogen source.

Soy hull, a material produced in large amounts during soybean processing was utilized for cellulase production through SSF. Of five known fungi [*Chaetomium globosum* (NCIM 874)] *Coriolus versicolor* (R-106), *Phanerochaete chrysoporium* (HHB-1037375 S), *Trichoderma reesei* (QM 9114) and *Neurospora sitophila* (NRRL 2884), *Phanerochaete chysoporium* gave maximum yields [CMCase (27.5 IU/gds)} and FPAse (12.2 IU/gds) of the enzyme (Jha *et al.*, 1995).

There are several reports describing co-culturing of two or more cultures for enhanced cellulase production. Gupte and Madamwar (1997 a, b) cultivated two strains of *Aspergillus ellipticus* and *A. fumigates,* and reported improved hydrolytic and o-culturing of two or more cultures for enhanced cellulase production. Gupte and Madamwar (1997 a, b) cultivated two strains of xylanase (5070 IU/g) and β- xwhen a mutant of *Trichoderma reesei* was co-cultured with a strain of *Pleurotus sajor-caju* with wheat straw as the substrate. The media constituents too play an important role in mixed culturing. In another study, *T. reesei* was co-cultured with *Aspergillus phoenicis* using dairy manure as a substrate to produce cellulase with a high level of g. In another study, nhanced cellulas*T. reesei* and *A. phoenicis*, the optimal media compositions were same (10 g/l manure supplemented with 2 g/l KH_2PO_4, 2 ml/l Tween 80 and 2 mg/l $COCl_2$) while the optimal temperature and pH were similar (25.50 C and pH 5.76 for *T. reseei*, 28.20 C and pH 5.4 for *A. phoenicis*). The mixed culture was therefore completed at 270 C and pH 5.5, which was close to the optimal values of both fungi. The mixed culture resulted in relatively high levels of total cellulase and alues of both fungi*et al.,* 2005b).

Comparatively less numbers of studies have been made with bacteria for cellulase production. Krishna (1999) studied the cellulase production by *Bacillus subtilis* (CBTK 106) through SSF using banana fruit stalk waste as the substrate. The optimum FPAse (filter paper activity) of 2.8 IU/gds, CMC (carboxymethylcellulase)ase activity of 9.6 IU/gds and cellobiase activity of 4.5 IU/gds were obtained at 72 h incubation with media containing banana fruit stock (autoclaved at 1210 C for 60 min, particles at 4.0 mm size). The total enzyme production was 12- folds higher in SSF than in SMF. Production of FPAse, endo-β-glucanase and β-glucosidase by *Cellulomonas biazotea* was investigated during growth on *Laptochloa fusca* (kallar grass). The organism produced 37.5, 17.5 and 6.1 IU/l/h for CMCase, FPAse and β-glucosidase, respectively, with cell mass productivity of 0.235 g/l/h (Rajoka and Malik, 1997).

5. Environmental Factors affecting Microbial Cellulase Production in SSF Systems

Environmental factors such as water activity and moisture content, temperature, pH, oxygen levels and concentrations of nutrients and products significantly affect microbial growth and cellulase production. In SmF, environmental monitoring is relatively simple because of the homogeneity of microbial cell suspensions and of the solutions of nutrients and enzyme in the liquid phase. Due to complex nature and heterogeneity of substrates environmental monitoring is more challenging in SSF.

5.1 Water Activity/Moisture Content

Moisture content is a critical factor for SSF processes because this variable influences growth and biosynthesis of cellulase (Tengerdy and Szakacs, 2003). Lower moisture content causes reduction in solubility of nutrients of the substrate, low degree of swelling and high water tension. On the other hand, higher moisture levels can cause a reduction in product yield due to steric hindrance of the growth of the producer strain by reduction in porosity (inter-particle spaces) of the solid matrix, thus interfering with oxygen transfer (Troquet *et al.*, 2003).

The moisture requirements of microorganisms must be better defined in terms of water activity (a_w) rather than moisture content of the solid substrate (Raimbault, 1998; Gervais and Molin, 2003). Water activity is defined as the relationship between the vapour pressure of water in a system and the vapour pressure of the pure water. From a microbiological point of view a_w indicates the available or accessible water for the growth of the microorganisms. The water activity affects the biomass development, metabolic reactions, and the mass transfer processes (Krishna, 2005).The optimum a_w for growth of a number of fungi used in SSF processes is at least 0.96 (Raimbault, 1998).

Water content of solid substrate between 55% and 70% is found optimum for the growth of most of the cellulase producing organisms, i.e. *Trichoderma* (Aiello *et al.*, 1996; Xia and Cen, 1998; Latifian *et al.*, 2007; Sun *et al.*, 2010), *Aspergillus* spp. (Jecu, 2000; Lee, 2007), *Penicillium* (Pericin *et al.*, 2008), etc. For example, Sun *et al.*, (2010) reported an initial moisture level of 70% was found optimum for cellulase production by *Trichoderma* sp. on apple pomace under SSF. Jecu (2000) reported a moisture content of 74% is optimum for endoglucanse production by *Aspergillus niger* when grown on a mixed substrate of wheat straw and wheat bran. Krishna (1999) reported a moisture content of 70% in banana wastes was optimal for production of FPAse and CMCase by *Bacillus subtilis* CBTK 106.

In contrast, moisture content of 50% was reported optimal for the growth and cellulase production by *Trichoderma koningii* AS3.4262 when waste from vinegar industry was chosen as substrate for SSF.

5.2 Temperature

The increase in temperature in SSF is a consequence of the metabolic activity of the microorganism when the heat removal is not sufficient. This affects directly spore germination, growth of the microorganisms and cellulase production. The temperature level reached is a function of the type of microorganism and the porosity, particle diameter and depth of the substrate (Gervais and Molin, 2003; Raghavarao *et al.*, 2003). Control of temperature is more difficult in SSF than in SmF. Thus, the control methods used in SmF are not suitable for SSF. In an industrial context, monitoring and controlling this variable is critical for scaling up (Bellon- Maurel *et al.*, 2003). Conventionally, aeration is the main method used to control the temperature of the substrate (Raimbault, 1998). Because high aeration rates can reduce the water activity of the substrate by evaporation, moisture-saturated air is usually used. The agitation of the fermentation mass can also help to control the temperature (Raghavarao *et al.*, 2003).

Usually a temperature range of 25-30^{0}C is found optimum for most mesophilic organisms including *Trichoderma* (Cen and Xia, 1999). For examples Latifian *et al.*, (2007) reported optimum temperature required for *Trichoderma reesei* for cellulase production in SSF was 25-30^0 C (Wen *et al.*, 2005a). Wen *et al* (2005b) reported a temperature range of 25.5-27^0 C for optimal production of cellulase (β -glucosidase) by the mixed fungal culture of *T. reesei* and *Aspergillus phoenicis* on dairy manure. Krishna (1999) reported incubation temperature of 35^0 C was optimum for *Bacillus subtilis* strain CBTK 106 for cellulase production using banana waste as the substrate. Soni *et al.* (2010) reported a temperature of 45^0 C was optimum for cellulase production by *Aspergillus fumigatus*.

5.3 Mass Transfer Processes: Aeration and Nutrient Diffusion

In SSF, the mass transfer processes related to gases and nutrient diffusion are strongly influenced by the physical structure of the matrix and by the liquid phase of the system (Raghavarao *et al.*, 2003). Raghavarao *et al.* (2003) described two kinds of phenomena of mass transfer: one at the micro-scale and other at the macro-scale outside the cells. The first one deals with the mass transfer into and out of the microbial cells. The second one includes several factors such as the bulk air flow into and from the bioreactor, natural convection, diffusion and conduction through the substrate, the materials of the bioreactor, the shear damage of the microorganism and the integrity of the substrate particles.

- Gas diffusion: Aeration essentially has two functions: (1) oxygen supply for aerobic metabolism and (2) removal of CO_2, heat, water vapour and volatile components produced during the metabolism (Gervais and Molin, 2003). The exchange of O_2 and CO_2 between the solid and the gas phase takes place at both inter-particulate and intra-particulate levels. This depends on those factors that increase the contact surface between the phases (Cannel and Moo- Young, 1980a; Fujian *et al.*, 2002): void fraction of the matrix, pore and diameter size of the particle, degree of mixture and depth of the matrix, additional aeration generated by forced step of sterile air and agitation and moisture level of the substrate. In general, the gas diffusion increases with the pore size and decreases with the reduction of the diameter of the particle due to substrate packaging (Cannel and Moo- Young, 1980a, b: Troquet *et al.*, 2003).

In SSF, the shear forces caused by rotation and agitation damage or disrupt fungal mycelia and reduce the porosity of the substrates, while the forces led by dynamic changes of air (including air pressure pulsation and internal circulation) are not shearing but normal. The substrates, which are more effectively utilized in the solid-state culture with periodically dynamic changes of air (dynamic culture), are looser and provide more room for fungal propagation than in the static culture. The maximum average filter paper enzyme activity (FPA) in the dynamic culture with the fermentation period 60 h is 20.4 IU/g at a bed height of 9.0 cm while the maximum average FPA is 10.8 IU/g in the static culture with the fermentation period 84 h (Fujian *et al.*, 2002). Under the optimum pressure amplitude, 1400 IU/g CMCase activity could be obtained in the system against 450 IU/gds compared with the tray fermenter (Tao *et al.*, 1999). Rocky Salimi and Hamidi-Esfahani (2009) also stressed that aeration rate had significant effect of cellulase of FPA.

- Nutrients diffusion: Nutrient diffusion occurs at an intra-particulate level and includes both the diffusion of nutrients toward the cells and the hydrolysis of solid substrates by the microbial enzymes (Cannel and Moo- Young, 1980b). This latter point is an important concept in SSF since a large part of the substrate is water insoluble (Raghavarao *et al.*, 2003). In substrates with a small pore size, the resistance to the intra-particle mass transfer increases with the diameter of the substrate particle (discussed further in Section 4.4) and the degradation of the substrate occurs mainly at the outer surface. Nutrient diffusion processes are especially important in bacterial and yeast SSF. They are not so critical for fungal cultures because the mycelium can better penetrate the solid matrix.

5.4 Substrate Particle Size

Particle size of the substrate play crucial role for enzyme production. Generally, smaller substrate particle size provides larger surface area for microbial attack and thus, is a desirable factor. However, too small substrate particles may result in substrate agumulation, which may interfere with microbial respiration/aeration, and therefore, result in poor growth. In contrast, larger particle provide better respiration/aeration efficiency (due to increased inter-particle space), but provide limited surface for microbial attack. This necessitates a compromised particle size for a particular process as well as substrate.

The particle size of 1.0-2.0 mm is usually considered suitable for cellulase production covering a good number of agricultural wastes, i.e. rice bran (Latifian *et al.*, 2007; Kang *et al.*, 2009; Rocky-Salimi and Hamidi-Esfahani, 2009), wheat bran (Jecu, 2000; Kang *et al.*, 2009), wheat straw (Awafo *et al.*, 2000), soy hull (Jha *et al.*, 1995), palm waste (Alam *et al.*, 2009), banana waste (Krishna, 1995), etc

5.5 pH

The pH of a culture may change in response to microbial metabolic activities. The most obvious reason is the secretion of organic acids such as citric, lactic and acetic acids particularly in fungal culture, which will cause the pH to decrease (Ray *et al.*, 2008).

Initial culture pH of 6.0 was found optimum for growth and cellulase production by *T. reesei* using wheat straw as carbon source (Awafo *et al.*, 2000). Similarly, the pH 5.5-5.7 was found optimum for cellulase production by *T. reesei* using dairy manure in SSF (Wen *et al.*, 2005 a, b). In contrast, Vintila *et al.,* (2009) reported cellulase reached maximum activity of 3.18 IU FPAse/ ml at 50^0 C and pH 4.8 when *T. viride* was employed in SSF of starch-based corn grain. Optimum pH of 4.0 for *Phanerochaete chrysosporium* was reported for cellulase production using soy hull in SSF (Jha *et al.,* 1995). The pH range of 4.5-5.5 on mixed substrate containing wheat straw: wheat bran of 9: 1 resulted in highest endo-glucanase (14.8 IU/ml) production (Jecu, 2000).

An optimal pH of 7.0 for *Bacillus subtilis* (CBTK 106) was reported for FPAse and CMCase activity using banana fruit stock waste as the solid substrate (Krishna, 1999). Similarly, Yang *et al* (2006) reported that for production of extra-cellular xylanase by the thermophilic fungus, *Paecilomyces thermophila* J18 on wheat straw, the ideal pH was 7.0-8.0.

5.6 Inoculation Size

The optimum size of inoculums was 1.0 x 10^4 viable spores/gds in case of cultivation of *Phanerochaete chysosporium* on soy hull (Jha *et al.*, 1995). Inoculum (*Bacillus subtilis* CBTK 106): substrate (banana feed stock wastes) ratio of 15% (v/w) resulted in highest bacterial cellulase production than other combinations (Krishna, 1999). Sun *et al.* (2010) reported inoculum size of 2 x 10^8 spores/flasks (500 ml) was used for cellulase production by *Trichoderma* species.

5.7 Surfactants

Many reports have shown the stimulatory effects of surfactants on enzyme production by microorganisms in SmF (Swain and Ray, 2010) or SSF (Jha *et al.,* 1995). Most of the surfactants used were chemically synthesized surfactants such as Tween 20, Tween 60, Tween 80, Triton X-100, polythelene glycol, sodium lauryl sulphate, sodium taurocholate, etc. Tween 80 is the most commonly used surfactant applied for cellulase production from several microorganisms such as *Aspergillus fumigatus* (Soni *et al.*, 2010). These surfactants had various effects in different enzymes. Recently, bio-surfactants, produced as metabolic by-products by bacteria, yeasts and fungi, are applied for enhancing enzyme production (Liu *et al.*, 2006). There are of special advantages over chemically synthesized surfactants owing to their biodegradability, low toxicity, solubilization of low solubility compounds and insensitivity to extreme temperature and pH (Mulligan, 2005).

The effects of rhamnolipid, a bio-surfactants were compared with those of Tween 80 on production of cellulase and xylnase by *Trichoderma viride* in SSF. Rhamnolipid at 0.018% (w/w) gave Avicelase and β-glucosidase activity 122.6 % and 157.6% higher than those of the control (no surfactant added). Further, the stimulatory effects of rhamnolipid were superior to those of Tween 80 (Liu *et al.*, 2006).

6. FERMENTER (BIOREACTOR) DESIGN FOR CELLULASE PRODUCTION IN SSF

Over the years, different types of fermenters (bioreactors) have been employed for various purposes, including cellulase production in SSF systems. Laboratory studies are generally carried out in Erlenmeyer flasks, Roux bottles, beakers, jars and glass tubes (as column bioreactor). Large scale fermentations have been carried out in tray-, drum- or deep through type fermenters. The developments of a simple and practical fermenter with automation, is yet to be achieved for the SSF processes. Most of the studies on cellulase production in SSF have been carried out using either Erlenmeyer flasks or Roux bottles. In the following paragraph, few examples have been cited in which tray -, drum- or deep through fermenters were used.

In some earlier studies, pan bioreactor, requiring a small capital investment, was developed for SSF of wheat straw (Awafo *et al.*, 1996). High yields of complete cellulase system were obtained in comparison to those in the SmF. A complete cellulase system is defined as one in which the ratio of the β, 1996). Hig activity to FPAse activity in the enzyme solution is close to 1:0. The prototype pan bioreactor however required further improvements so what optimum quantity of substrate could be fermented to obtain high yields of complete cellulase system per unit space.

Xia and Cen (1999) studied cellulase production by *T. reesei* ZU-02 in shallow tray- and deep through fermenters. In shallow tray fermenter, the solid substrate (lignocellulosic waste from the xylose industry) could be reused in at least three batches and the highest cellulase (FPAse) activity (158 IU/g *koji*) was obtained in the second fermentation batch. To produce cellulase on a larger scale, a deep through fermenter with forced aeration was used and 128 IU/g *koji* (~305 IU/g cellulose) was reached after 5 days of SSF. Alam *et al* (2009) studied solid- state bioconversion of oil palm empty fruit branches for cellulase production by *T. harzianum* T2008 using two SSF systems: Erlenmeyer flask (500 ml) and horizontal rotary drum bioreactor (50 l). The highest cellulase activity on the 4th day of fermentation in the Erlenmeyer flask was 8.2 FPAse/gds of empty fruit branch, while the activity from the rotary drum bioreactor was 10.1 FPAse/gds on the 2nd day of fermentation.

Liu and Yang (2007) reported cellulase production by *Trichoderma koningii* AS3.4263 in a deep through fermenter with forced aeration using vinegar industry waste as the substrate. FPAse activity of 5.87 IU/gds and CMCase activity of 12.98 IU/gds were achieved after 84 h of SSF.

7. Biomass Conversions and Application of Microbial Cellulases

Cellulases were initially investigated several decades back for the bioconversion of biomass which gave way to research in the industrial applications of the enzyme in animal feed, food, textiles and detergents and in the paper industry. With the shortage of fossil fuels and the arising need to find alternative source of renewable energy and fuels, there is renewal of interest in the bioconversion of lignocellulosic biomass using cellulases and other enzymes. In the other fields, however, the technologies and products using cellulases have reached the stage where these enzymes have become indispensable.

7.1 Textile Industry

Cellulases are used in the biostoning of denim garments for producing softness and the faded look of denim garments replacing the use of pumice stones which were traditionally employed in the industry (Bhat, 2000). They act on the cellulose fibre to release the indigo dye used for colouring the fabric, producing the faded look of denim (Sriram and Ray, 2005). *Humicola insolens* cellulase is most commonly employed in the biostoning, though use of acidic cellulase from *Trichoderma* along with proteases is found to be equally good (Bhat, 2000).

7.2 Laundry and Detergents

Cellulases, in particular EGIII and CBH I, are commonly used in detergents for cleaning textiles (Sriram and Ray, 2005). Several reports disclose that EG III variants, in particular from *T. reesei*, are suitable for the use in detergents (Clarkson *et al.*, 2000). *T. viride* and *T. harzianum* are also industrially utilized natural sources of cellulases, as *A. niger* (Kottwitz *et al.*, 2005). Cellulase preparations, mainly from species of *Humicola* (*H. insolens* and *H. grisea* var. *thermoide*a), which are active under mid alkaline conditions and at elevated temperatures are commonly added in washing powders and in detergents (Uhlig, 1998).

7.3 Food and Animal Feed

In food industry, cellulases are used in extraction and clarification of fruit and vegetable juices, production of fruit nectars and purees, and in the extraction of olive oil. Glucanases are added to improve the malting of barely in beer manufacturing, and in wine industry, better maceration and colour extraction is achieved by use of exogenous hemicellulases and glucanases (Saigal and Ray, 2008). Cellulases are also used in carotenoid extraction in the production of food colouring agents. Enzyme preparations containing hemicellulase and pectinase in addition to cellulases are used to improve the nutritive quality of forages. Improvements in feed digestibility and animal performance are reported with the use of cellulases in feed processing (Sriram and Ray, 2005).

7.4 Pulp and Paper Industry

In the pulp and paper industries, cellulases and hemicellulases have been employed for biochemical pulping for modification of the coarse mechanical pulp and hand sheet strength properties, deinking of recycled fibers and for improving drainage and sewerage of paper mills (Bhat, 2000). Cellulases are employed in the removing inks, coating and toners from paper. Bio-characterization of pulp fibres is another application where microbial cellulases are employed (Bhat, 2000).

7.5 Bio-Fuel

Perhaps the most important application currently being investigated actively is in the utilization of lignocellulosic wastes for the production of bio-fuel. The lignocellulosic residues represent the most abundant renewable resource available to mankind but their use is limited only due to lack of cost effective technologies. A potential application of cellulase is the conversion of cellulosic materials to glucose and other fermentable sugars, which in turn can be, used as microbial substrates for the production of single- cell proteins (SCP) or variety of fermentation products like ethanol (Sriram and Ray, 2005) (Figure 2). Organisms with cellulase systems that are capable of converting biomass to alcohol directly are already reported. For example, *Trichoderma* (designated strain A10), isolated from cow dung directly fermented cellulosic biomass to ethanol in SmF and the yield of ethanol was 2g/l (Stevenson and Weimer, 2002). The ethanol yield and productivity obtained during fermentation of lignocellulosic hydrolysates is decreased due to the presence of inhibiting compounds, such as weak acids, furans and phenolic compounds formed or released during hydrolysis.

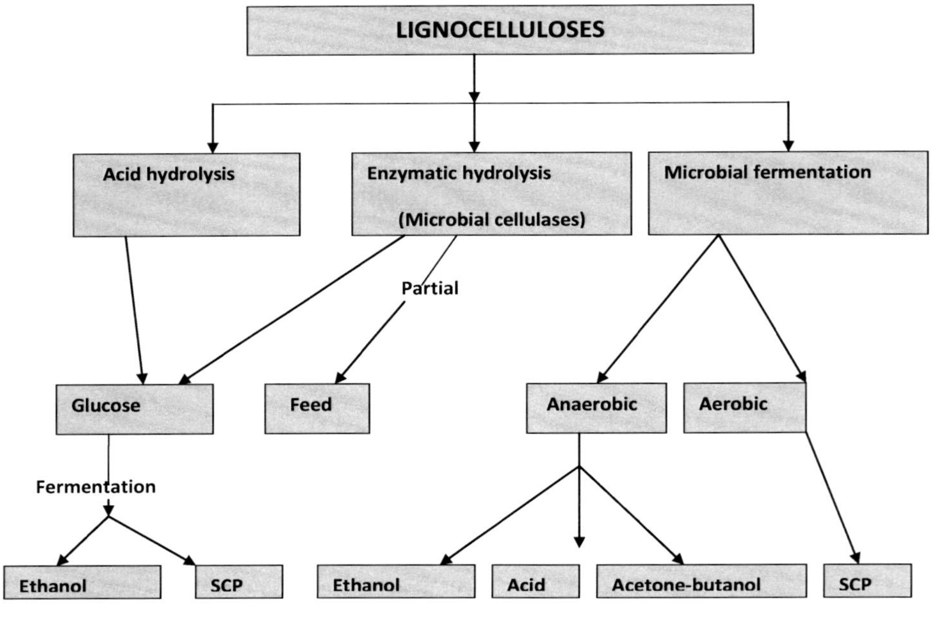

Figure 2. Possible uses of microbial cellulases in biotechnological processes.

Trichoderma reesei cellulase complex degraded the inhibitors found in the acid hydrolysis, resulting in the increase in ethanol productivity and yield (Palmqvist and Hahn-Hàgerdal, 2000). But none of these systems described is effective alone to yield a commercially viable process. The strategy employed currently in bio-ethanol production from lignocellulosic residues is a multi-step process involving pre-treatment of the residue to remove lignin and hemi-cellulose fraction, cellulase treatment at 50^0 C to hydrolyze the cellulosic residue to generate fermentable sugars, and finally use of fermentative microorganism such as *Saccharomyces cerevisiae* or *Zymomonas mobilis* to produce alcohol from the hydrolyzed cellulosic material (Ward and Singh, 2002, 2005). To develop efficient technologies for bio-ethanol production, significant research have been directed at the biotechnological and genetic improvement of the existing organisms utilized in the process. The use of pure enzymes in the conversion of biomass to ethanol or to fermentation products is currently uneconomical due to the high costs of commercial cellulases. Effective strategies are yet to resolve and active research has to be taken up in this direction (Ray and Edison, 2005).

Conclusion and Future Perspective

The biological aspects of solid state bio-processing of cellulosic biomass become the crux of future research involving cellulases and cellulolytic microorganisms. Critical analysis of literature shows that production of cellulases by SSF offers several advantages such as easy enzyme recovery, low cost of production, high product concentration and reducing energy requirement. It has been well established that cellulase titres produced in SSF systems are several-folds higher than in SmF systems. However, the problems which warrant attention is not limited to cellulase production alone, but a concerted effort in understanding the basic physiology of cellulolytic microorganisms coupled with engineering principles applied to SSF. The aspects open to consideration include cheaper technologies of pre-treatment of cellulosic biomass for a better microbial attack, designing of bioreactors and better system for process optimization for higher and qualitative cellulase yield, treatment of biomass for production of hydrolytic products, which can then serve as feedstock for downstream fermentative production of valuable primary and secondary metabolites and protein engineering to improve cellulase qualities..

References

Aiello, C., Ferrer, A. and Ledesma, A. (1996). *Effect of alkaline treatments at various temperatures on cellulase and biomass production using submerged sugarcane bagasse fermentation with* Trichoderma reesei *QM 9414*. Bioresour.Technol.57: 13-18.

Alam, M. Z., Mamun, A. A., Qudsieh, I., Muyibi, S. A., Salleh, H. M. and Omar, N. M. (2009). *Solid state bioconversion of oil palm fruit branches for cellulase enzyme production using a rotary drum bioreactor*. Biochem. Eng. J. 46: 61-64.

Awafo, V. A., Chahal, D. S. and Simpson, B. K. (1996). *Production of cellulase systems by selected mutants of* Trichoderma reesei *in solid-state fermentation and their hydrolytic potentials*. Symposium on Biotechnology for Fuels and Chemicals, N°17, Vail, Colorado, ETATS-UNIS, vol. 57-58, pp. 461-470

Awafo, V. A., Chahal, D. S. and Simpson, B. K. (2000). *Evaluation of combination treatments of sodium hydroxide and stream explosion for the production of cellulase systems by two* Trichoderma reesei *mutants under solid state fermentation conditions.* Bioresour. Technol.73: 235-245.

Asha Poorna, C. and Prema, P. (2007). *Production of cellulase-free endo-xylanase from novel alkaophilic thermotolerant* Bacillus pumilus *by solid state fermentation and its application in waste paper recycling.* Bioresour. Technol. 98: 485-490.

Bellon-Maurel, V., Orliac, O. and Christen, P. (2003). *Sensors and measurements in solid state fermentation: a review.* Process Biochem. 38: 881- 896.

Bhat, M.K. (2000). *Cellulases and related enzymes in biotechnology.* Biotechnol. Adv. 18: 355- 383.

Bhat, M.K. and Bhat, S., (1997). *Cellulose degrading enzymes and their potential industrial applications.* Biotechnol. Adv. 15 (3/4): 583- 620.

Canel, E. and Moo-Young, M. (1980a). *Solid-state fermentation systems.* Process Biochem. 15:2-7.

Canel, E. and Moo-Young, M. (1980b). *Solid-state fermentation systems.* Process Biochem. 15: 24-28.

Cen, P. and Xia, L. (1999). *Production of cellulase by solid state fermentation.* In: Advances in Biochemical Engineering/Biotechnology. Vol. 63, Recent Progress in Bioconversion of Lignocellulosic, (Ed., T. Scheper), Springer-Verlag, Berlin, Heidelberg, pp. 68-92.

Clarkson, K.A., Weiss, G.L. and Larenas, E.A. (2000). *Detergent compositions containing substantially pure EG III Cellulase*, Eur. Patent EP 0586 375 BI (to Genencore International, Inc.) 3s July 2001.

Detroy, R.W. and Julian, G. S. (1983). *Biomass conversion: Fermentation chemicals and fuels.* Crit. Rev. Microbiol. 10: 203- 228.

Elshafei, A. M., Vega, J. L., Klassen, K. T., Claussen, E. C. and Gaddy, J. L. (1990). *Cellulase and hemicellulase formation by fungi using corn stover as a substrate.* Biol.Wastes. 32, 209- 218.

Elisashvilli, V., Kachlishvili, E. and Bakradze, M. (2002). *Dependence of activities of polysaccharide hydrolases and oxidases from* Cerrena unicolor *on the source of carbon and aromatic acids in culture media.* Prikl. Biochim. Mikrobiol. 38: 243- 247.

Fujian, X., Hongzhang, C. and Zuohu, Li (2002*). Effect of periodically dynamic changes of air on cellulase production in solid state fermentation.* Enzyme Microb. Technol. 30: 45-48.

Gallo, B.I., Andreotti, R., Roche, C., Ryu, D. and Mandels, M. (1978). *Cellulase production by a new mutant strain of* Trichoderma reesei *MCG 77.* Biotechnol. Bioeng. Symp. 8 : 89-101

Gervais, P. and Molin, P. (2003). *The role of water in solid-state fermentation.* Biochem. Eng. J. 13: 85- 101.

Gupte, A. and Madaawar, D. (1997a). *Solid state fermentation of lignocellulosic waste for cellulase and beta-glucosidase production by co-cultivation of* Aspergillus ellipticus *and* Aspergillus fumigatus. Biotechnol. Prog. 13: 166- 169.

Gupte, A. and Madaawar, D. (1997b). *Production of cellulolytic enzymes by coculturing of* Aspergillus ellipticus *and* Aspergillus fumigatus *grown on bagasse under solid state fermentation.* Appl. Biochem. Biotechnol. 62: 267- 274.

Han, Y.W., Timpa, J., Ciegler, A., Courtney, J., Curry, W.F. and Lembremont, A. (1981). *Gamma-ray induced degradation of lignocellulosic materials.* Biotechnol. Bioeng. 23: 2525- 2535.

Heerah, A. S., Mudhoo, A., Mohee, R. and Sharma, S. K. (2008). *Steam pre-treatment of lingocellulosic wastes for biomethanogenesis : a preliminary study.* Rasayan J. Chem. 1: 503-514.

Himmel, M.E., Ruth, M.F. and Wyman, C. E. (1994). *Cellulase for commodity products from cellulosic biomass.* Curr. Opin. Biotechnol. 10: 358- 364.

Jecu, L. (2000). *Solid state fermentation of agricultural wastes for endo-glucanase production.* Industrial Crops Products 11: 1-5.

Jha, K., Khare, S. K. and Gandhi, A. P. (1995). *Solid state fermentation of soyhull for the production of cellulase.* Bioresour. Technol.54: 321-322.

Kang, S. W., Park, Y. S., Lee, T. S., Hong, S. I. and Kim, S. W. (2009). *Production of cellulases and hemicellulases by* Aspergillus niger *KK2 from lignocellulosic biomass.* Bioresour. Technol. 91: 153-156.

Kanotra, S. and Mathur, M.S. (1995). *Isolation and partial characterization of a mutant of* Trichoderma reesei *and its application in solid-state fermentation of paddy straw alone or in combination with* Pleurotus sajor-caju . J. Environ. Sci. Health A30: 1339- 1360.

Kottwitz, B. and Schambill, F. (2005). *Cellulase and cellulase containing detergent.* US Pat. 20050020472, 27 January, 2005.

Krishna, C. (1999). *Production of bacterial cellulose by solid state bioprocessing of banana wastes.* Bioresour. Technol.69: 231-239.

Krishna, C. (2005). *Solid state fermentation systems- an overview.* Crit. Rev. Biotechnol. 25: 1- 30.

Latifian, M., Hamidi-Esfahani, Z. and Barzegar, M. (2007). *Evaluation of culture conditions for cellulase production by two Trichoderma reesei mutants under solid state fermentation conditions.* Bioresour. Technol. 94: 3634-3637.

Lee, N. S. (2007). *The production of fungal mannanase, cellulase and xylanase using palm kernel meal as a substrate.* Walailak J. Sci. Technol. 4: 62-82.

Liu, J., Yuan, X., Zeng, G., Shi, J. and Chen, S. (2006).*Effect of biosurfactant on cellulase and xylanase production by* Trichoderma viride *in solid substrate fermentation.* Process Biochem. 41: 2347-2351.

Liu, J. and Yang, J. (2007). *Cellulase production by* Trichoderma koningii *A 53.4262 in solid state fermentation using lingocellulosic waste from the vinegar industry.* Food Technol. Biotechnol. 45: 420-425.

Lonsane, B.K., Ghildyal, N.P., Budiatman, S. and Ramakrishna, S.V. (1985). *Engineering aspects of solid state fermentation.* Enzyme Microb. Technol. 7: 258- 261.

Mekala, N. K., Singhania, R. R., Sukumaran, R. K. and Pandey, A. (2008). *Cellulase production under solid state fermentation by* Trichoderma reesei *RUT C 30: Statistical optimization of process parameters.* Appl. Biochem. Biotechnol. 151: 122-131.

Mitchell, D.A., von Meien, O.F. and Krieger, N. (2003). *Recent developments in modeling of solid- state fermentation: Heat and mass transfer in bioreactors.* Biochem. Eng. J. 13: 137- 147.

Mulligan, C. N. (2005). *Environmental applications for bio-surfactants*. Environ Pollut. 133: 183–98.

Ooijkaas, L. P., Weber, F. J., Buitelaar, R. M., Tramper, J. and Rinzema, A. (2000). *Defined media and inert supports: their potential as solid-state fermentation production systems.* Trends Biotechnol. 18: 356-360.

Palmqvist, E. and Hahn-Hägerdal, B. (2000*). Fermentation of lignocellulosic hydrolysates. I: inhibition and detoxification.* Bioresour. Technol. 74: 17- 24.

Pamment, N., Robinson, C., Hilton, J. and Moo-Young, M. (1978). *Solid state cultivation of* Chaetomium cellulolyticum *on alkali pretreated sawdust*. Biotechnol. Bioeng. 20 : 1735-1744.

Pandey, A., Selvakumar, P., Soccol, C. R. and Nigam, P. (1999). *Solid state fermentation for the production of industrial enzymes*. Curr. Sci. 77: 149-162.

Pandey, A., Soccol, C.R. and Mitchell, D. (2000). *New developments in solid state fermentation: I- bioprocesses and products*. Process biochem. 35: 1153- 1169.

Pericin, D., Madarev-Popovic, S., Radulevi-Popvic, L. and Skrinjar, M. (2008). *Evaluate of pumpkin oil cake as substrate for cellulase production by* Penicillium roquefortii *in solid state fermentation.* Roman. Biotecnol. Lett. 13 : 3815-3820.

Pêrez- Guerra, N., Torrado- Agrasar, A., López- Macias, C. and Pastrana, L. (2001). *Main characteristics and applications of solid substrate fermentation.* Electronic J. Environ. Agric. Food Chem. 1: 1-20.

Pothiraj, C., Balajj, P. and Eyini, M. (2006). *Enhanced production of cellulases by various fungal cultures in solid state fermentation of cassava waste.* Afr. J. Biotechnol. 5: 1882-1885.

Raghavarao, K.S.M.S., Ranganathan, T.V. and Karanth, N.G. (2003). *Some engineering aspects of solid-state fermentation*. Biochem. Eng. J. 13: 127- 135.

Raimbault, M. (1998). *General and microbiological aspects of solid substrate fermentation.* Electron. J. Biotechnol. 1: 1- 22.

Rajoka, M. I. and Malik, K. A. (1997). *Cellulase production by* Cellulomonas biazotea *cultured in media containing different cellulosic substrates.* Bioresour. Technol. 59: 21-27.

Ray, R.C. and Edison, S. (2005). *Microbial biotechnology in agriculture and aquaculture- an overview.* In: Microbial Biotechnology in Agriculture and Aquaculture, Volume1 (Ed., Ramesh C. Ray), Science Publishers, New Hampshire, pp.1- 28.

Ray, R.C., Shetty, K. and Ward, O.P. (2008). *Solid-state fermentation and value-added utilization of horticultural processing wastes*. In: Microbial Biotechnology in Horticulture, Volume 3 (Eds., Ramesh C. Ray and O.P. Ward), Science Publishers, New Hampshire, pp.231- 272.

Rocky-Salimi, K. and Hamidi-Esfahani, Z. (2009). *Evaluation of the effect of particle size, aeration rate and harvest time on the production of cellulase by* Trichoderma reesei *QM 9414 using response surface methodology.* Food Bioproduct Process. Doi. 10.1016/j.fbp.2009.06.006.

Ryu, D.D.Y. and. Mandels, M. (1980). *Cellulases: Biosynthesis and applications.* Enzyme Microb. Technol. 2 91-102.

Saigal, D. and Ray, R.C. (2007). *Wine making: Microbiology, biochemistry and biotechnology*. In: Microbial Biotechnology in Horticulture, Volume 3 (Eds., Ramesh C. Ray and O.P. Ward), Science Publishers, New Hampshire, pp. 1-33.

Soni,R., Nazir, A. and Chadha, B. S. (2010). *Optimization of cellulase production by a versatile* Aspergillus fumigatus *fresenius strain (AMA) capable of efficient deinking and enzymatic hydrolysis of Solkafloc and bagasse.* Industrial Crops Products 31: 277-283.

Sriram, S. and Ray, R.C. (2005). Trichoderma*: Systematics, molecular taxonomy and agricultural and industrial applications.* In: Microbial Biotechnology in Agriculture and Aquaculture, Volume I (Ed., R.C. Ray), Science Publishers, Enfield, New Hampshire, pp. 335- 376.

Stevenson, D.M. and Weimer, P. J. (2002). *Isolation and characterization of a* Trichoderma *strain capable of fermenting cellulose to ethanol.* Appl. Microbiol. Biotechnol. 59: 721-726.

Sukumaran, R., Singhania, R. R. and Pandey, A. (2005). *Microbial cellulases-production, applications and challenges.* J. Sci. Ind. Res. 641 : 832-844.

Sun, H., Ge, X., Hao, Z. and Peng, M. (2010). *Cellulase production by* Trichoderma *sp. on apple pomace under solid state fermentation.* Afr. J. Biotechnol. 9: 163- 166.

Suryanarayan, S. (2003). *Current industrial practice in solid state fermentation for secondary metabolite production: The Biocon India experience.* Biochem. Eng. J. 13: 189- 195.

Swain M. R., and Ray R. C. (2010). *Production, characterization and application of a thermostable exo-polygalacturonase by* Bacillus subtilis *CM5.* Food Biotechnol. 24: 37-50.

Tao, S., Beihui, L., Zuohu, L. and Deming, L. (1999*). Effects of air pressure amplitude on cellulase productivity by* Trichoderma viride *SL-1 in periodic pressure solid state fermenter,* Process Biochem. 34: 25-29.

Tengerdy, R.P. (1996). *Production by solid-state fermentation.* J.Sci. Ind. Res.55: 313- 316.

Tengerdy, R.P. and Szakacs, G. (2003). *Bioconversion of lignocellulose in solid substrate fermentation.* Biochem. Eng. J. 13: 169–179.

Troquet, J., Larroche, C. and Dussap, C.G. (2003). *Evidence for the occurrence of an oxygen limitation during soil bioremediation by solid- state fermentation.* Biochem. Eng. J. 13: 103- 112.

Uhlig, H. (1998). *Industrial Enzymes and Their Applications.* John Wiley & Sons, Inc. New York, 435pp.

Urbánszki, K., Szakács, G. and Tengerdy, R. P. (2000). *Standardization of the filter paper activity assay for solid substrate fermentation.* Biotechnol. Lett. 22: 65- 69.

Vandevoorde, L. and Verstraete, W. (1987). *Anaerobic solid state fermentation of cellulosic substrates with possible application to cellulase production*. Appl. Microbiol. Biotechnol. 26: 479 - 484.

Vintila, T., Dragomireseu, M., Jurcoane, S., Vintila, D., Caprita, R. and Maniu, M. (2009). *Production of cellulase by submerged and solid state cultures and yeasts selection for conversion of lignocelluloses to ethanol.* Roman. Biotechnol. Lett. 14 : 4275-4281.

Ward, O.P. and Singh, A. (2002). *Bioethanol technology: Development and perspectives.* Adv. Appl. Microbiol. 51: 53-79.

Ward, O.P. and Singh, A. (2005). *Microbial technology for bioethanol production from agricultural and forestry wastes.* In: Microbial Biotechnology in Agriculture and Aquaculture, Volume I (Ed., R.C. Ray), Science Publishers, New Hampshire, pp.449-479.

Wen, Z., Liao, W. and Chen, S. (2005a). *Production of cellulase by* Trichoderma reesei *from dairy manure.* Bioresour. Technol. 967: 491-499.

Wen, Z., Liao, W. and Chen, S. (2005b). *Production of cellulose/β-glucosidase by the mixed fungal culture* Trichoderma reesei *and* Aspergillus phoenicis *on dairy manure*. Process Biochem. 40: 3057-3094.

Wither, S.G., (2001). *Mechanism of glycosyl transferase and hydrolases*. Carbohydr. Polym. 44: 325–337.

Xia, L. and Cen, P. (1999). *Cellulase production by solid state fermentation on lingocellulosic waste from the xylose industry*. Process Biochem.34: 909-912.

Yang, S. Q., Yan, Q, J., Jiang, Z. Q., Li, L. T., Tian, H. M. and Wang, Y. Z. (2006*). High level of xylanases production by the thermophilic* Paecilomyces thermophila *J18 on wheat straw in solid state fermentation. Bioresour.* Technol.97: 1794-1800.

Zang, Y-H.P., Lynd, L.R., (2005). *Determination of the number-average degree of polymerization of cellodextrins and cellulose with application to enzymatic hydrolysis.* Biomacromolecules 6: 1510–1515.

Zheng, Z. and Shetty, K. (1999). *Solid-state fermentation and value- added utilization of fruit and vegetable processing by-products*. In: Encyclopaedia of Food Science and Technology, 2nd Edn. (Ed., F.J. Francis), Wiley Publishers, New York, pp. 2165- 2174.

In: Cellulase: Types and Action, Mechanism, and Uses
Editor: Adam E. Golan
ISBN: 978-1-61761-983-0

Chapter 5

Enhanced Enzyme Saccharification of Cereal Crop Residues using Dilute Alkali Pretreatment

T. Vancov[a,b,*] ***and S. McIntosh***[a]
Industry & Investment NSW, Wollongbar Primary Industries,NSW, Australia[a] Primary Industries Innovation Centre, University of New England Armidale, NSW, Australia[b]

Abstract

Mild alkali pretreatment of lignocellulosic biomass is an effective pretreatment method which improves enzymatic saccharification. Alkaline pretreatment successfully delignifies biomass by disrupting the ester bonds cross-linking lignin and xylan, resulting in cellulose and hemicellulose enriched fractions. Here we report the use of dilute alkaline (NaOH) pretreatment followed by enzyme saccharification of cereal crop residues for their potential to serve as feedstock in the production of next-gen biofuels in Australia. Specifically, we discuss the impacts of varying pretreatment parameters on enzymatic digestion of residual solid materials. Following pretreatment, both solids and lignin content were found to be inversely proportional to the severity of the pretreatment process. Higher temperatures and alkali strength were also shown to be quintessential for maximising sugar recoveries from enzyme saccharifications. Essentially, pretreatment at elevated temperatures led to highly digestible material enriched in both cellulose and hemicellulose fractions. Increasing cellulase loadings and tailoring enzyme activities with additional β-glucosidases and xylanases delivered greater rates of monosaccharide sugar release and yields during saccharification. Sugar conversion efficiency of alkali treated sorghum and wheat straw residues following enzyme saccharification, approached 80 and 85%, respectively. Considering their abundance and apparent ease of conversion with high sugar yield, cereal crop residues are ideally suited for the production of second generation biofuels and/or use as feedstock for future biorefineries.

Keywords: alkaline pretreatment, cellulases, enzyme saccharification, lignocellulose, cereal residues, wheat straw, sorghum straw.

* Corresponding author: Tony Vancov, Industry & Investment NSW, 1243 Bruxner Highway, Wollongbar, 2477 NSW, Australia. Tel: +61 2 6626 1359; Fax: +61 2 6628 3264; E -mail: tony.vancov@industry.nsw.gov.au

INTRODUCTION

Interest in commercial scale production of alternative transportation fuels chiefly emanates from issues relating to the use, impacts and rising demand of traditional fossil fuels. Growing dependency on oil, inability to protect supply lines from global political intrigue, projected declines in worldwide petroleum reserves and record crude oil prices (US$145/bbl in June '08) are major drivers for the development of alternative fuels. Global petroleum demands have steadily increased from 57 x 10^6 barrels/day in 1973 to 82 x 10^6 barrels/day in 2004 and are anticipated to rise another 50% by 2025 [1]. Allowing for current rates of production and existing reserves, we will soon approach Hubbert's predicted 'peak oil' levels [2]. Since the industrial revolution atmospheric CO_2 levels have increased from ~275 to ~380 ppm owing to the burning of fossil fuels. Consequently, atmospheric temperatures have risen by 0.6 ± 0.2°C during the twentieth century. If left unchecked, CO_2 levels could easily surpass 550 ppm by the middle of this century [3].

Biofuels, fuels derived from plant biomass are currently the only sustainable class of liquid fuels [4]. First-generation biofuels such as ethanol are currently produced from plants rich in carbohydrates (i.e. sugar and starch). However, as demand for the feedstock intensifies so does the debate between 'food vs fuel'. Moreover, 1st generation ethanol produced from crop starch (corn, wheat) is unsustainable and does not significantly diminish green house gas (GHG) emissions [5, 6]. These shortcomings can be addressed by producing ethanol from lignocellulosic material (next generation biofuels), such as agricultural and forest waste residues. Second-generation biofuels are derived from the inedible and/or unexploited part of the plant (lignocellulose) and can be sourced from plant residues or organic waste such as crop straw, forestry thinnings or contents of landfill.

Accelerating lignocellulosic ethanol research is essential in achieving next generation biofuel production. Various governments and corporations throughout the US, Europe and Asia have heavily invested in emerging lignocellulosic technologies in anticipation of approaching commercial reality (the US has committed over US$1 billion). For example, various joint ventures such as Iogen/Royal Dutch Shell (Canada), Abengoa (Spain), CRAC (China) and DuPont/Danisco (USA) are in the process of delivering pilot scale second generation ethanol facilities based on a range of agricultural feedstocks. These process developments are principally designed for biomass feedstocks at hand. Only two small scale commercial ventures are being developed in Australia, both narrowly focusing on producing sugar streams and ethanol from sugarcane bagasse. If Australia is serious about developing a second generation biofuels industry it will need to consider other feedstocks besides bagasse, such as crop and forestry waste residues.

Australia has approximately 500, 000 km^2 of arable land for producing vast amounts of lignocellulosic biomass, particularly drought tolerant plants [7]. Conservative estimates place agricultural biomass residues at about 65 million dry tonnes per year [8], of which about 25% could be made available for ethanol conversion after accounting for soil management practices and livestock feed [9]. Sorghum straw, wheat straw, sugarcane bagasse, eucalyptus, pine and a number of potential energy crops (e.g. oil mallee) are also among the biomass residues identified as potential renewable resources for ethanol production. Of particular interest is sorghum and wheat as they are two of the largest crop species cultivated in Australia covering both winter and summer cropping cycles. Wheat is Australia's largest broad acre cereal crop with plantings of more than 14 million ha in 2008-2009 seasons [10]. Sorghum is a drought tolerant summer grain crop grown specifically for animal fodder and is expected to increase in plantings to 1 million ha [10, 11]. It's an attractive cropping species because it tolerates heat, moisture and nutrient stresses and is well suited to Australia's changing climatic conditions. Furthermore, by late 2008 the first of two ethanol refineries based on sorghum grain (1st generation ethanol facility) was commissioned by Dalby Bio-Refinery, with a second facility (Pinkenbar Biofuel Project) in the planning stage [12].

Lignocellulose forms the structural framework of plant cell walls and comprises cellulose, hemicellulose and lignin, in proportions varying with the source of the material [13, 14]. Cellulose is a linear polymer composed of 100 to 10,000 anhydroglucose subunits linked by β-1,4-glucoside bonds [15]. In its native state, cellulose molecules form fibres which are largely composed of compact crystalline domains separated by more amorphous regions. Inside plant cell walls, the fibres are embedded in a matrix composed of lignin and hemicellulose. The crystallinity of cellulose in association with hemicellulose and lignin makes lignocellulosic substances highly recalcitrant to decomposition [15]. Hemicellulose is a major constituent of plant cell wall material which makes up 30-40% of many agricultural residues [16]. Xylose is the most abundant sugar in the hemicellulose of hardwoods and crop residues, while mannose is more abundant in softwoods [16, 17]. Lignin, the third major component of lignocellulose, is a heterogeneous aromatic polymer with high molecular weight, and is the second most profuse renewable carbon source on earth [18]. Together with hemicellulose, lignin's key function is to bond cellulose fibres and cells together in plants.

Three key R&D areas which greatly influence lignocellulosic to ethanol conversion efficiency are pretreatment, enzymatic hydrolysis and fermentation. All three stages must be fine-tuned and optimised for a particular feedstock. Efficient utilization of lignocellulosic biomass requires pretreatment to liberate cellulose from its lignin seal and disrupt its recalcitrant structure before effective enzymatic hydrolysis to simple sugars can take place [19]. A range of chemical, physical and biological processes to release these sugars have been configured, yet all face challenges of cost, technological breakthroughs and infrastructure needs [20-22]. In recent years, alkali-based processes have become prominent in pretreatment of straw and stover-type residues, mainly because they operate under lower temperatures, pressures and residence times compared to other pretreatment technologies. The extent of these savings depends on the nature of the biomass feedstock and in particular the lignin content [23]. Sodium hydroxide and lime pretreatments have received a great deal of attention [24, 25], owing in part to the incorporation of cost-effective practises such as chemical and water recycling, and partly because lower enzyme loads are required in converting cellulose to glucose [26, 27].

Alkali pretreatment successfully delignifies biomass by disrupting the ester bonds cross-linking lignin and xylan, resulting in cellulose and hemicellulose enriched fractions; a mechanism of action similar to soda or kraft pulping [28]. Numerous studies have evaluated the use of alkaline pretreatment and enzyme saccharification on a range of lignocellulosic material with varying degrees of success [30-32]. For the most part, the studies examine pretreating and saccharifying the lignocellulosic feedstock in the one vessel, without prior separation or removal of inhibitory compounds. Although most authors report near theoretical sugar yields and demonstrate the fermentation potential of resulting hydrolysates with a range of microorganisms, they fail to highlight the need for high enzyme dosages and extended reaction and fermentation times beyond the cost-benefit threshold. They also neglect to discuss the impacts of individual treatment parameters on saccharification yields and changes in monosaccharide compositions. Overcoming toxicity associated with the release of sugar and lignin decomposition products (furans, phenols etc) and other inhibitors, factor quiet highly in making the process economically unviable.

The feedstock's physicochemical characteristics and the nature of the pretreatment process strongly impacts on the success of the downstream enzymatic hydrolysis. As mentioned above, pretreatment processes usually liberate and/or generate biomass inhibitors which interfere with enzymatic hydrolysis. Other significant issues affecting hydrolysis include: nature of substrate, cellulase loading, and reaction conditions such as temperature and pH, and end-product inhibitors. Efficient solubilisation of cellulose and hemicellulose component requires synergistic action between different enzymes, which incidentally need to be tailored for individual biomass and pretreatment processes [29]. Commercial cellulase R&D is currently focused on overcoming some of these shortcomings and in ultimately creating enzyme blends with significantly reduced processing costs and boarder application conditions [30].

Despite a plethora of studies reporting the use of dilute alkali as an effective lignocellulosic pretreatment option, few have reported using Australian biomass as a feedstock. Moreover, apart from our studies, no one has previously reported using sorghum residues. This chapter reports sugar yields and profiles from post-grain harvested sorghum and wheat straw residues using mild alkali process parameters and low enzyme dose saccharifications. Specifically, we examine and describe three characteristic phases: 1) the function of key pretreatment parameters (alkali strength, temperature, and residence time) and their impact on sugar solubilisation, lignin reduction and solid losses; 2) enzymatic hydrolysis efficacy of pretreated solid residues and variations in sugar composition with respect to pretreatment parameters; and 3) the role of individual and combined enzyme activities and their impact on the rates and yields of sugar release. Due to the impact of phenolic compounds on downstream processes we also discuss their release during pretreatment and saccharification. Understanding these key elements will enable further process optimisation of wheat residues and assist in determining the efficacy of the conversion strategy.

METHODS AND MATERIALS

Materials

Post-grain harvested sorghum straw (*Sorghum bicolour var. MR Buster*) and wheat straw (*Triticum aestivum*) was sourced from the Liverpool plains Northern NSW, Australia. The residues were dried at 55-60°C for 48 h, ground in a rotary mill (Thomas Wiley Laboratory Mill) and passed through a 1.5 mm screen. Milled material was stored at room temperature in sealed containers. All chemicals used were of reagent grade or analytical grade and purchased from Sigma Chemical Co. (St. Louis, MO).

Pretreatment

Biomass feedstock pretreatment was conducted as described by McIntosh and Vancov [31]. To evaluate the effect of pretreatment parameters, a 2x3x4 factorial design was applied for individual biomass samples. Sodium hydroxide (NaOH) at concentrations of 0- 2.0% (w/v) was used to pretreat milled samples at a solid loading of 10% (w/v). Treatments were performed in triplicate at either 60°C in a static water bath and/or in an autoclave at 121°C (15psi) with residence times of 30, 60 and 90 min. The pretreated material was separated into solid and liquor (prehydrolysate) fractions using a Buchner funnel fitted with glass fibre filters (GF-A, Whatman). Pretreated solids were washed with water until the filtrate registered a neutral pH, sealed in plastic bags to retain moisture and stored at -20°C.

Enzyme Assays

Cellulase (NS50013), β-glucosidase (NS50010) and xylanase (NS50030) preparations were kindly supplied by Novozyme (Bagsvaerd, Denmark). Enzyme activities as described by supplier are 70 filter paper unit (FPU) /g, 250 cellobiase units (CBU) /g and 500 fungal xylanase units (FXU) /g respectively. Total cellulase activity of NS50013 was confirmed using the filter paper assay as described by the National Renewable Energy Laboratory (NREL) laboratory procedure LAP006 [32]. Protein content of liquid enzyme preparations was determined using a commercial bicinchoninic acid (BCA) protein assay reagent kit (Pierce Products, USA) and reported in table 1.

Table 1. Specific activity of the commercial enzymes used in mild alkaline pretreated wheat and sorghum straw saccharification.

Enzymes	Specific Activity (U/mg protein)[a]		
	NS50013	NS50010	NS50030
Endoglucanase	14.20	0.11	0.02
Exoglucanase	1.51	0.07	0.05
Xylanase	7.05	75.00	129.50
β-glucosidase	1.07	10.08	0.04
Pectinase	0.03	0.4	ND
Cellulase*	70.00	ND	ND
Protein (mg/ml)[†]	135	150	33

[a] At pH 5.0 and 50°C
*Measured as filter paper units/g protein
ND Not determined
[†] Concentration of Novozymes preparations

Endoglucanase, exoglucanase and xylanase activities were individually determined in reaction mixtures (10 mL) containing 1% (w/v) carboxymethyl cellulose (CMC), 0.5% (w/v) avicel® and 0.5% (w/v) oat spelt xylan and 0.5% (w/v) citrus pectin, respectively, in 50 mM citrate buffer (pH 5.2), and appropriately diluted enzyme solutions as described by McIntosh and Vancov [31]. After 30 min incubation at 50°C, the reducing sugar liberated in the reaction mixture was measured by the dinitrosalicylic acid (DNS) method. One unit (U) of enzyme activity is defined as the amount of enzyme that produces 1 μmol of reducing sugar as glucose (xylose for xylanase) or galacturonic acid (for pectinase) in the reaction mixture per minute, per mg protein under the above specified conditions (Table 1).

β-Glucosidase activities were assayed in reaction mixtures (1 mL) containing 4 mM *p*-nitrophenyl *β*-D-glucoside, 50 mM acetate buffer (pH 5.0) and appropriately diluted enzyme solutions as described by McIntosh and Vancov [31]. After incubation at 50°C for 30 min, the reaction was stopped by adding 100 uL of ice-cold 100mM NaOH, and the colour that developed as a result of *p*-nitrophenol liberation was measured at 405 nm. A unit (U) of enzyme activity is defined as the amount of enzyme that releases 1 μmol of *p*-nitrophenol per minute, per mg of protein in the reaction mixture under these assay conditions. *β*-glucosidase activities present in commercial preparations are reported in Table 1.

Enzymatic Saccharification

Enzymatic saccharifications were performed according to the method described by McIntosh and Vancov [31]. Essentially, solid residues at a 5% (w/v) loading were resuspended in flasks containing 50 mM citrate buffer (pH 5.2) and appropriately diluted enzymes (as specified in the text). Hydrolysis was performed in a shaking water bath at 50°C and 150 rpm for up to 72 h and in the presence of 10 mM sodium azide to prevent microbial contaminant growth. Samples were withdrawn at time points specified in the text and immediately chilled on ice, centrifuged at 8000g for 5 min, filtered and stored at -20°C awaiting sugar analysis. For reproducibility, units of enzyme activity throughout the manuscript are those reported by the manufacturer. Net enzyme saccharifications were calculated by averaging values for sample triplicates and subtracting average values for the respective controls.

Analytical Methods

Neutral detergent fibre (NDF), acid detergent fibre (ADF), acid detergent lignin (ADL) and acid insoluble ash (AIA) were determined for untreated sorghum straw by Industry & Investment NSW's Diagnostic and Analytical Services (Wagga Wagga, Australia) using ANKOM Technology Methods as reported by McIntosh and Vancov [31]. The difference between NDF and ADF provides an estimate of detergent hemicellulose. Detergent cellulose is calculated by subtracting the values for ADL plus AIA from ADF. Carbohydrate content of untreated material was also determined by measuring the hemicellulose (xylan and araban) and cellulose (glucan) derived sugars in supernatants following concentrated acid hydrolysis as described by NREL method [33]. Acid insoluble lignin content of untreated straws and bagasse and the solid fraction remaining after pretreatment was determined according to the NREL methods [33]. Likewise, water and ethanol soluble sugars were extracted from untreated samples and quantified according to NREL methods [34] (see Table 2).

Sugar composition of prehydrolysate and enzymatic saccharification liquors were determined using high performance liquid chromatography (HPLC) according to the procedure described by McIntosh and Vancov [31]. The separation system consisted of a solvent delivery system (Controller 600 Waters, Milford, MA) equipped with an autosampler (717, Waters), a refractive index detector (410 differential refractometer, Waters) and a computer software based integration system (Empower, Waters).

Sugars, acetic acid and ethanol were analysed using either a Sugar-Pak 1 (6.5 x 300mm, Waters) or an IC-Pak Ion-Exclusion 50Ao 7μm (7.8 x 300mm, Waters), both fitted with the IC-Pak Ion Exclusion guard-Pak (Waters). The Sugar–Pak 1 column was maintained at 70°C, and sugars were eluted with degassed Milli-Q filtered water containing 50mg/l Ca-EDTA at a flow rate of 0.5 mL/min. The IC-Pak Ion-Exclusion column was maintained at 60°C, and sugars, acetic acid and ethanol were eluted with degassed Milli-Q filtered water containing 2mM H2SO4 at a flow rate of 0.8 mL/min. The refractive index detector was maintained at 50°C for all applications. Peaks were detected by refractive index and were identified and quantified by comparison to retention times of authentic standards (glucose, xylose, galactose, arabinose, mannose, fructose, sucrose, cellobiose, acetic acid, and ethanol).

Total reducing sugars were determined using the dinitrosalicylic acid (DNS) method as described by NREL [32].

Xylan Extraction

The method was performed according to the procedure described by McIntosh and Vancov [31]. Briefly, ground biomass samples were pretreated at a solid loading of 10% (w/v) in NaOH at a concentration of 0.75%, 1.0% and 2% (w/v) for 60 min at 121°C (15psi). The pretreatment hydrolysate was separated from remaining solids using a Buchner funnel and glass fibre filters (GF-A, Whatman) and was centrifuged (20°C; 10 min; 10000g) to pellet particulates. The hydrolysate was adjusted to $\leq$ pH 4.0 with 6N HCL with rapid stirring. After 10 min of continual stirring the precipitate was sedimented by centrifugation as before. Three volumes of cold (4°C), 96% ethanol were added to the remaining supernatant whilst stirring for 15 min at ambient temperature. The xylan precipitate was collected by centrifugation as before, dried and weighed.

Acid-Insoluble Lignin Extraction

The method is as described by McIntosh and Vancov [31]. Particulates were removed from liquors (described above) were separated from remaining solids using a Buchner funnel and glass fibre filters (GF-A, Whatman) and were centrifuged (20°C; 10 min; 10000g) to pellet particulates. The hydrolysate was heated $\geq$ 60°C and adjusted to $\approx$ pH 2.0 with conc. H_2SO_4 with rapid stirring. After 5 min of continual stirring the samples were cooled to ambient temperature and the precipitate was sedimented by centrifugation as before. The acid-insoluble lignin precipitates were washed with water (pH 2.0) by gently inversion, collected by centrifuged as before, dried and their weight recorded.

Total Phenolic Determination

The enzymatic method described by McIntosh and Vancov [31] was used to determine total phenolic content of hydrolysates. Samples to be analysed were centrifuged and filtered (0.45 micron) prior to assaying. 25 µl aliquot of appropriately diluted phenolic sample were mixed with 225 µl of enzyme-reagent working solution into 96 well microtitre plate (clear F-bottom Greiner Bio-one). The enzyme-reagent working solution was freshly prepared with 0.1 M potassium phosphate buffer solution (pH 8.0) containing 30 mM 4-aminoantipyrine (4-AP), 20 mM hydrogen peroxide (H_2O_2) and 6.6 µM HRP). After 15 minutes at room temperature, the absorbance was read at 540 nm, using a Flurostar (BMG labtechnologies, GmbH) plate reader. Vanillic acid standards (0 to 500 ng / mL) were subjected to the same assay conditions as the samples. Total phenolics were reported as vanillic acid equivalents.

Statistical Methods

Each set of observations was modelled as a response to the classifying factors generated by the experimental design. The data was analysed using analysis of variance which enabled partitioning of total variation in the data into components due to temperature, time, alkaline strength and interactions between those terms.

The modelling process enabled prediction of the expected (average) response at each combination of the experimental factors and a measure of the experimental error. Estimated experimental error was used to calculate the "Least significant difference" (l.s.d., $p = 0.05$) between three averages required to indicate a statistically important effect. Statistical analysis and graphical presentation were conducted using software provided by the R Development Core Team [35].

RESULTS

Compositional Analysis of Straw Residues

The chemical composition of wheat and sorghum straw (presented in table 2) is generally attributed to and reflects a number of factors such as cultivar type, farming inputs and practises, geographical location, seasonal conditions, stage of harvest and analytical procedures. The hybrid sorghum variety (MR Buster) used in this study is principally a grain variety as opposed to forage hybrids which have been selected for decreased lignin contents (*bmr*). The holocellulose fraction totalled 59.4 and 62% of dry sorghum and wheat straw biomass, respectively, with cellulose being the major component at 32.4 and 36 % whilst the remaining 27 and 26 % derived from hemicelluloses. Both acid detergent and acid insoluble lignin levels for sorghum and wheat straw were 2.9 and 5.9 % and 7.0 and 7.6%, respectively. Water extractive compounds accounted for approximately 210 and 130 mg/g dry sorghum and wheat straw, respectively, of which 14 and 35 mg was identified as the non-structural disaccharide, sucrose. Further solvent extraction of sorghum and wheat straws with ethanol resulted in 87 and 55 mg of material, respectively, presumably composed of oils, pigments and waxes. The profile and size of individual wheat and sorghum straw components are comparable to reported values in the literature [36, 37].

Table 2. Composition of untreated cereal straws.

Component	Straw residue	
	Sorghum	Wheat
Neutral Detergent Fibre	63.0	69.0
Acid Detergent Fibre	36.0	43.0
Acid Detergent Lignin	2.9	5.9
Acid Insoluble Ash	0.7	0.9
Cellulose	32.4	36.0
Hemicellulose	27.0	26.0
Acid Insoluble Lignin	7.0	7.6
Water Extractives	21.1	13.0
Ethanol Extractives	8.7	5.5

[a] Composition percentages are on dry-weight basis.

Optimising Enzymatic Hydrolysis of Mild Alkali Treated Wheat Straw

For any individual biomass feedstock and pretreatment strategy it is essential to tailor the saccharification process (enzyme mixture and conditions) to maximise sugar yields [38]. Others reasons for optimization is to compensate for imbalances and/or shortfalls in commercial available preparations. Commercial cellulase mixtures maybe abundant in β-endoglucanase and cellobiohydrolyase, but are generally low in β-glucosidase and xylanase activity. They have been shown to be particularly inadequate for efficient monomeric sugar release from substrates containing higher amounts of arabinoxylan [39]. As shown in table 1, the Novozymes cellulase preparation (NS50013) has 10-fold and 18-fold less β-glucosidase and xylanase activities, respectively, than NS50010 and NS50030 enzyme preparations, hence necessitating enzyme blending. The rate and extent of saccharification in response to differing enzyme combinations and dosages from NaOH (1.0% NaOH; 60 min; 121°C) pretreated wheat straw was examined and the data plotted in Figure 1. The pretreatment regime was employed to evaluate pretreated material that has been substantially delignified yet retained most of its xylan fraction.

The combination of cellulase with β-glucosidase substantially promoted sugar release and was greater than the individual preparations. A supplementary experiment (unreported data) revealed that increasing the ratio of NS50010 to NS50013 from 1:1 to 4:1 (a 4-fold increase in β-glucosidase activity) lead to a corresponding rise in saccharification. However, beyond the ratio of 1:1 the gains were neither statistically significant nor cost-effective for cellulose conversion, and this ratio was subsequently used in following enzyme trials, including sorghum straw saccharifications. Tengborg and co-workers [40] also described similar benefits and limitations of β-glucosidases in enzymatic saccharifications of lignocellulosics in their work on softwoods.

Rates of sugar release over 62 h and yield improved as cellulase and β-glucosidase dosage was raised 4 and 6-fold for sorghum and wheat straw, respectively. For example, in wheat straw trials total sugar release increased 1.3 fold in the presence of 10 FPU cellulase and 10 CBU β-glucosidase, mainly benefiting glucose release (Figure 1a). Final total sugar yields were lower that anticipated. We assumed that presence of xylanase activity in NS50010 (refer to table 1), would be adequate for hydrolysing the hemicellulose fraction. Hemicellulose (xylan) is known to act as a physical barrier, restricting enzyme access to cellulose fibres [41-43]. The hydrolytic efficiency was improved by supplementing the cellulase and β-glucosidase mixture with xylanase (NS50030). The final combined glucose and xylose yields increased by 180 mg following the addition of 1.5 FXU xylanase to the enzyme cocktail. As shown in Figure 1, approximately 90% of glucose and xylose was released within 14 h and hydrolysis was completed inside 24 h.

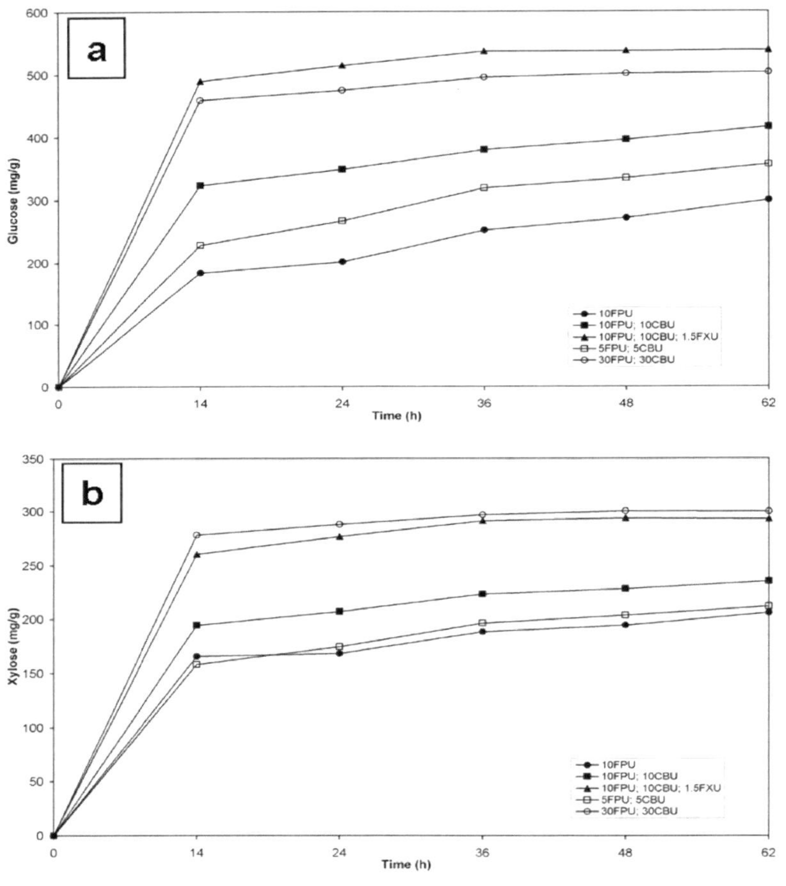

Figure 1. Time course of glucose (a) and xylose (b) release from enzymatic saccharification of alkali pretreated wheat straw. Enzyme combinations and dosage expressed as units of activity per gram of pretreated material with cellulase, β-glucosidase and xylanase activity measured in FPU, CBU and FXU respectively. Glucose and xylose yields are presented as mg/g pretreated material. Data represents averages of three separate experiments. Average l.s.d. ($p = 0.05$) = 12.9 (glucose) and 9.6 (xylose).

There was no advantage in extending saccharification after this time; albeit in the absence of additional xylanase, extended saccharification time was found to be necessary. Both glucose and xylose yields increased by 1.3 and 1.24 fold respectively, in the presence of the tri-enzyme mix. More importantly, this combination was capable of delivering similar rates of sugar release and total yields to saccharifications dosed with 3 fold higher cellulase/β-glucosidase (30 FPU/ 30 CBU) blends. In essence, addition of xylanase in the saccharification reaction had the effect of reducing the cellulase enzyme loadings 3-fold. Further increases in xylanases (3.0 FXU) failed to promote greater sugar gains (unreported data).

In a complementary study (data not shown), various enzymatic mixtures were used to saccharify wheat straw exposed to harsh pretreatment conditions (2% NaOH/ 90 min/ 121°C). Optimal mixtures (10 FPU/ 10 CBU/ 1.5 FXU) released total sugar yields which peaked at 940 mg /g pretreated material, with glucose to xylose ratios approaching 3:1 within the first 14 h. Increasing the enzyme load 3-fold (30 FPU/ 30 CBU/ 1.5FXU), failed to improve final sugar yield suggesting that cellulase levels may have reached saturation point. Disparities in cellulase saturation loading amongst wheat and other herbaceous straw saccharifications are well documented and reported to result from variations in enzyme activities and substrate composition/structure [46-48]. The highly reactive nature of alkali pretreated straw, demonstrated by its near complete digestion to monomeric sugars and rapid reduction in volume (80% within 8 h), could prove to be advantageous in overcoming limitations in solid to liquid load ratios and limitations of dilute sugar streams for fermentation.

Enzymatic Hydrolysis of Mild Alkali Treated Sorghum Straw

Like the wheat straw samples, combing cellulase with β-glucosidase not only increased sugar release but surpassed individual preparations (Figure 2). The rate of sugar release over 63 h and total sugar yield improved as the cellulase and β-glucosidase dosage was raised 4-fold. Once again the final sugar yields were lower than anticipated. This deficit was overcome by supplementing the cellulase and β-glucosidases mixture with xylanase, i.e. final sugar yields increased by 45% (63h incubation) following the addition of 1.5 FXU xylanase to the 2.5 FPU / 3.75 CBU enzyme cocktail. Doubling the load of cellulase/ β-glucosidase (5.0 FPU / 7.5 CBU) whilst maintaining xylanase at 1.5 FXU, resulted in an additional 25% gain. Further xylanases (3.0 FXU) dosage increase failed to promote or deliver greater sugar gains (data not shown). Several authors have reported mixed responses to raising cellulase loads whilst maintaining uniform xylanase activity in saccharification trials, especially in the presence of non-cellulolytic enzymes such as xylanase and pectinase [38, 49, 50].

Despite the low and high enzyme dosage combinations (2.5 FPU / 3.75 CBU / 1.5 FXU; 5 FPU / 7.5 CBU /1.5 FXU) delivering similar end point yields (900 and 950mg/g, respectively), the latter dose is probably more cost-effective because of its faster rates of saccharification. Approximately 88% of the pretreated material was digested within 14 h and hydrolysis was virtually completed inside 24 h. Like the wheat straw samples, no benefits where realised by prolonging saccharification beyond this point; though in the absence of added xylanase, extended times were required. Basically, supplementing enzyme saccharification mixtures with xylanase allowed for a 4-fold reduction of cellulase loadings and increased the initial rates of sugar release.

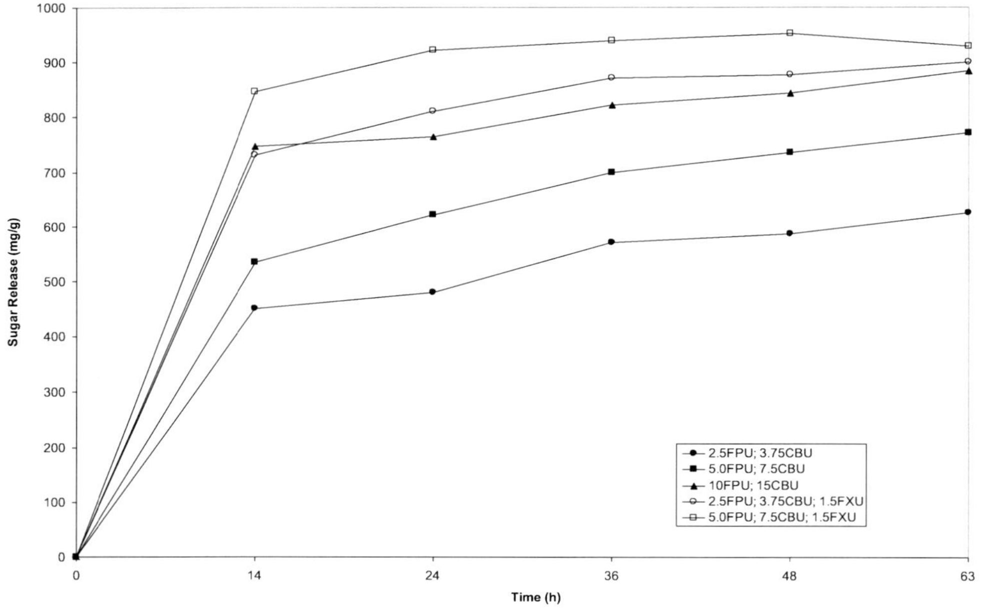

Figure 2. Enzyme digest time trials of alkali pretreated sorghum straw. Enzyme dosage expressed as units of activity per g of pretreated material with cellulase, β-glucosidase and xylanase activity measured in FPU, CBU and FXU respectively. Total sugar release is presented as mg/g pretreated material. Data represents averages of three separate experiments.

Alkali Pretreatment Reduces Straw Mass

Treating cereal straws with dilute NaOH resulted in dark coloured liquors containing solid residue material. We found that the colour intensity of the liquor generally increased with pretreatment severity. Conversely, mass and colour of remaining solids in prehydrolysate liquors decreased with severity. Others have reported similar reductions in solids during alkali pretreatment and attribute the degree of solubilisation with the severity in temperature, residence time and alkali concentrations. Under mild pretreatment conditions (1% NaOH; 60min; 60°C), solid losses were 25% (w/w) compared to 63% when pretreated at harsher conditions (2% NaOH at 121°C). Although each variable under study contributed to solid loss, we found that temperature had the greatest impact followed by alkalinity and then residence time. Comparable solid losses and treatment parameter trends have been reported in related studies on wheat straws [44]. However, a survey of the literature reveals some disparity in alkali pretreatment susceptibility between different crop residues [44-46]. These solid losses represent solubilisation of the hemicellulose fraction and other components into prehydrolysate liquors. Aside from lignin (discussed later), several studies have reported hydrolysis of hemicellulose and release of oligoxylans (polyoses) of mixed molecular weights following exposure to alkali-based chemicals during the pretreatment process [47-49]. Once considered a drawback of alkaline chemistry (i.e. reduction in total fermentable sugar yield), current biorefinery platforms are exploiting alkali-based processes for recovery of valuable high molecular weight oligoxylans/ arabinoxylans [50-52].

We initially attempted to quantify liberated pentose sugars (xylose and arabinose), in order to determine extent of hemicellulose solubilisation. However, HPLC analysis of prehydrolysate liquors revealed a complex profile of monosaccharides and high molecular weight oligosaccharides, levels of which were found to be proportional with the strength of alkali pretreatment at 121°C. Pretreating sorghum and wheat straw at 121°C for 30 min in 2.0% and 0.75% NaOH solutions resulted in 18.5 and 20 % and 33 and 33% solubilisation, respectively, of the hemicellulose fraction. This fraction was correspondingly isolated as a crude xylan precipitant from prehydrolysate liquors as outlined in the mass balance section. In addition to alkaline strength, variation of temperature and time may also impact on the yield of isolated arabinoxylans [52].

Total Sugar Yields in Enzyme Saccharified Hydrolysates

Hydrolysis of both cellulose and hemicellulose in pretreated lignocellulosics via enzymatic action is critical in releasing monomeric sugars for fermentation to bioethanol. The rate and extent of enzymatic saccharification of the polysaccharide provides a measure (indicator) of the pretreatment's effectiveness. This section of work reports on enzyme saccharification of both wheat and sorghum straw pretreated under varying conditions. Specifically, we examine whether a relationship between pretreatment severity and enzyme saccharification of the pretreated material exists, and if so attempt to identify and describe the key variables. Twenty four pretreatment combinations derived from varying test parameters such as alkaline concentration (4 levels), time (3 levels) and temperature (2 levels), were trialled in triplicate on wheat and sorghum straw.

To avert large rapid sugar releases, which could potentially mask the identity of important pretreatment variable(s), saccharifications were dosed with low enzyme activities. That is, enzymatic saccharification strategies used in this study were not intended to maximise sugar recoveries but to augment and draw out those variables critical to the success of the pretreatment process. Pretreated solids were subject to enzymatic saccharification using the following set of conditions: substrate load 5% (w/v) in citrate buffer (pH5.0) at 50°C for 48hrs. The enzyme mixture consisted of 2.5 FPU cellulase, 2.5 CBU β-glucosidase and 1.5 FXU xylanase per gram of pretreated solids. Sugar yields were quantified by HPLC analysis, and total sugar release was modelled as a response to pretreatment parameters and expressed as a function of alkaline strength, temperature and residence time (Figure 3).

The data in Figure 3 demonstrates that increases in pretreatment temperature, residence time and alkali concentration improved enzymatic saccharification efficiency of the test material. In both cases, temperature had the greatest ($p < 0.05$) impact on enzyme saccharification, above alkaline strength and/or time. That is, pretreatment at 121°C was more acquiescent to enzymatic hydrolysis than at 60°C. Within the 121°C sorghum straw trials, increasing alkaline strength from 0 to 2% led to a 465% improvement in total sugar release. Likewise, wheat straw trials showed a 520% increase in total sugar yield. Pretreating wheat and sorghum straw with 2% NaOH for 30 and 60min, respectively, at 121°C followed by enzyme saccharification yielded the highest recorded total sugar release of 850 and 799 mg/g pretreated material. Raising reaction time to 90 min under the same conditions failed to liberate further monomeric sugars, however, reducing treatment time to 30 min for the sorghum straw slightly diminished sugar yields. Conversely, an 8% increase in total sugar recovery was noted by raising the alkali strength from 1% to 2%. Total sugar release from pretreated wheat straw was also found to slightly decline (~ 8%) when the reaction time was reduced to 60 min. Hu and Wen [53] and Wang et al [54] reported similar responses to temperature and alkaline concentrations, albeit, they recovered significantly less total sugars at elevated NaOH strengths.

At the lower pretreatment temperatures of 60°C, sugar release was found to generally rise with increasing NaOH concentration. A maximum yield of 667 and 621mg/g for wheat and sorghum straw, respectively, was attained in 2% NaOH followed by saccharification. Under these conditions, statistically similar ($p < 0.05$) yields were obtained from solid material exposed to elevated temperature and reduced hydroxide combinations (121°C / 0.75% NaOH). This raises the possibility that under mild alkaline conditions the optimal pretreatment temperature may be lower than 121°C, offering potential power and cost savings in an industrial process. No discernable differences between the 30 and 60 minute treatments were observed at 60°C, however, extending the time to 90 min improved total sugar yields for all combinations of alkalinity. In the absence of NaOH, increasing time did not influence sugar yields but raising the temperature to 121°C slightly improved saccharification.

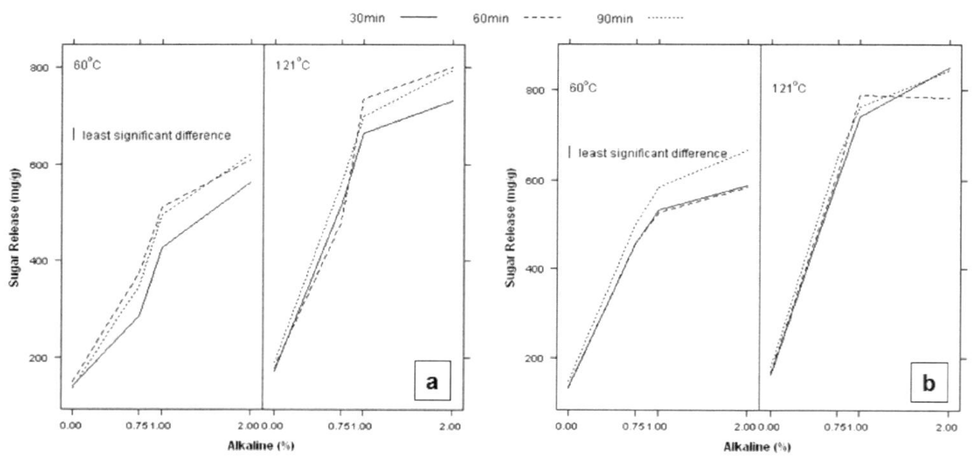

Figure 3. Total sugar release from NaOH pretreated and enzyme saccharified sorghum (a) and wheat (b) straw presented as a function of alkaline strength, temperature and residence time. Sugar yields are expressed as mg/g pretreated material. Data represents averages of three independent experiments. The average l.s.d. ($p = 0.05$) = 24.4 & 25.0 for sorghum and wheat straw, respectively.

Sugar Composition of Enzyme Saccharified Hydrolysates

The effectiveness of enzymatic saccharification on pretreated material is principally evaluated by the degree of conversion of cellulose to glucose monomers. For alkaline based pretreatment processes, this also includes the release of monomeric pentose (xylose and arabinose) sugars from hemicellulose. Alkaline pretreatment partially solubilises the hemicellulose fraction leaving a material enriched in cellulose [45, 48, 54]. Thus quantifying individual sugar components in enzyme treated hydrolysates permits rapid appraisal of its fermentation potential and assists in determining the best possible conversion strategy. Constituent monosaccharides of sorghum and wheat straw enzyme saccharified hydrolysates were quantified and expressed as a function of alkaline strength, temperature and residence time in Figures 4a and b, respectively.

Generally, enzymatic hydrolysis of cellulose correspondingly increased with pretreatment temperature, residence time and alkali concentration. Temperature had the greatest significant ($p < 0.05$) impact, with 121°C delivering greater cellulose saccharification than 60°C. Maximum glucose yields were recorded (540 and 567 mg/g) for sorghum and wheat straw, respectively, when samples were pretreated at 121°C for 90 min in 2% NaOH. Comparable glucose yields were observed with a pretreatment time of 60 and 30 min for sorghum (532 mg) and wheat straw (552mg), respectively. Within the 121°C treatments, elevating alkaline strengths resulted in a significant ($p < 0.05$) increase in glucose recovery for all pretreatment times. Similar trends were noted amongst samples treated at the lower temperature. Glucose release from sorghum (390 mg/g) and wheat (410 mg/g) straw exposed to 2% NaOH at 60°C for 90 min, were found to rival and surpass glucose levels resulting from straws treated with 0.75% NaOH at 121°C. This suggests that increasing alkaline strength may potentially act as a trade-off to reducing reaction temperatures.

Increasing pretreatment severity also improved hemicellulose saccharification and xylose release. Temperature had a significant ($p < 0.05$) effect with 121°C producing greater xylose release than 60°C. Maximum xylose yields were attained when sorghum and wheat straw samples were exposed to 2 and 1% NaOH at 121°C, giving a peak yield of up to 235 and 275 mg, respectively, after 60 min of treatment time. As observed for glucose yields, reducing alkali strength (0.75%) for both samples resulted in significantly ($p > 0.05$) lower xylose release, implying modest expose of the lignocellulosic structure. When both straw types were pretreated at conditions optimal for glucose recovery (i.e. 2% NaOH /121°C / 90 min), significantly lower xylose yields were obtained. Others have reported similar declines in xylose yield, which incidentally correlates with elevated xylan levels in prehydrolysate liquors and pretreatment settings [44, 46, 52]. In the control samples, xylose release was very small (25 mg to 45 mg/g), irrespective of temperature settings.

Lowering the pretreatment temperature to 60°C led to a reduction in the maximum xylose yield for both sorghum (205 mg/g) and wheat straw (227 mg/g). However, we noted that xylose levels from sorghum straw exposed to 1-2% NaOH at 60°C for 60 and 90 min exceeded xylose release from enzyme digested solids pretreated with 0.75% NaOH at 121°C. Xylose concentrations from comparable wheat straw samples were found to be similar. Inadequate hemicellulose hydrolysis at this lower temperature has probably physically constrained and impeded cellulase breakdown. Supplementing the enzyme mixture with additional hemicellulase/xylanase activity should improve hydrolysis of mildly treated substrates containing higher amounts of xylan [55]. Pretreatment conditions for maximum arabinose sugar release correlated with those observed for xylose sugars at both temperatures. Maximum yields of up to 33 mg/g pretreated material were attained under these conditions. Arabinose yields from solids pretreated in 2% NaOH at 121°C were also significantly ($p < 0.05$) reduced. Glucose and xylose yields from controls (water treated materials) were approximately 4 and 6-fold, respectively, lower than yields resulting from alkali catalysed pretreatment, confirming the need for an alkali catalyst.

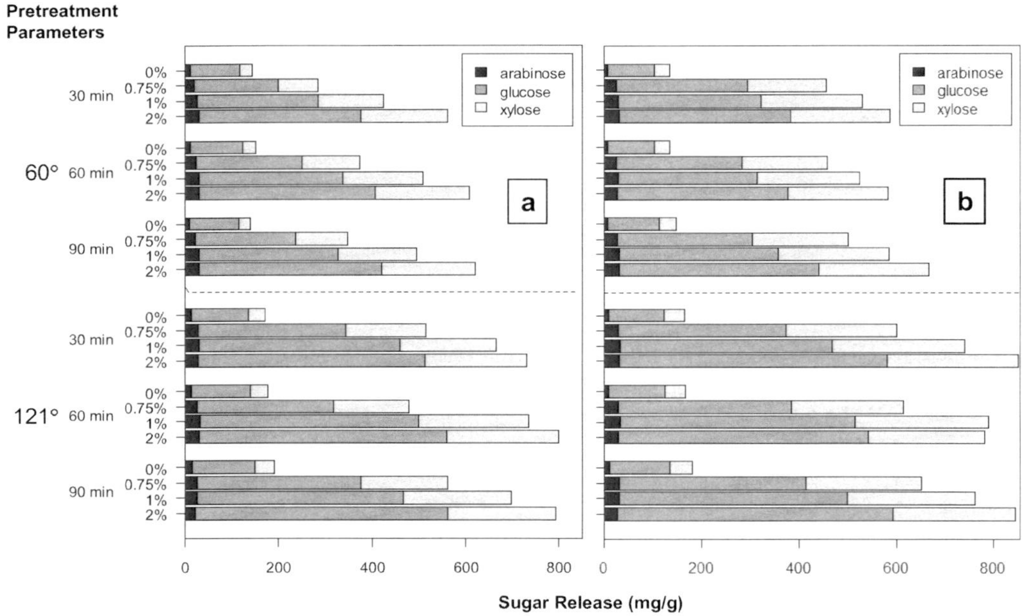

Figure 4. Monosaccharide profile of sugars released from alkali pretreatedand enzyme saccharified sorghum (a) and wheat (b) straw presented as a function of alkaline strength, temperature and residence time. Sugar yields are expressed as mg/g pretreated material. Data represents averages of three independent experiments. The average l.s.d. ($p = 0.05$) are: 15.2 & 13.2 (glucose), 10.7 & 11.1 (xylose), 3.6 & 2.7 (arabinose) and 24.4 & 25.0 (total yields) for sorghum and wheat straw, respectively.

Delignification during Mild-Alkaline Pretreatment

The degree of delignification reflects the effectiveness of the alkaline pretreatment process. Moreover, it is critical in improving enzymatic degradation of lignocellulosics and is ultimately influenced by pretreatment severity [56]. The effect of NaOH pretreatment on the delignification of sorghum and wheat straw were quantified by determining the decline of acid-insoluble lignin in pretreated solids and is presented in Figure 5. Of all parameters tested, temperature had the most significant ($p < 0.05$) impact on delignification. At 121°C, delignification of sorghum straw ranged from 18% (0.75% NaOH/ 30 min/ 121°C) to a maximum of 77.3% (2.0% NaOH/ 90 min/ 121°C). Slightly lower lignin losses (70%) were also achieved with shorter 30 min reaction time.

For wheat straw, delignification extended from 33% (0.75% NaOH/ 30 min/ 121°C) to a maximum of 72% (2.0% NaOH/ 90 min/ 121°C). At 2% NaOH/ 121°C, similar reductions in lignin content were attained irrespective of reaction time. At reduced alkaline strengths, the degree of delignification between 30min and 90 min was significant ($p < 0.05$). Generally, we found that increasing alkaline concentration significantly ($p < 0.05$) improved delignification at 121°C, whereas, responses to increasing time were less pronounced. Under similar reaction conditions, Varga and co-workers [46] reported almost complete delignification (>95%) when alkaline concentrations were raised to 10%, though total recoverable carbohydrate levels were drastically diminished.

At 60°C delignification was substantially reduced and ranged from 10.2% (0.75% NaOH/ 30 min) to 45% (1.0% NaOH/ 60 min) for sorghum straw. There were no significant ($p <$

0.05) gains in delignification by raising the NaOH to 2% and residence time to 90 min. Irrespective of reaction temperature and time, alkaline treatments at 0.75% NaOH were ineffective in delignifying sorghum straw. Similarly, delignification was substantially reduced at the lower temperature and was found to extend from 15% (0.75% NaOH/ 30 min) to 42% (2.0% NaOH/ 90 min) for wheat straw. However, the maximum delignification achieved at 60°C in 2% NaOH surpasses that at 121°C in 0.75% NaOH. Alkaline treatment at 0.75% NaOH failed to effectively delignify both straw residue types, irrespective of reaction temperature and time. Incidentally, this coincided with reduced sugar yields in saccharified hydrolysates.

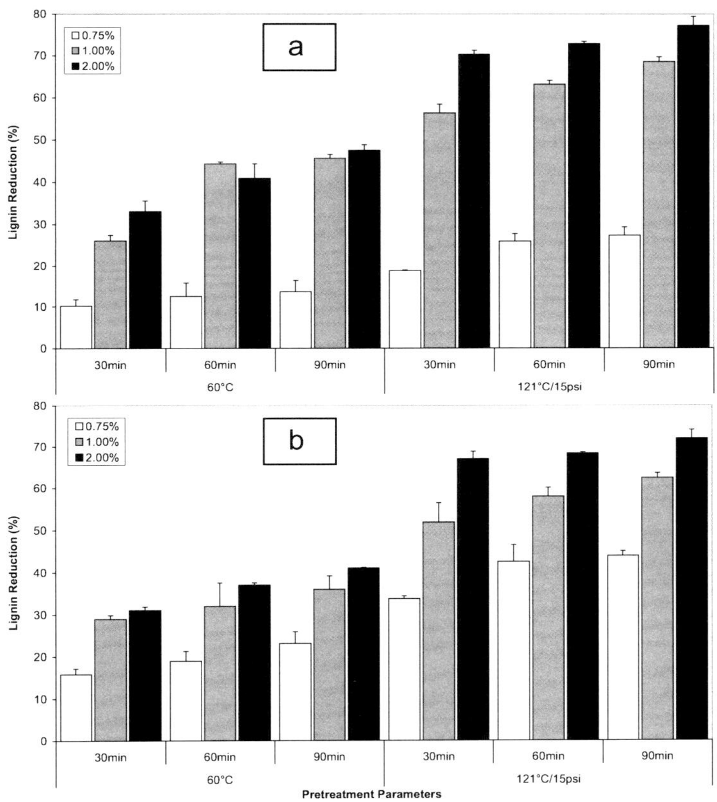

Figure 5. Reduction of acid-insoluble lignin in sorghum and wheat straw pretreated in 0.75% (white) 1.0% (grey) and 2.0% (black). NaOH expressed as a function of temperature and residence time. Results are presented as percent reduction of untreated straws. The data presented are averages of three separate experiments. The average l.s.d. ($p = 0.05$) = 3.44 and 4.77 for sorghum and wheat straw, respectively.

These findings imply a correlation between delignification and enhanced enzyme saccharification of pretreated sorghum and wheat straw. In fact several studies have demonstrated strong negative correlations between lignin content and sugar released by enzymatic hydrolysis [57, 58]. Various researchers have confirmed that lignin directly acts as a physical barrier, restricting cellulase access to cellulose, and reduces the enzyme's activity through non-productive binding [41].

Release of Phenolics into Pretreated Liquors and Saccharified Hydrolysates

Phenols, furans, carboxylic acids and inorganic salts formed or released during pretreatment of lignocellulosic materials are known to have an inhibitory effect on downstream processes [59, 60]. Sorghum and wheat, like most monocotyledons, are typically rich in phenolic acid esters associated with hemicellulose and lignin [61]. Total phenolics from both pretreated liquors and enzyme saccharified hydrolysates were quantified and the data presented as a function of change in pretreatment temperature, residence time and alkaline strength in Figure 6.

Excluding the control samples, approximately 5 and 7.4-fold more total phenolics were found in pretreated sorghum and wheat hydrolysates (2200 &1486 µg/g) respectively, compared to enzyme saccharification mixtures (450 and 200 µg/g). With respect to sorghum pre-hydrolysates, temperature had a larger impact on phenolic levels than reaction time or alkalinity (Figure 6a).

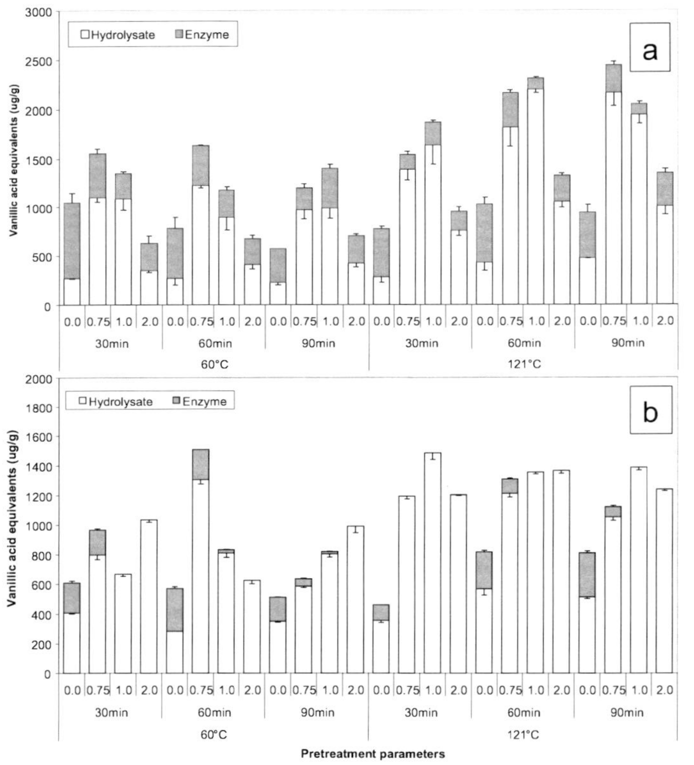

Figure 6. Total phenolics present in pretreated liquors and enzyme saccharified hydrolysates of sorghum (a) and wheat (b) straw samples as a function of alkaline strength, temperature and residence time. Data represents averages of three separate experiments.

Samples treated at 121°C released almost double the amount (2200 µg/g) of phenolics than those pretreated at 60°C (1200 µg/g). At 121°C, increasing pretreatment reaction time and alkaline strength enhanced phenolic release from sorghum samples. However, at 2% NaOH, total phenolic levels diminished, either because of product decomposition or alteration. Sorghum produces pigmented phenolic compounds (e.g. anthocyanins) which are reportedly susceptibility to degradation and/or changes to their oxidative state at elevated pH's [62]. Decline in phenolic release was more pronounced at the lower (60°C) pretreatment temperature and/or residency time. With respect to wheat straw samples, increasing pretreatment reaction time and alkaline strength >0.75% at 121°C did not generally enhance phenolic release. At 60°C responses to changes in time and alkalinity were varied, though a net decrease in total yields was observed.

Total phenolic content in enzyme saccharified hydrolysates were substantially lower, especially following higher temperature treatments. Most of the phenolics were recovered in prehydrolysate liquors. Conversely, saccharification mixtures of samples treated at 60°C and all the water controls contained higher phenolic content. Total phenolics in their respective hydrolysates were comparatively low. Overall these results suggest that harsher pretreatment conditions should provide saccharified hydrolysates with reduced phenolic content and greater fermentation potential.

Mass Balance

An overall mass balance diagram describing the process stages from pretreatment to enzymatic hydrolysis is presented in Figure 7. Both sorghum and wheat straw at a solid loading of 10% (w/v) were pretreated under conditions optimised for maximum sugar recovery (2% NaOH / 121°C / 60 and 30 min for sorghum and wheat, respectively). The remaining insoluble fraction was separated from the pretreatment hydrolysate prior to enzymatic saccharification. The amount of recovered material corresponded to ≈ 45 and 51% (w/w) of the original (sorghum and wheat straw respectively) starting material and was subjected to saccharification.

Enzyme saccharification was achieved using low dosages of cellulase (2.5 FPU), β-glucosidase (2.5 CBU) and 1.5 FXU xylanase (per gram of pretreated solids) and incubated at 50°C for up to 48 h. Sugar yields were recorded at 240 & 279 mg of glucose, 94 &136 mg of xylose and 13 &15 mg of arabinose per gram of original starting material (sorghum and wheat straw, respectively). Recovered prehydrolysate liquors were further fractionated through titration with 6N H_2SO_4. At pH 4.0, 135 & 162.6 mg/g of acid insoluble lignin was recovered as a precipitate from sorghum and wheat straw, respectively. Addition of 3 volumes of cold ethanol to the aqueous phase led to the precipitation of 90 & 86 mg/g crude xylan from sorghum and wheat straw, respectively. Prehydrolysate liquors also contained approximately 145 & 35 mg/g of water extractive storage carbohydrate and other unquantified polysaccharides, phenolics and degradation compounds from sorghum and wheat straw, respectively.

Conclusion

In closing, we find that dilute alkali treatment satisfies some of the more important requisites of an effective pretreatment process, namely; it is an excellent delignifying agent, produces a cellulose enriched fraction that is responsive to enzyme digestion with high and rapid sugar release, and generates low levels of phenolic compounds. Our study also shows that there may be opportunities for further process optimisation, particularly in pretreatment temperatures, enzyme combinations and dosages. Using alkaline pretreatment to extract oligoxylans and lignins while simultaneously improving cellulose hydrolysis can be a means of advancing the viability and grounds for an integrated biorefinery. However, selecting an appropriate pretreatment regime requires a degree of compromise between maximising glucose yield and minimising creation of inhibitors. Considering their abundance and high sugar potential, sorghum and wheat straw are an excellent feedstock for biorefineries.

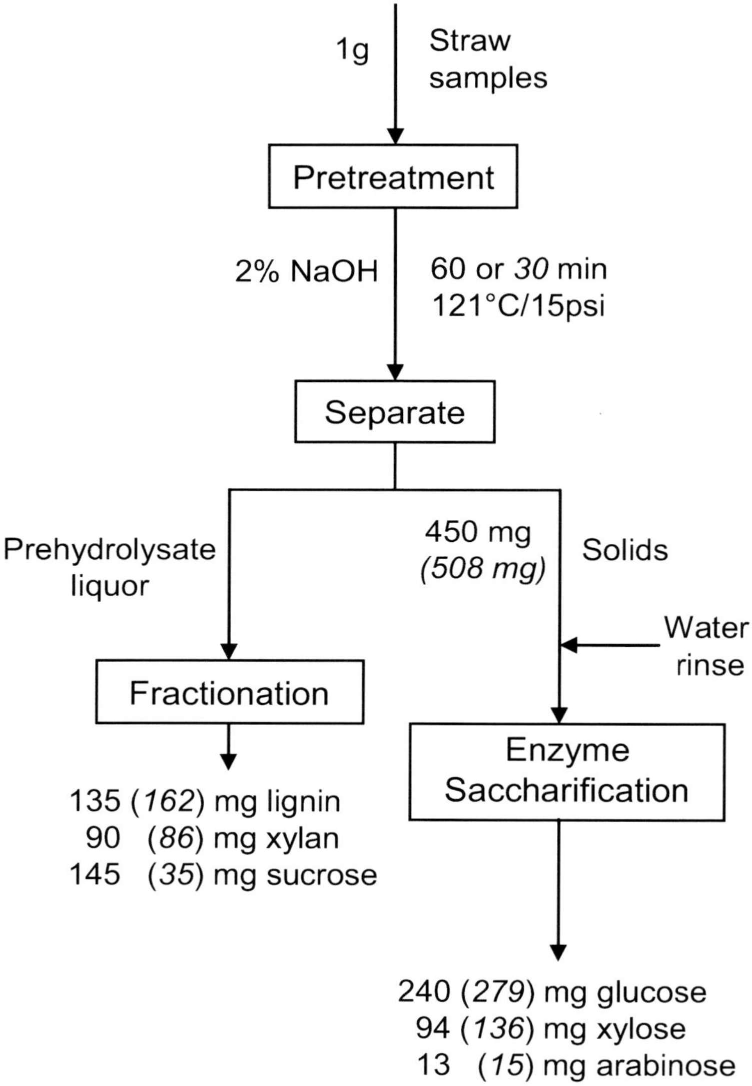

Figure 7. Mass balance of pretreatment and enzymatic hydrolysis process steps. Numbers in brackets represent data for wheat straw.

ACKNOWLEDGMENTS

We gratefully acknowledge the financial support provided by Climate Action Grant (TOC/CAG/013-2007) for this work and the support of Industry and Investment NSW, Australia. We express our gratitude to Mr Steve Pepper for technical assistance and Mr Steve Morris for providing advice and assistance in the presentation of the data.

REFERENCES

[1] Ragauskas, A.J., Nagy, M., Kim, D., Eckert, C.A., Hallett, J.P., and Liotta, C.L., (2006) *From wood to fuels: integrating biofuels and pulp production.* Industrial biotechnology. 2(1): 55-65.

[2] Campbell, C.J. and Laherrere, J.H., *The End of Cheap Oil. Global production of convention oil will begin to decline sooner than most people think, probably within 10 years*. 1998, Scientific American. p. 78-83.

[3] Zhang, Y.H.P., (2008) *Reviving the carbohydrate economy via multi-product lignocellulose biorefineries.* Journal of Industrial Microbiology & Biotechnology. 35(5): 367-375.

[4] Ragauskas, A.J., Williams, C.K., Davison, B.H., Britovsek, G., Cairney, J., Eckert, C.A., Frederick, W.J., Hallett, J.P., Leak, D.J., Liotta, C.L., Mielenz, J.R., Murphy, R., Templer, R., and Tschaplinski, T., (2006) *The path forward for biofuels and biomaterials.* Science. 311(5760): 484-489.

[5] Spatari, S., Bagley, D.M., and Maclean, H.L., (2010) *Life cycle evaluation of emerging lignocellulosic ethanol conversion technologies.* Bioresource Technology. 101(2): 654-667.

[6] Johansson, D. and Azar, C., (2007) *A scenario based analysis of land competition between food and bioenergy production in the US.* Climatic Change. 82(3): 267-291.

[7] Warden, A. and Haritos, V., (2008) Report: *Future Biofuels for Australia. Issues and opportunities for conversion of second generation lignocellulosics.* Rural Industries research and Development Corporation.

[8] Fung, P.Y.H., Kirschbaum, M.U.F., Raison, R.J., and Stucley, C., (2002) *The potential for bioenergy production from Australian forests, its contribution to national greenhouse targets and recent developments in conversion processes.* Biomass and Bioenergy. 22(4): 223-236.

[9] Salleh, A., *Biofuels made from wheat stubble could lead to a greener future.* 2008. [cited; Available from: http://www.abc.net.au/science/articles/2008/09/22/2368009.htm?site= science&topic=latest].

[10] *Australian Crop Report*, in *Australian Bureau of Agricultural and Resource Economics*. 2008: Canberra.

[11] *Queensland Department of Primary Industries and Fisheries_Sorghum Grain Report.* Queensland Department of Primary Industries and Fisheries 2008 [cited; Available from: http://www.dpi.qld.gov.au]

[12] *Biofuels Association of Australia*. 2007[cited; http://biofuelsassociation.com.au.]

[13] Marsden, W.L. and Gray, P.P., (1986) *Enzymatic hydrolysis of cellulose in lignocellulosic material.* . CRC Crit. Rev. Biotechnol. 3: 235-276.

[14] Tsao, G.T., Ladisch, M.R., and Bungay, H.R., *Biomass refining*. In Advanced Biochemical Engineering, ed. H.R. Bungay and G. Belfort. 1987 John Wiley & Sons. 79-101.

[15] Bisaria, V.S. and Ghose, T.K., (1981) *Biodegradation of cellulosic materials: substrates, microorganisms, enzymes and products.* Enzyme Microb. Technol. 3: 90-104.

[16] Krull, L.H. and Inglet, G.I., (1980) *Analysis of neutral carbohydrates in agricultural residues by gas-liquid chromatography.* J. Agric. Food. Chem. 28: 917-919.

[17] Beck, M.J., (1989) *Factors affecting efficiency of biomass fermentation to ethanol.* . Biotechnol. Bioeng. Symp. 17: 617-627.

[18] Gosselink, R.J.A., De Jong, E., Guran, B., and Abacherli, A., (2004) *Co-ordination network for lignin--standardisation, production and applications adapted to market requirements (EUROLIGNIN).* Industrial Crops and Products. 20(2): 121-129.

[19] Eggeman, T. and Elander, R.T., (2005) *Process and economic analysis of pretreatment technologies.* Bioresource Technology. 96(18): 2019-2025.

[20] Lloyd, T.A. and Wyman, C.E., (2005) *Combined sugar yields for dilute sulfuric acid pretreatment of corn stover followed by enzymatic hydrolysis of the remaining solids.* Bioresource Technology. 96(18): 1967-1977.

[21] Kim, S. and Holtzapple, M.T., (2005) *Lime pretreatment and enzymatic hydrolysis of corn stover.* Bioresource Technology. 96(18): 1994-2006.

[22] Blanca Antizar-Ladislao, J.L.T.-G., (2008) *Second-generation biofuels and local bioenergy systems.* Biofuels, Bioproducts and Biorefining. 2(5): 455-469.

[23] Mcmillan, J., *Enzymatic Conversion of Biomass for Fuels production.*, in *Pretreatment of lignocellulosic biomass*, M. HImmel, J. Baker, and R. Overend, Editors. 1994, Americal Chemical Society: Washington, DC. p. 292-324.

[24] Chang, V., Nagwani, M., and Holtzapple, M., (1998) *Lime pretreatment of crop residues bagasse and wheat straw.* Applied Biochemistry and Biotechnology. 74(3): 135-159.

[25] Chen, Y., Sharma-Shivappa, R., Keshwani, D., and Chen, C., (2007) *Potential of Agricultural Residues and Hay for Bioethanol Production.* Applied Biochemistry and Biotechnology. 142(3): 276-290.

[26] Sendich, E., Laser, M., Kim, S., Alizadeh, H., Laureano-Perez, L., Dale, B., and Lynd, L., (2008) *Recent process improvements for the ammonia fiber expansion (AFEX) process and resulting reductions in minimum ethanol selling price.* Bioresource Technology. 99(17): 8429-8435.

[27] Dale, B.E., Leong, C.K., Pham, T.K., Esquivel, V.M., Rios, I., and Latimer, V.M., (1996) *Hydrolysis of lignocellulosics at low enzyme levels: Application of the AFEX process.* Bioresource Technology. 56(1): 111-116.

[28] Smook, G., ed. *Handbook for Pulp and Paper Technologists*. 1982, Joint Textbook Committee of The Paper Industry: Montreal, Canada.

[29] Doi, R.H. and Kosugi, A., (2004) *Cellulosomes: plant-cell-wall-degrading enzyme complexes.* Nat Rev Micro. 2(7): 541-551.

[30] Merino, S. and Cherry, J., *Progress and Challenges in Enzyme Development for Biomass Utilization.*, in *Biofuels*. 2007, Springer Berlin. p. 95-120.

[31] Mcintosh, S. and Vancov, T., (2010) *Enhanced enzyme saccharification of Sorghum bicolor straw using dilute alkali pretreatment* Bioresource Technology. 101: 6718-6727.

[32] Adney, B. and Baker, J., *Measurement of Cellulase activity*. 2008. NREL Technical Report NREL/TP-510-42628.

[33] Sluiter, A., Hames, B., Ruiz, R., Scarlata, C., Sluiter, J., Templeton, D., and Crocker, D., (2008) *Determination of Structural Carbohydrates and Lignin in Biomass*. NREL Technical Report NREL/TP-510-42618. [http://www.nrel.gov/biomass/pdfs/42618.pdf]

[34] Sluiter, A., Ruiz, R., Scarlata, C., Sluiter, J., and Templeton, D., (2008) *Determination of Extractives in Biomass*. NREL Technical Report NREL/TP-510-42619. [http://www.nrel.gov/biomass/pdfs/42619.pdf].

[35] R Development Core Team, *R: A language and environment for statistical computing*. 2009, Vienna, Austria: R Foundation for Statistical Computing.

[36] Saha, B.C. and Cotta, M.A., (2006) *Ethanol Production from Alkaline Peroxide Pretreated Enzymatically Saccharified Wheat Straw*. Biotechnology Progress. 22(2): 449-453.

[37] Mehmood, S., Gulfraz, M., Rana, N., Ahmad, A., Ahring, B., Minhas, N., and Malik, M., (2009) *Ethanol production from Sorghum bicolor using both separate and simultaneous saccharification and fermentation in batch and fed batch systems*. African Journal of Biotechnology. 8: 2857-2865.

[38] Chang, V.S. and Holtzapple, M.T., (2000) *Fundamental factors affecting biomass enzymatic reactivity*. Applied Biochemistry and Biotechnology. 84-86: 5-37.

[39] Berlin, A., Maximenko, V., Gilkes, N., and Saddler, J., (2007) *Optimization of enzyme complexes for lignocellulose hydrolysis*. Biotechnology and Bioengineering. 97(2): 287-296.

[40] Tengborg, C., Galbe, M., and Zacchi, G., (2001) *Reduced inhibition of enzymatic hydrolysis of steam-pretreated softwood*. Enzyme Microb Technol. 28(9-10): 835-844.

[41] Jeoh, T., Ishizawa, C., Davis, M., Himmel, M., Adney, W., and Johnson, D., (2007) *Cellulase digestibility of pretreated biomass is limited by cellulose accessibility*. Biotechnology and Bioengineering. 98(1): 112-122.

[42] Kabel, M.A., Bos, G., Zeevalking, J., Voragen, A.G.J., and Schols, H.A., (2007) *Effect of pretreatment severity on xylan solubility and enzymatic breakdown of the remaining cellulose from wheat straw*. Bioresource Technology. 98(10): 2034-2042.

[43] Yang, B. and Wyman, C.E., (2004) *Effect of xylan and lignin removal by batch and flowthrough pretreatment on the enzymatic digestibility of corn stover cellulose*. Biotechnology and Bioengineering. 86(1): 88-98.

[44] Chen, Y., Sharma-Shivappa, R., Keshwani, D., and Chen, C., (2007) *Potential of agricultural residues and hay for bioethanol production*. Applied Biochemistry and Biotechnology. 142(3): 276-290.

[45] , R.A., Chen, Y., Sharma-Shivappa, R.R., Boyette, M.D., and Osborne, J., (2007) *A comparison of chemical pretreatment methods for improving saccharification of cotton stalks*. Bioresource Technology. 98(16): 3000-3011.

[46] Varga, E., Szengyel, Z., and Réczey, K., (2002) *Chemical pretreatments of corn stover for enhancing enzymatic digestibility*. Applied Biochemistry and Biotechnology. 98-100(1): 73-87.

[47] Cheng, K.-K., Zhang, J.-A., Ping, W.-X., Ge, J.-P., Zhou, Y.-J., Ling, H.-Z., and Xu, J.-M., (2008) *Sugarcane Bagasse Mild Alkaline/Oxidative Pretreatment for Ethanol Production by Alkaline Recycle Process.* Applied Biochemistry and Biotechnology. 151(1): 43-50.

[48] Varga, E., Schmidt, A., Reczey, K., and Thomsen, A., (2003) *Pretreatment of corn stover using wet oxidation to enhance enzymatic digestibility.* Applied Biochemistry and Biotechnology. 104(1): 37-50.

[49] Zhang, Q. and Cai, W., (2008) *Enzymatic hydrolysis of alkali-pretreated rice straw by Trichoderma reesei ZM4-F3.* Biomass and Bioenergy. 32(12): 1130-1135.

[50] Huang, H.-J., Ramaswamy, S., Tschirner, U.W., and Ramarao, B.V., (2008) *A review of separation technologies in current and future biorefineries.* Separation and Purification Technology. 62(1): 1-21.

[51] Moure, A., Gullón, P., Domínguez, H., and Parajó, J.C., (2006) *Advances in the manufacture, purification and applications of xylo-oligosaccharides as food additives and nutraceuticals.* Process Biochemistry. 41(9): 1913-1923.

[52] Persson, T., Ren, J.L., Joelsson, E., and Jönsson, A.-S., (2009) *Fractionation of wheat and barley straw to access high-molecular-mass hemicelluloses prior to ethanol production.* Bioresource Technology. 100(17): 3906-3913.

[53] Hu, Z. and Wen, Z., (2008) *Enhancing enzymatic digestibility of switchgrass by microwave-assisted alkali pretreatment.* Biochemical Engineering Journal. 38(3): 369-378.

[54] Wang, Z., Keshwani, D.R., Redding, A.P., and Cheng, J.J., *Alkaline Pretreatment of Coastal Bermudagrass for Bioethanol Production.* 2008, DigitalCommons@University of Nebraska - Lincoln.

[55] Bura, R., Chandra, R., and Saddler, J., (2009) *Influence of xylan on the enzymatic hydrolysis of steam-pretreated corn stover and hybrid poplar.* Biotechnology Progress. 25(2): 315-322.

[56] Sun, Y. and Cheng, J., (2002) *Hydrolysis of lignocellulosic materials for ethanol production: a review.* Bioresource Technology. 83(1): 1-11.

[57] Chen, F. and Dixon, R.A., (2007) *Lignin modification improves fermentable sugar yields for biofuel production.* Nat Biotech. 25(7): 759-761.

[58] Guo, G.-L., Hsu, D.-C., Chen, W.-H., Chen, W.-H., and Hwang, W.-S., (2009) *Characterization of enzymatic saccharification for acid-pretreated lignocellulosic materials with different lignin composition.* Enzyme and Microbial Technology. 45(2): 80-87.

[59] Klinke, H.B., Thomsen, A.B., and Ahring, B.K., (2004) *Inhibition of ethanol-producing yeast and bacteria by degradation products produced during pre-treatment of biomass.* Applied Microbiology and Biotechnology. 66(1): 10-26.

[60] Palmqvist, E. and Hahn-Hägerdal, B., (2000) *Fermentation of lignocellulosic hydrolysates. I: inhibition and detoxification.* Bioresource Technology. 74(1): 17-24.

[61] Akin, D., (2008) *Plant cell wall aromatics: influence on degradation of biomass.* Biofuels, Bioproducts and Biorefining. 2(4): 288-303.

[62] Bjorøy, Ø. Rayyan, S., Fossen, T., and Andersen, Ø., (2009) *Structural Properties of Anthocyanins: Rearrangement of C-Glycosyl-3-deoxyanthocyanidins in Acidic Aqueous Solutions.* Journal of Agricultural and Food Chemistry. 5

In: Cellulase: Types and Action, Mechanism, and Uses
Editor: Adam E. Golan
ISBN: 978-1-61761-983-0

Chapter 6

CELLULOLYTIC ENZYMES ISOLATED FROM BRAZILIAN AREAS: PRODUCTION, CHARACTERIZATION AND APPLICATIONS

Heloiza Ferreira Alves do Prado[1], Rodrigo Simões Ribeiro Leite[2], Daniela Alonso Bocchini Martins[3], Eleni Gomes[4] and Roberto da Silva[4]

[1]Univ. Estadual Paulista – UNESP, FE - Ilha Solteira Campus, Phytotechnology, Food Technology and Social Economy Department, Ilha Solteira, SP, Brazil
[2]Federal University of Grande Dourados – UFGD, Biological and Enviromental Science Faculty, Dourados, MS, Brazil
[3]Univ. Estadual Paulista – UNESP, IQ - Araraquara Campus, Biochemistry and Chemical Technology Department, Araraquara, SP, Brazil
[4]Univ. Estadual Paulista – UNESP, IBILCE – São José do Rio Preto Campus, Laboratory of Biochemistry and Applied Microbiology, São José do Rio Preto, SP, Brazil

The plant cell wall consists of cellulose, hemicelluloses and pectin as well as the phenolic polymer lignin. Cellulose is the most abundant polysaccharide in nature and the major constituent of a plant cell wall providing its rigidity. Cellulose consists of β-1,4 linked D-glucose units that form linear polymeric chains of about from 8000 to 12000 glucose units. In crystalline cellulose, these polymeric chains are packed together by hydrogen bonds to form highly insoluble structures. Hemicelluloses, the second most abundant polysaccharides in nature, have a heterogeneous composition of various sugar units. Hemicelluloses are usually classified according to the main sugar residues in the backbone of the polymer such as xylan, (galacto)glucomannan, arabinan, galactan found in cereals and hardwood, softwood and hardwood, The main chain sugars of hemicelluloses are modified by various side groups such as 4-O-methylglucuronic acid, arabinose, galactose, and acetyl, making hemicelluloses branched and variable in structure. Pectins are a family of complex polysaccharides containing a backbone of α-1,4 linked D-galacturonic acid. Pectins contain two different types of regions. In the region of pectin classified as a smooth region, D-galacturonic acid residues can be methylated or acetylated, whereas the region classified as a hairy one consists of two different structures, D-xylose substituted galacturonan and rhamnogalacturonan to which long arabinan and galactan chains are linked via rhamnose. The cellulose wall is strengthened by lignin, a highly insoluble complex branched polymer of substituted phenylpropane units joined together by carbon–carbon and ether linkages forming an extensive cross-linked network within the cell wall.

The hydrolytic action of cellulases and hemicellulases has a fundamental importance to obtain fermentable sugars from lignocellulosic biomass. The enzymatic hydrolysis of cellulose into glucose involves the synergistic action of at least three different enzymes: endoglucanase or endo-β-1,4-glucanase (EC 3.2.1.4), exoglucanase or exocellobiohydrolase (EC 3.2.1.91), and β-1,4-glucosidase or cellobiase (EC 3.2.1.21). Endoglucanases hydrolyze the polymers internally, resulting in a reduction of the degree of polymerization, whereas the exoglucanases act by removing units of cellobiose from either the reducing or the nonreducing ends of the molecule. B-glucosidase hydrolyzes cellobiose and other cellodextrins into glucose. B-glucosidase is responsible for the control of the entire speed of the reaction exerting a crucial effect on the enzymatic degradation of the cellulose, preventing the accumulation of cellobiose. Because of hemicellulose heterogeneity, the enzymatic hydrolysis of xylan requires different enzymatic activities. Two enzymes, β-1,4-endoxylanase (EC 3.2.1.8) and β-xylosidase (EC 3.2.1.37), are responsible for hydrolysis of the main chain, the former attacking the internal main chain xylosidic linkages and the latter releasing xylosyl residues by means of endwise attack of xylooligosaccharides. However, for complete hydrolysis of hemicellulose, side chain cleaving enzyme activities are also necessary, such as, α-L-arabinofuranosidases (EC 3.2.1.55), endomannanases (EC 3.2.1.78), β-mannosidases (EC 3.2.1.25), and α-galactosidases (EC 3.2.1.22).

These enzymes also have applications in maceration of vegetables, clarification of juices and wines, extraction of juices, juice scents, juice pigments, and biobleaching of pulp. It can also be used as fermentation substrates to produce liquid fuels, food products, or other chemicals of interest. Specifically, β-glucosidase can also be used by the food industry to increase the bioavailability of the isoflavones in the human intestine, and by the beverage industry to stabilize the coloration of juices and wines.

Brazil is an agroindustrial country known for its production of soy, corn, sugar cane, cassava, coffee, and so on, and for its high consumption of wheat, which generates large amounts of residues that have considerable potential for solid state fermentation (SSF) applications. SSF is a well-known process for enzyme production and is defined as fermentation involving solids in the absence (or near absence) of free water. However, the substrate must possess enough moisture to support growth and metabolism of microorganisms. The ability of some microorganisms to metabolize cellulose and hemicelluloses makes them potentially important to take advantage of vegetable residues. Agricultural and agro-industrial waste, like sugarcane bagasse, wheat bran, rice peel, corn straw, corncob, fruit peels and seeds, effluents from the paper industry and orange bagasse, have increased as a result of industrialization, becoming a problem regarding space for disposal and environmental pollution. These residues represent an alternative source for microbial growth aiming at the production of biomass or enzymes.

Cellulose and hemicelluloses represent more than 50% of the dry weight of agricultural residues. They can be converted into soluble sugars either by acid or enzymatic hydrolysis. There is a current tendency to apply the SSF process in the development of bioprocesses to attain products with higher added values, such as antibiotics, alkaloids, organic acids, biopesticides, biofuel, aromatic compounds, and enzymes. So the agroindustrial residues can be used as a plentiful and cheap source of renewable energy in the world.

On the other hand, most of the processes of industrial application of enzymes occur at high temperatures. So, the use of thermostable enzymes appears to be appropriated because they preserve their catalytic activity at high temperatures. A series of advantages such as faster reaction, decreased viscosity in processed fluid, increased solubility of the substrate, and reduced contamination risk by undesired organisms have been proposed for use of thermostable enzymes in biotechnology processes.

The purpose of this chapter is to show some isolated Brazilian strains with high potential to produce cellulases and hemicellulases in solid state fermentation. In this case, we analyzed the following enzymes isolated from Brazilian areas such as agricultural, Amazon florest and Cerrado areas: CMCases, β-glicosidases and xylanases. We reviewed the analyses carried out by Brazilian researchers that have used different agricultural residues as substrates for solid state fermentations producing cellulases and hemicellulases. It has also isolated different microorganisms that can be potentially interesting for industrial processes. Finally, the properties and potential applications on biotechnological processes of the isolated enzymes are also analyzed.

I. Cellulose

The plant's cell wall contains cellulose, hemicellulose and lignin. The cellulose microfibrils of the plants cell wall are embedded in an amorphous matrix of lignin and hemicellulose. These three types of polymers are strongly linked by noncovalent interactions as well as covalent bonds, compounding the material known as lignocellulose (Fig. 1). The lignocellulose material represents 90% of dry weight of vegetal cells (STICKLEN, 2008). The structure, configuration and composition of cell walls vary depending on plant taxa, tissue, age and cell type, and also within each cell wall layer (GLAZER; NIKAIDO, 1995).

In general, softwoods (gymnosperms such as pine) have higher lignin content than hardwoods (angiosperms such as eucalyptus and oak). The content of hemicelluloses, in turn, is greater in grasses. On average, the lignocellulose consists of about 45% cellulose, 30% hemicellulose and 25% lignin.

Plants produce about 180 billion tons of cellulose per year globally, making this polysaccharide the largest organic carbon reservoir on earth. Cellulose in the plant cell wall is found in the form of 30 nm diameter microfibrils. Each microfibril is an unbranched polymer with about 15,000 anhydrous glucose molecules that are organized in β-1,4 linkages (that is, each unit is attached to another glucose molecule at 180° orientation) (Fig.2). The coupling of adjacent cellulose chains and sheets of cellulose by hydrogen bonds and van der Waal's forces results in a parallel alignment and a crystalline structure with straight, stable supra-molecular fibers of great tensile strength and low accessibility (DEMAIN et al., 2005; KRASSIG, 1993; NISHIYAMA et al., 2003; NOTLEY et al., 2004; ZHANG and LYND, 2004b; ZHBANKOV, 1992). Cellulose also has amorphous or soluble regions, in which the molecules are less compact, but these regions are staggered, making the overall cellulose structure strong. So far, cellulose is the only polysaccharide that has been used for commercial cellulosic ethanol production, probably because it is the only one for which there are commercially available deconstructing enzyme mixtures.

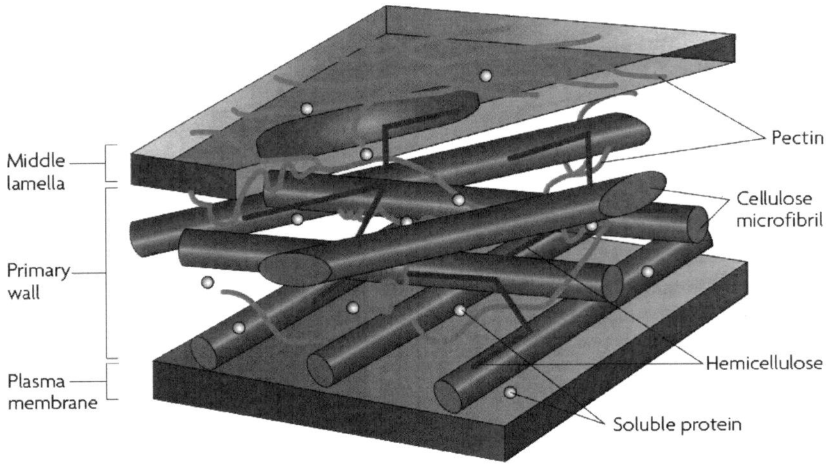

Figure 1. Plant cell wall structure containing cellulose microfibrils, hemicellulose, pectin, lignin and soluble proteins. STICKLEN (2008).

The cellulose molecule is very stable, with a half life of 5–8 million years for β-glucosidic bond cleavage at 25 °C (WOLFENDEN and SNIDER, 2001), while the much faster enzyme-driven cellulose biodegradation process is vital to return the carbon in sediments to the atmosphere (BERNER, 2003; COX et al., 2000; FALKOWSKI et al., 2000; SCHLAMADINGER and MARLAND, 1996).

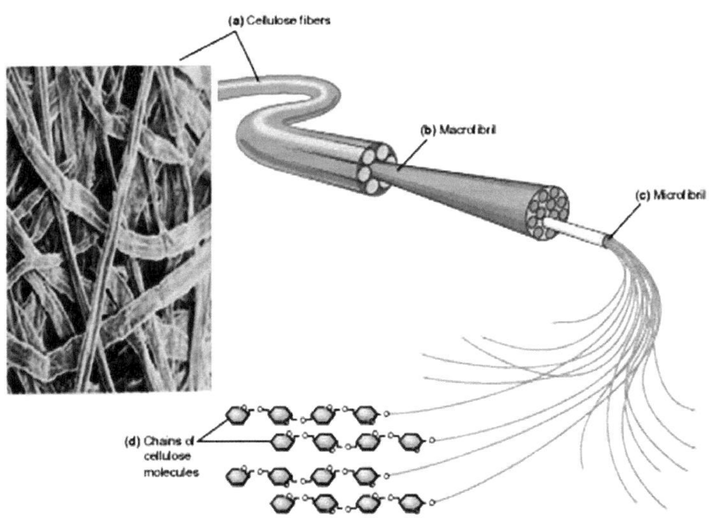

Figure 2. Stucture of cellulose. (a) Cellulose fibers from a ponderosa pine. (b) Macrofibrils compose each fiber. (c) Each macrofibril is composed of bundles of microfibrils. (d) Microfibrils, in turn, are composed of bundles of cellulose chains. Cellulose fibers can be very strong; this is one reason why wood is such a good building material. (http://nutrition.jbpub.com/resources/chemistryreview9.cfm)

II. Hemicellulose

Cellulose microfibrils are coated with other polysaccharides such as hemicellulose or xyloglucans. All dicotyledonous cell walls and about half of monocotyledonous ones consist mainly of hemicellulose. However, in the monocotyledons, such as cereals and other grasses, cell walls mostly consist of glucuronoarabinoxylans. Depending on the plant species, 20–40% of the plant cell-wall polysaccharides are hemicellulose.

The hemicellulose fraction consists of highly branched heteropolysaccharide and generally not crystalline. The sugar residues found in hemicellulose include pentoses (D-xylose, L-arabinose), hexoses (D-galactose, L-galactose, D-mannose, L-raminose, L-fucose) and uronic acids (D-galacturonic acid). These sugar residues are modified by acetylation and methylation. Thus, the classification of hemicelluloses depends on the type monomers united to form of heteropolymers, and the same may be called xylans, mannans, galactans or arabinan (GLAZER, NIKA, 1995).

The hemicellulose term was first introduced by Schulze (1891) to describe fractions of plant cell wall polysaccharides easily hydrolysable. Since then, this term has been used to describe different groups of non-cellulosic complex heteropolysaccharide, classified according to the main monosaccharide (WILLIAMS, 1989). Hemicelluloses can also be defined as polysaccharides present in plant cell wall and on the middle lamella extracted from tissues of superior plants by alkaline treatment. It can be extracted of certain carbohydrates from endosperm of cereals, non-starch polysaccharides, that are described as gums or cereal pentosans (TIMELINE, 1964; WILKIE, 1979). Subsequently, the hemicelluloses were redefined, with bases in their chemical properties, to include only those cell wall polysaccharides linked noncovalently to cellulose (BAUER et al., 1973; KENNEDY, WHITE, 1988).

The hemicelluloses are major constituents of lignocellulosic materials. It are important structural components found in close association with lignin and cellulose, yet interacting, covalently, with pectin (WILLIAMS, 1989; ZIMMERMANN, 1989).

Most hemicellulose molecules are relatively small, containing between 70 and 200 residues in monosaccharides hemicelluloses of hardwoods larger molecules, with 150-200 units (COUGHLAN, 1992; KENNEDY, WHITE, 1988). The composition of the hemicelluloses in plants can be influenced by many factors, such as growth, maturity, type of soil, climate, day length, geographical location and type of fertilizer used (WILKIE, 1979).

III. Lignocellulolitic Enzymes

The hydrolytic action of cellulases and hemicellulases is of fundamental importance to obtain fermentable sugars from lignocellulosic biomass. These can be used as fermentation substrates to produce liquid fuels, food products, or other chemicals of interest (ROMERO et. al, 1999; KANG et. al, 2004).

III.1. Cellulase

Cellulases are enzymes that form a complex capable of acting on cellulosic materials, promoting its hydrolysis. These biocatalysts are highly specific enzymes that act in synergy to release sugars, including glucose.

These enzymes have been studied during the Second World War. The deterioration of uniforms, tents, bags and other objects of the camps, made of cotton, drew the attention of army soldiers American, installed in the Solomon Islands, South Pacific. Some organizations, such as the Quartermaster Corps, along with the military, set up labs for explanations and immediate solutions to this problem, which included the detection of agents spoilage organisms, their mechanisms of action and control methods. The working group, consisting of eight researchers, led by Dr. Elwyn T. Reese went on to conduct their experiments in the laboratory of the military, in Natick, Massachusetts, USA. As a result of the investigations was isolated a strain, coded QM6a, a filamentous fungus, later identified as Trichoderma viride. This feature was attributed t excrete enzymes capable of degrading cellulose. Until 1953, Dr. Reese and his working group had determined that natural enzymes, named cellulases, are complexes of several molecules with different abilities in the degradation of the substrate. In 1956, Dr. Reese allied to his knowledge of Dr. Mary Mandels which, together, went to work in the lab in Natick. Since then, the research focus is no longer preventing the hydrolysis of cellulose and is now improving the production of enzymes responsible for this phenomenon, using the microorganism strain isolated previously. Mandels and Reese have published several studies on the influence of the main factors affecting the production of enzymes and formulations of ideal culture medium for growth of T. viride. Later, selected mutants of Trichoderma with high volumetric productivity of cellulase expression, already initially reaching 3 IU L-1 h-1. Since then, each decade was marked by significant advances in studies on the enzymes of the cellulolytic complex. The beginning of the century was marked by large investments in the production of cellulases, especially focused on its application for production of ethanol fuel (CASTRO and PEREIRA Jr, 2010).

Advances in research on cellulases occurred in several areas of knowledge. Over the years, and until the present day, scientific contributions have been generated continuously, as the screening of microorganisms producer cellulases, the increased expression of cellulases by genetic mutations, the purification and characterization of components of this enzymatic complex, the understanding of the mechanisms of attack on cellulose, the cloning and expression of genes, the determination of three-dimensional structures of cellulases and the demonstration of the industrial potential of these enzymes (BATH and BATH, 1997).

The enzymatic hydrolysis of cellulose into glucose involves the synergistic action of at least three different enzymes: endoglucanase, exoglucanase, and β-1,4-glucosidase (PALMA-FERNANDEZ et al. 2002; BHATIA et al. 2002; ZHANG and LYND, 2004; LEITE et al. 2007; CASTRO and PEREIRA Jr, 2010).

Endoglucanase or endo-β-1,4-glucan-4-glucanohydrolases (EC 3.2.1.4) is the cellulolytic enzyme complex responsible for initiating the hydrolysis. This enzyme hydrolyzes randomly the inner regions of the amorphous structure of cellulose fiber, resulting in a reduction of the degree of polymerization.

Exoglucanases act by removing units of cellobiose from either the reducing or the non-reducing ends of the molecule. This cellulose group consists of cellobiohydrolase (CBH) and glucanohidrolase (GH). Glucanohidrolase or β-1,4-D-glucan-glucanohydrolase (EC 3.2.1.74), also known as cellodextrinases, is little reported, but the strategy has to hydrolyze cellulose fiber of high importance because it is capable of releasing glucose polymer directly. The cellobiohydrolase or β-1,4-D-glucan-cellobiohydrolase (EC 3.2.1.91), is known to hydrolyze only the terminal non-reducing fiber and cellulose oligosaccharides with degree of polymerization (DP) > 3 releasing cellobiose, although there are reports of attacks by reducing end this enzyme. CBH participates in the primary hydrolysis of the fiber and is responsible for ''amorphogenesis'', which is a phenomenon not yet fully elucidated, but it is known that involves a physical disruption of the substrate, resulting in destratification or segmentation of the cellulose fibers. The amorphogenesis promotes increases in the rate of hydrolysis of cellulose for making amorphous regions of crystalline polymer, leaving it more exposed to celulases. The CBH also can be divided into two types: CBH I, which hydrolyzes terminal reducing (R), whereas CBH II hydrolyzes terminal non-reducing (NR). These enzymes usually are inhibited by its hydrolysis product (cellobiose) (ZHANG and LYND, 2004).

The third and last enzyme of the cellulolytic complex is β-glucosidase or β-glucoside glucohydrolase (EC 3.2.1.21). β-glucosidase hydrolyzes cellobiose and other cellodextrins soluble (DP <7) into glucose. B-glucosidase is responsible for the control of the entire speed of the reaction exerting a crucial effect on the enzymatic degradation of the cellulose, preventing the accumulation of cellobiose. Just as CBH is also reported it can be inhibited by their hydrolysis product. (CASTRO and PEREIRA Jr, 2002; LEITE et al. 2007; LEITE et al, 2008).

Cellulase systems exhibit higher collective activity than the sum of the activities of individual enzymes, a phenomenon known as synergism. There are at least three forms of synergism have been reported: (i) endo-exo synergy between endoglucanases and exoglucanases, acting in the amorphous regions of fiber, provides reducing and non-reducing end for the action of CBH I and CBH II, respectively (ii) exo-exo synergy between exoglucanases processing from the reducing and non-reducing ends of cellulose chains, (iii) synergy between exoglucanases and β-glucosidases that remove cellobiose (and cellodextrins) as end products of the first two enzymes (CASTRO and PEREIRA Jr, 2010; LYND et al, 2002)

So, when cellulase enzyme systems act in vitro on insoluble cellulosic substrates, three processes (ZHANG and LYND, 2004) occur simultaneously:

1) chemical and physical changes in the residual solid-phase cellulose (not yet solubilized);
2) primary hydrolysis: involving the release of soluble intermediates with a degree of polymerization (DP) up to 6 from the surface cellulose molecules. The enzymatic depolymerization step performed by endoglucanases and exoglucanases is the rate-limiting step for the whole cellulose hydrolysis process;
3) secondary hydrolysis: involving hydrolysis of soluble intermediates to lower molecular weight and of cellobiose to glucose by β-glucosidases, although some β-glucosidases also hydrolyze longer cellodextrins.

Figure 3 illustrates the synergistic action between exoglucanases, endoglucanase And β-glucosidase in the hydrolysis of cellulose fiber.

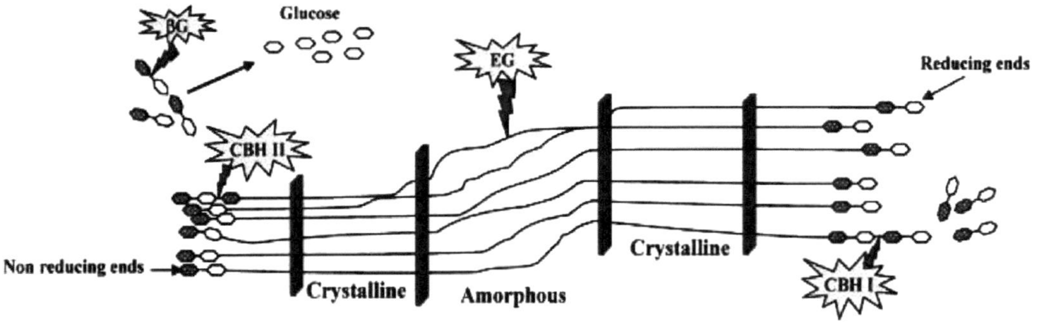

Figure 3. Enzymatic degradation of cellulose to glucose (PERÉZ et al, 2002). CBHI Cellobiohydrolase I acts on the reducing ends; CBHII cellobiohydrolase II acts on the non-reducing ends; EG endoglucanases hydrolyze internal bonds. βG β-Glucosidase cleaves the cellobiose disaccharide to glucose.

IV. Cellulases Application

Already in the 90s, cellulases, together with hemicellulases accounted for more than 20% of the global market enzyms. In 2008, considering only the Brazilian imports and exports, the cellulases mobilized an amount USD 1.35 millions.

The cellulases production in industrial scale began in the mid 80's, for their application as an additive to animal feed in order to increase the digestibility of feeds for ruminants and monogastric animals. Then these enzymes began to be used as an input for the food industry, whose aim was to improve sensory properties of pasta. In this sector, the cellulases have also begun to act in the processing of beverages, promoting the clarification of fruit juices and wines and on maintenance of a stable rheology of the final product. Subsequently, the cellulolytic enzymes began to be used on a large scale in the following industries: textile then deployed in Biopolis procedures (defibrillation of fabrics like cotton, linen, wool and viscose) and biowashing (washing process using cellulolytic enzymes) for production of jeans to look worn, known for *stone wash*. The abrasive power of these enzymes on the fibers of the fabric, resulting in increased softness and decolorize the same (SAID and PIETRO, 2002). Previously to achieve this result, the pumice stone was used in industrial washing machines. The pumice stone use resulted in serious damage to health workers and the environment. Due to the large amount of dust emitted and the need for a greater number of rinses, to remove the excess stone impregnated in the fabric. This process results in greater spending for water and energy (DILLON, 2004). Beyond the textile industry, these enzyme was applied in industries of paper and pulp for the control mechanical modification of pulp; on release of paint from the surface of the fibers to be recycled; on laundry, in order to increase the brightness, removing dirt and softness of fabrics, also ease the wear of clothes, noted by the fluff and pellets that appear after repeated washings (CASTRO and PEREIRA Jr, 2010).

Within the cellulase group, β-glucosidase is an enzyme with special properties on different sectors of application which is show in sequence.

The β-glucosidases are a heterogenous group of hydrolytic enzymes, which can be classified according to substrate specificity, being divided into three groups: (1) aryl β-gluosidases, enzymes that act on substrates aryl glycosides, (2) cellobiase, which have high specificity for hydrolysis of cellobiose, (3) enzymes that have broad specificity, there are that act on different substrates glycosides (VILLENA et al., 2006). According to BHATIA et al. (2002) most of the microbial β-glucosidases are classified in the latter category.

All three cellulases of cellulolitic complex suffer catabolite repression by the final product of hydrolysis. The cellobiose accumulation is preventing by β-glucosidase action. The β-glucosidase is responsible for controlling the overall rate of cellulolytic hydrolysis reaction, thereby playing a crucial effect on the enzymatic degradation of cellulose (LEITE et al. 2007a; PARRY et al., 2001). The β-glucosidase function on enzymatic process of cellulose, makes this enzyme present a great application potential for the ethanol industry. Thus, microbial β-glucosidase research associated with other cellulolytic enzymes can help to make the obtaining fuels from agro-industrial waste rich in cellulose.

The β-glucosidase can be applied in the juice and beverage industry. Cellulolytic enzymes disrupted the cell wall matrix of plants, which facilitates the breakdown of plant cell by increasing the juice extraction. During the winemaking process the cellulase addition helps the extraction of anthocyanins and terpenes. These compounds are present in grape skin, and are responsible for color and aroma of wine, respectively (TRAONA-MASSON and PELLERIN, 1998). Anthocyanins are natural pigments that provide the color blue, red, violet and purple of many plant species (COUTINHO, 2002). β-glycosidase with high specificity can hydrolyze anthocyanins, producing anthocyanidins (aglycone) and sugar free (LEITE et al., 2007b). The anthocyanidins have a lower solubility and pigmentation. This characteristic permits the removal of them, by precipitation and filtration. This effect can be used for depigmentation of juices with a high concentration of anthocyanins, or in smoothing of the red wine color to obtain the rose wine (PALM-FERNANDES, 2002).

The presence of β-glycosidases in the vinitification process, collaborate to the flavor compounds release from their glycosidic precursors (Fig. 4). Their compounds are collectively called terpenes, among them are nerol, α-terpineol, geraniol, linalool, citronellolis and others. When these compounds are glycosylated, it shown a low volatility that contributes to the little aroma of wine (BHATIA et al., 2002). The endogenous β-glucosidase from grape are unable to act during the vinitification process, due to low stability of them (BARBAGALLO et al., 2004). The use of microbial enzymes actuating on the conditions of vinitification process can be an alternative to increasing wines flavor. Thus, several studies have been conducted in order to find enzymes that can be used in vinitification process (BARBAGALLO et al., 2004, GUEGUEN et al., 1997, HERNÁNDEZ et al., 2003; SPAGNA et al., 2002).

The β-glycosidase action can be hinder on the non-volatile terpene glycosides that show other sugars linked in glucose molecule, such as rhamnose and arabinose (VILLENA et al., 2007). It is reported about the synergistic action of rhamnosidases, arabinosidases and β-glucosidase in the release of volatile terpenes with greater efficiency (BARBAGALLO et al., 2004; BELANCIC et al., 2003; HEMINGWAY et al., 1999).

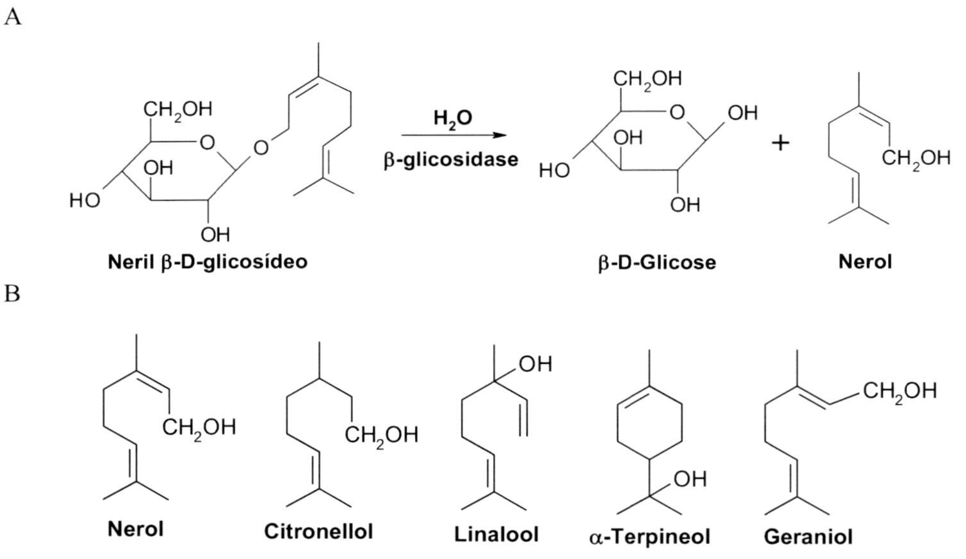

Figure 4. (A) Hydrolysis by β-glucosidase of Neril β-D-glucoside to nerol and β-D-glucose (HEMINGWAY et. Al., 1999). (B) Free terpenes that contribute to wine flavor (MAIC and MATEO, 2005).

The plants produce a wide variety of low molecular weight compounds called secondary metabolites, which include isoflavones, which are predominantly found in leguminous, especially soybeans. In plants, the isoflavones are related to the defense against attack of predators and pathogens (GENOVESE and LAJOLO, 2001). So, the flavonoids group is subdivided into different classes: flavanones, flavones, flavonols, anthocyanins and isoflavones (LOPES et al., 2000).

Isoflavones or isoflavonoids are phenolic compounds widely distributed in the plant kingdom. These compounds occur naturally in fruits, vegetables, tea, red wine, and soyabeans. Their concentration is relatively larger in leguminous, particularly soybeans (ESTEVES E MONTEIRO, 2001). Most of the soy isoflavones are found in glycoside form as daidzin, genistin and glycitin (Fig. 5). The isoflavones free of sugar molecule (deglycoside form) are called of aglycones. The aglycone moiety, released as a result of hydrolytic activity of β-glucosidases, has potent biological activity (PARK et al., 2001; RIELD et al., 2005). Because, only aglycones are absorbed by the human intestinal tract since the isoflavones glycosilated can not cross the epithelial barrier of the intestine. So, this letter isoflavone has not functional effect on the human organism (OTIENO and SHAH, 2007). So, the aglycones can be several uses in the field of medicine as antitumor agents, in general biomedical research, and in the food industry. The hydrolysis of diadzin and genistin to diadzein and genistein, respectively, with the release of glucose was demonstrated with the *Lactobacillus casei* subspecies *rhamnosus* β-glucosidase, thus reducing the undesirable bitter and astringent isoflavones glucosides from soyabean cooked syrup (SCS) (MATSUDA et al., 1994; BHATIA et al, 2002). Similarily, phloridzin was hydrolyzed to liberate the aglycone moiety, which is a precursor of melanin. The latter is known to reduce the risk of skin cancer and promotes dark color of hair (RIDGWAY et al., 1997).

Glycosides Isoflavones

R1	R2	Compound
H	H	Daidzin
OH	H	Genistin
H	OCH_3	Glycitin

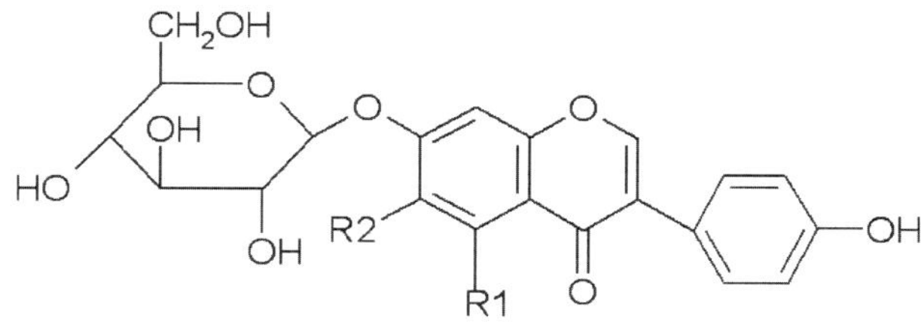

Deglycosides Isoflavones - Aglycones

R1	R2	Compound
H	H	Daidzein
OH	H	Genistein
H	OCH_3	Glycitein

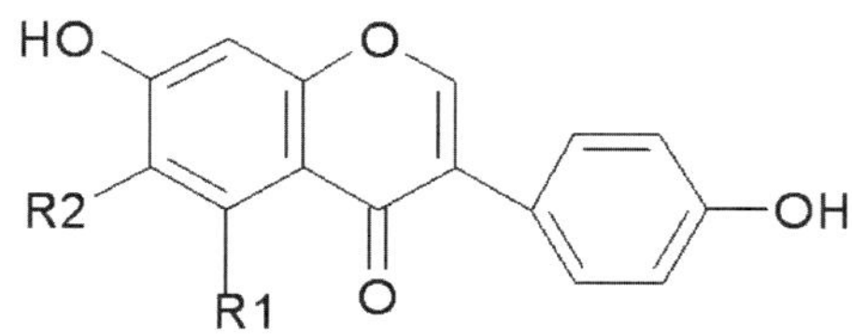

Figure 5. Structures of the isoflavones found in soy (SONG et al., 1998).

Aglycones, especially genistein, exhibit estrogenic properties due to structural similarity with the female hormone 17-β-estradiol. Aglycones also act as an inhibitor of topoisomerase I and II enzymes and histidine kinases. These enzymes are involved in cell proliferation and tumor formation. Antioxidant activity is another property evident in the presence of aglycones (LIMA, 2003). Because of these characteristics, in recent decades has increased interest in the benefits it can provide a diet rich in isoflavones, contributing to the control and prevention of many chronic diseases such as cancer (breast, prostate and colon), osteoporosis, menopausal symptoms, cardiovascular and other diseases (CARRÃO-PANIZZI and BORDINGNON, 2000).

The isoflavone glycosides can be hydrolysed by the action of β-glucosidase produced by microorganisms living in the human intestine. However, we know that intestinal flora is highly variable and may be influenced by age, ethnicity, diet, drug intake, and generally by person lifestyle. Thus, hydrolysis and consequently the absorption of isoflavones, may vary greatly from one person to another (RAFI et al., 2003). Within this context, it is possible to see in the literature several papers that aim to increase the concentration of free isoflavones using plant or microbial β-glucosidase in foods derived from soybeans. (PETERSON and BARNES, 1993; CARRÃO-PANIZZI and BORDINGNON, 2000; PARK et al. 2001; RIELD et al., 2005; OTIENO and SHAH, 2007). Brazil is one of major soybeans producer and exporter in world trade, it is of great national interest to develop technology to add value to this product.

V. Cellulase Production

The lignocellulosic raw materials are renewable most abundantly found in nature, mainly, by agro-industrial materials, the urban waste and the wood of angiosperms and gimnospermas. Among these, the agro-industrial materials are distinguished by the character of agro-industrial waste, conferred for his achievement after processing of raw materials with higher added value. Brazilian economy is based heavily on agricultural production, with the main products being soy, corn, cotton, sugarcane, coffee, cassava, and several fruits. In addition to the commercialization of agricultural products *in natura*, the agro-industrial sector involved with the processing of fruit juices and production of alcohol and sugar has become an important export. Thus, agro-industrial waste such as sugarcane bagasse (a fibrous residue of cane stalks left over after the crushing and extraction of the juice from the sugar cane), wheat bran, rice peel, corn straw, corncob, fruit peels and seeds, effluents from paper industry and orange bagasse are also produced in large quantities in Brazil. Disposing these abundant wastes is a hard problem for the agro-industries and these wastes can also become a great environmental pollution problem if its disposing is not adequately carried out. (PANDEY et al, 2000, SOCCOL 2003; DaSILVA eta 2005, GRAMINHA et al, 2007, LEITE et al 2008). Figure 6 shows historical series for generating agro-industrial waste in Brazil, represented with basis on wet mass. Sugarcane bagasse is predominant among the agro-industrial waste estimated. In 2007 were generated 147 million tones of sugarcane bagasse. Brazil produced about a total of 606 million tones of agro-industrial waste in 2007 (CASTRO and PEREIRA Jr 2010). These total about 105 million tones correspond to cellulosic fraction that is available to use in different process such as an alternative source for the microbial growth aiming the production of biomass or enzymes.

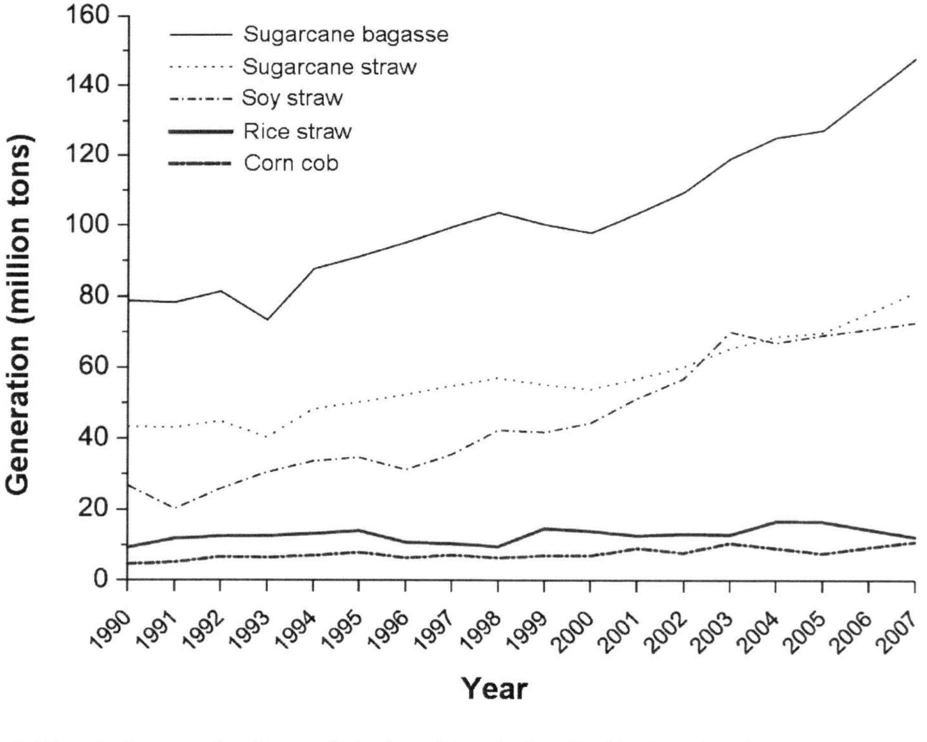

Figure 6. Historical generation lignocellulosic residues in Brazil. (CASTRO and PEREIRA Jr 2010).

The ability of some microorganisms to metabolize lignocellulosic material makes them potentially important to take advantage of vegetable residues. For cellulase production, the microbial growth can be carried out using solid state fermentation (SSF) or submerged fermentation (SF).

The differentiation these two types of processes is the water content present in the reaction medium. In SSF there is absence or near absence of free water. The water in these systems is complexed with the solid matrix or substrate as a thin layer absorbed by the particle surface. In general, these processes the moisture content rage between 30-85% and the water activity ranges from 0.40 to 0.90. These characteristics are similar the conditions found in nature and even allowing to be conducted without prior sterilization, since the contamination is remote (RAIMBAULT, 1998; ROBINSON and NIGAM, 2003). Water has many functions in a bioprocess, such as the diffusion of nutrients in the reaction medium; absorption of nutrients by microorganisms of the medium, as well as the metabolites removal; maintenance of function and stability of biological structures such as proteins, nucleotides and carbohydrates; stability of the layered structure and conservation of the permeability of the plasmatic membrane (RAIMBAULT, 1998).

In SF, the nutrients and substrate of culture medium are dissolved in water. So, there is a increased on energy demand associated with the sterilization of the medium and the removal of product from the fermented medium. There is a lower concentration of products obtained by FS, so, the downstream processing of the protein molecules is facilitated by the absence or low concentration of the substrate particles or compounds of secondary metabolites. The high water content and the nature of the medium diluted, facilitate control of the growth temperature, reducing degradation of the molecules, in particular enzymes with low thermostability. On the other hand, when operated with high substrate concentrations, rheological problems can occur in the system. But, diffusional processes and of the mixture are facilitated by the homogeneous character of the system. Currently, there is monitoring technologies of the variables in the fermentative medium during fermentation process.

There is a big trend for the application of SSF process in the of development of bioprocesses, such as biodegradation and bioremediation of toxic compounds, nutritional enrichment of agricultural residues and obtainment of products with high added values as organic acids, biopesticides, biofuels, aromatics and enzymes (PANDEY, 2003). The slope for the microbial enzymes production using the SSF, is related to the following characteristics: this type of process is more adequated to use of the agro-industrial waste as substrates, and it reduce the final product costs. The growth conditions of SSF are similar to the natural habitat, when of the filamentous fungi cultivated, resulting in better adaptation to environmental conditions and consequently higher production. The risk of bacterial contamination are significantly reduces, because the absence of free water and the spaces between the substrate particles allows easy aeration with reduction in energy consumption. The formed products, after fungi fermentation, are more concentrated and usually extracted with the addition of small amounts of water, making it possible to obtain enzymatic extracts with a high concentration of interest enzyme (BIANCHI et al. 2001; MARTINS, 2003).

The selection of a suitable substrate is a key factor for success of the SSF. Besides the substrate composition suitable for the induction of the desired product, the particle size of the substrate also be a factor with strong influence on the FES. Small particles provide more surface contact, allowing greater access to micro-nutrients, but depending on the type of substrate and the moisture content, can compress itself, hindering aeration and oxygen availability, as well as heat dissipation, limiting microbial growth. Already large particles to facilitate aeration of the medium, but may hinder access of the microorganism, limiting the contact surface of the substrate and hindering the mass transfer (PALM, 2003). Other parameters should also be evaluated and optimized for process efficiency, such as moisture and initial pH, incubation temperature, aeration, amount of inoculum, supplementation with nutrients, extraction and purification of the product (PANDEY et al., 2003).

The fermented material from SSF for cellulases and hemicellulases production can be applied to the animal feed industry. The microbial growth on plant biomass, for produce these enzymes, cause a partial disruption of cell walls of the biomass, increasing the digestibility of these materials for animal consumption (SILVA, 2006). Filho (2004) reports that the addition of cellulases and hemicellulases increases the digestibility of cereals (oats, rye, barley and wheat) those are rich in non-starch polysaccharides, mainly for monogastric animals. These animals have a reduced capacity for digestion of food lignocellulosic.

The filamentous fungi are considered the most suitable microorganisms for FES because their hyphae can grow on the particle surface and penetrate into the intra-particle spaces, and thus colonize the solid substrate (Santos et al, 2004). In general, the requirement of high water activity for the development of bacteria causes they do not adapt to FES processes. However, scientific reports show that some bacterial cultures can be adapted to this type of process (NAMPOOTHIRI; PANDEY, 1996; OROZCO, 2008).

Thus, some microorganisms were isolated from different regions of Brazil, including the Cerrado and Amazon biomes as well as cultivated soils (Table 1). These microorganisms were cultured in different culture media. The culture media were formulated using agro-industrial waste. The enzyme production was analyzed for each microbial species.

V.1. ENZYMATIC PRODUCTION

The enzymatic production can be analyzed by solid state fermentation or submerged fermentation as explain above. In this case, solid state fermentation was selected for waste agro-industrial application as substrate.

Table 1. Microorganism cellulose producers isolated on different areas of the Brazilian Biomes

Strain	Collect place	Temperature (°C)	Reference
Aureobasidium pullulans ER-16	orange juice industry residues, Catanduva, São Paulo State	28	LEITE et al, 2007a
Thermoascus aurantiacus 179-5	decayed wood, Amazon area, Manaus, Amazonas State	50	DaSILVA et al, 2005
Neosartorya spinosa P2D19	decomposition soil material, Cerrado area, Selviria, Mato Grosso do Sul State	35	ALVES PRADO et al, 2010

In order to evaluate enzymatic production, the microorganisms were cultivated using four different types of substrates such as: wheat bran, soy bran, soy peel, corn cob, corn straw, sugarcane bagasse, orange bagasse (a fibrous residue of orange peel left over after the extraction of the juice from the orange fruit), green grass, dried grass, cassava bran, and eucalyptus sawdust.

So, the SSF was carried out in 500 mL Erlenmeyer flasks containing 5 g of moistened substrates (grounded to 2–3 mm size) with 10 mL of mineral solution aiming an initial humidity content of 75%. The mineral solution was made up of 0.1% $(NH_4)_2SO_4$, 0.1% $MgSO_4.7H_2O$, and 0.1% NH_4NO_3 (w/v). After the inoculation of the microorganism, the fermentation was incubated at correspondent growth temperature. Enzyme extraction was achieved by adding 50 mL of distilled water to each flask followed by 2 h on a rotary shaker at 80 rpm. Crude extracts were centrifuged (10,000g/20 min), and then the supernatant was used for enzyme activities assays.

Xylanase, endoglucanase, and avicelase enzymes activities were measured by determining the release of reducing sugars by the 3,5-dinitrosalicylic acid method (14). The 100 mM sodium acetate buffer pH 5.0 was used containing 0.5% of xylan (Birchwood-Sigma), 0.5% of carboxymethylcellulose (C5768 Sigma), and 0.5% of avicel (Co Sigma) as substrates for xylanase, endoglucanase, and avicelase enzymes, respectively. β-glucosidase activity was determined using 50 μL of the extract, 250 μL of 100 mM sodium acetate buffer pH 5.0, and 250 μL of 4 mM 4-nitrophenyl β-D-glucopyranoside (PNPG, Sigma). After 10 min, the reaction was stopped by the addition of 2 mL of 2 M sodium carbonate. The activities were measured at 410 nm and expressed in international units, defined as the amount of enzyme required to produce 1 μmole of nitrophenol (β-glucosidase), xylose (xylanase), and glucose (CMCase and avicelase) per minute, under assay conditions (DaSILVA et al, 2005; LEITE eta al, 2007a; ALVES PRADO et al, 2010)

Aureobasidium pullulans ER-16 produced β-glucosidase in all substrates tested, but the production in wheat bran was found to 1.10 U/mL after 72 h (Table 2). The endoglucanase and xylanase production was obtained in wheat bran 1.0 and 4.7 U/mL, respectively. No endoglucanase production was detected in soy bran and corn cob cultivation. In addition, no xylanase activity was detected in soy bran cultivation (Table 2). Avicelase activity was not detected from any of the substrates. Avicelase is the activity responsible for crystalline cellulose degradations and it is frequently found in low concentrations or absent on the cellulolytic system of many microorganisms (GOMES et al, 2000). These assays confirm that the *A. pullulans* is among these.

Table 2. Effect of different substrate on enzyme production by *Aureobasidium pullulans* ER-16. After 72 hours of incubation the clear filtrate was used for assays of enzyme activities. (LEITE et al., 2007a)

Substrates	Xylanase (U/mL)	CMCase (U/mL)	β-Glucosidase (U/mL)
Corn cob	0.79	ND	0.23
Soy bran	ND	ND	0.14
Soy peel	0.66	0.50	0.17
Wheat bran	4.7	1.0	1.10

ND: not determined

The *Thermoascus aurantiacus*, grew and produced enzymes more efficiently at 50°C, even with growth at temperatures 30°C to 60°C. As is shown in Table 3, the *T. aurantiacus* produced xylanase and CMCase independently of the material used as nutrient source (as it was used only distilled water to moisten the substrate all the nutrients necessary to the growth of fungus came from the raw material used). Orange bagasse, sugar cane bagasse and sawdust were very poor substrate to the growth of the fungus, consequently the enzyme production on these substrates were very low too. Avicelase activities were significantly lower among all the lignocellulosic materials tested as carbon sources (Table 3). β-glucosidase activity was obtained in wheat bran, soy bra, corn cob and soy peel substrates. Wheat bran was a good substrate for avicelase and β-glucosidase activities.

Table 3. Effect of different substrate on enzyme production by *Thermoascus aurantiacus*. After 4 days of incubation the clear filtrate was used for assays of enzyme activities. (DaSILVA et al, 2005; LEITE et al., 2008)

Substrates	Xylanase (U/mL)	CMCase (U/mL)	Avicelase (U/mL)	β-Glucosidase (U/mL)
Corn straw	97	59	0.65	ND
Corn cob	107	60	0.86	1.8
Dried grass	99	60	0.65	ND
Green grass	102	59	0.65	ND
Eucalyptus Sawdust	37	9	0.30	ND
Orange bagasse	8	4	0.30	ND
Soy bran	ND	ND	ND	4.2
Soy peel	ND	ND	ND	1.3
Sugarcane bagasse	11	3	0.60	ND
Wheat bran	64	30	1.50	5.8

ND: not determined

Six different media compositions with solid substrates were analyzed for xylanase production from *Neosartorya spinosa* P2D19, (Table 4). All six analyzed media presented microbial growth, measured as protein content and micelial growth visualization. The media composed of wheat bran and corn cob presented good xylanase production. The enzymatic activity was 15.10 U/mL for wheat bran and 8.50 U/mL for corn cob after 72 h of fermentation. For CMCase activity was observed 3.60 U/mL with wheat bran as substrate. For other substrates the level of activity was low. So, the β-glucosidase activity was better using wheat bran as substrate and in other analyzed substrates the activity was lower (ALVES PRADO et al, 2010).

As observed on analyzes above, the agro-industrial waste is a good alternative substrate for enzyme production. The results of these assay demonstrated that wheat bran as a potential agro-industrial waste. Wheat bran is a complex substrate rich in proteins (14%), carbohydrates (27%), minerals (5%), fat (6%), and B-vitamin (HAQUE et al, 2002), this probably favored the growth and the production of enzymes for the microorganism. Previous works (IEMBO et al, 2002; KALOGERIS et al 2003) report the production of cellulases and hemicelulases, using derived wheat (bran and straw) as substrate.

Table 4. Xylanase and CMCase activity on crude enzyme from *Neosartorya spinosa* strain P2D19 obtained by solid-state fermentation using different substrates, after 72 hour fermentation (ALVES PRADO et al, 2010).

Substrates	Xylanase activity (U/mL)	CMCase activity (U/mL)	β-Glucosidase (U/mL)
Cassava bran	3.45	0.09	0.10
Corn cob	8.50	0.20	0.23
Corn straw	0.95	0.16	ND
Sugar cane bagasse	0.50	0.05	ND
Wheat bran	15.1	3.60	1.45
Wheat bran + Sawdust	2.10	0.11	ND

ND: not determined

In order to evaluate β-glucosidase production through fermentation process, the wheat bran was used in a new cultivation and samples were removed every 24 h throughout a period of 144 h. The highest β-glucosidase production, about 7.0 U/mL (or 70 U/g), was obtained in 48–72 h of SSF with the *fungus T. aurantiacus* CBMAI-756 (Fig. 7). KALOGERIS et al. 2003 presented similar results to those obtained in the present study, with a strain of *T. aurantiacus* grown on wheat straw as a carbon source with 80% moisture.

The microorganism *A. pullulans* ER-16 showed maximum production of enzyme, about 1.3 U/mL (or 13 U/g), at 120 h of cultivation (Fig. 7). This value was higher than other values described in the literature. IEMBO et al. 2002 reported a maximum production of 0.5 U/mL at 168 h of growth, using wheat bran as carbon source with 67% moisture content on SSC for enzyme production by *Aureobasidium* sp.

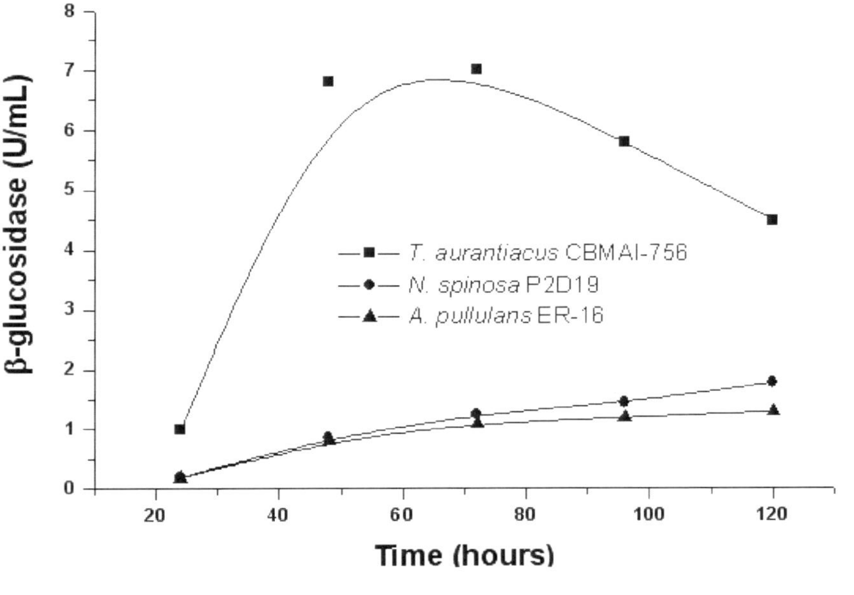

Figure 7. Time course of β-glucosidase production by *A. pullulans*, *N. spinosa* and *T. aurantiacus* in wheat bran.

The *Neosartorya spinosa* P2D19 showed maximum level of β-glucosidase activity of 1.78 U/mL (or 17.8 U/g), at 120 h of cultivation (Fig. 7). This activity value was higher than values described in the literature and the *A. pullulans* ER-16 analyzed n this study(LEITE et al, 2007a) .SAHA et al. 1994, using *A. pullulans* in submerged cultivation with a mixture of wheat bran and corn bran at 1% (w/v) as carbon source, reported 0.27 U/ml after 96 h.

So, these isolated microorganisms, shown a great potential to produce β-glucosidase from agro-industrial waste. Thus, the use of agro-industrial waste-for enzymes production may be a good alternative for lignocellulolytic enzyme production. Likewise this practice reduces the emission of agro-industrial waste directly into the environment. On the other hand, it is possible verify the capacity, of the investigated environment, for the isolation of news microbial strains capable of producer enzymes.

VI. Cellulase Properties

In some cases, the use of enzymes of mesophilic organisms can be a disadvantage, once they generally undergo denaturation in temperatures higher than 55°C, resulting in low efficiency of hydrolysis, demanding large quantities of enzyme and increased cost to conduct the hydrolysis under aseptic conditions. The employment of thermostable enzymes to carry out hydrolysis at higher temperatures is advantageous because it increases the speed of reaction and avoids microbial contamination contributing to increase technical and economical viability of the process. Other difficulty in using the β-glucosidases or others cellulose in industrial process is its strong enzyme inhibition by product. So, the search for β-glucosidases insensitive to product inhibition, and of high-thermal stability, has increased recently. Enzymes with these characteristics would improve the process of saccharification of lignocellulosic materials (ZANOELO et al, 2004). An alternative to bypass the problem of inhibition, is to associate the cellulose's enzymatic hydrolysis to a simultaneous alcoholic fermentation, where the glucose will be microbiologically converted to ethanol (SAHA et al, 1994; CARDONA and SÁNCHEZ, 2007). Nevertheless, the enzymes in this process need to be relatively stable due to ethanol present in the reaction medium. Most of the processes of industrial application of enzymes occur at high temperature, so the use of thermostable enzymes appears to be appropriated because they preserve their catalytic activity at high temperatures. A series of advantages such as faster reaction, decreased viscosity in processed fluid, increased solubility of the substrate, and reduced contamination risk by undesired organisms have been proposed for use of thermostable enzymes in biotechnology processes (BRUINS et al, 2001; GROMIHA, 2001; DaSILVA et al 2005; LEITE et al, 2008).

Crude and purified enzymes produced by *A. pullulans* ER-16, *N. spinosa* P2D19 and *T. aurantiacus* CBMAI-756 were studied and some iterating characteristics were observed.

Neosartorya spinosa P2D19 has only xylanase characteristics studied up till now. Thus, enzymatic properties were determined on crude xylanase produced by solid state fermentation using wheat bran as carbon source for *N. spinosa* P2D19. The optimal temperature activity was 60°C and this xylanase was stable until 55°C after 1 hour of heat (Fig. 8a). The optimal pH was pH 5.0-5.5, but 87% of activity was maintained on pH 7.5 (Fig. 8b). So, in optimal pH activity, it was seen at two peaks of pH. It may suggest that there are two proteins with distinct catalytic activities or that the same enzyme is capable of acting at different pH values.

On the other hand, these two peaks of pH can be related to isoenzymes. BADHAN et al., (2004), studying xylanase produced by *Myceliophthora* sp, found different xylanase isoforms using an isoelectric foccusing and these isoforms are depend on the substrate type used on solid state fermentation. More studies are necessary for accurate conclusions about this proposed work, because it is obtained of crude enzyme and purification assays were not realized yet. Based on performed analyses, This xylanase was stable between pH 5.5 to pH 8.5 after 24 h (Fig. 8b).

a) **b)**

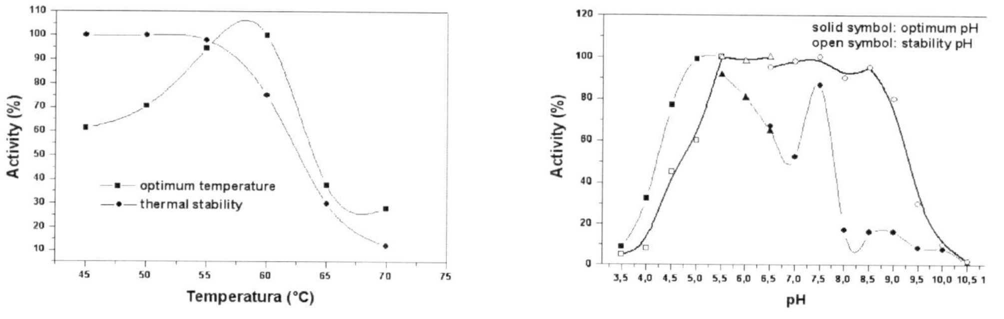

Figure 8. Effect of pH (a) and temperature (b) on xylanase activity and xylanase stability from *Neosartorya spinosa* strain P2D19. The buffers used were: Acetate (—■— pH 3.5-5.5), MES (—▲— pH 5.5-6.5) and Glycine-NaOH (—●— pH 7.0-10.5).

Effect of the temperature and pH in the activity and stability of crude enzymes from *T. aurantiacus* CBMAI-756 is shown in TABLE 5. The crude enzymatic extract was obtained by SSF using corn cob. Crude xylanase from *T. aurantiacus* shown its optimum pH at 5.0-5.5 and its activity fell to 50% when pH ranged from 5.5 to 6.5. This optimum pH was a little higher than that reported by GOMES et al (1994) to his strain of *T. aurantiacus*, which was pH 4.5. This xylanase was 100% stable within a wide range of pH, which is from pH 3.5 pH to 8.0. This result for stability is close to the result obtained by GOMES et al (1994) which reported stability over a broad pH range, exhibiting more than 80% of its total activity between pH 3.0 and 9.0. The optimum temperature of crude xylanase was 75°C. By comparing this value with the available reports on activity of crude xylanase in literature, it was verified that *T. aurantiacus* under study has an optimum temperature for xylanase activity higher than other mesophyllic fungus, i.e, 70°C for *T. terrestris* and *S. cellulophilum,* 55°C for *T. Reesei*, and 55°C for *Penicillium* sp DURAN et al (1984). Our result for optimum temperature is smaller than that reported by GOMES et al (1994) for his strain of *T. aurantiacus*, which was at 80°C. This xylanase remained 100% stable until the treatment at 60°C, starting to undergo denaturation from this temperature, and at 70°C only 22% of the initial activity was recovered after the treatment (DaSILVA et al, 2005).

The optimum pH for the crude CMCase activity was 5.5. However in pH 4.0 the activity was almost completely lost and in pH 7.0 the enzyme still retained 58% of its activity. The optimum pH found for the CMCase is comparable to those for the other *T aurantiacus* strains (YOSHIOKA et al, 1982; GOMES et al, 2000). The result obtained with this CMCase was 100% stable on a wide range of pH values, from 3.0 the 8.0. Our result on pH stability has shown CMCase less stable than others related by GOMES et al (2000) and KAWAMORI et al (1987). The optimum temperature of the CMCase was 75ºC and at 85ºC the crude enzyme still had 60% of the original activity. The optimum temperature found for the CMCase is in good agreement with the reports for the other *T. aurantiacus* strains (YOSHIOKA et al, 1982; GOMES et al, 2000). This crude CMCase was 100% stable to the treatment at 60ºC and started to undergo denaturation from this temperature. At 70ºC, 68% of the initial activity was still recovered after the treatment. (DaSILVA et al, 2005)

The data have shown that the xylanases and CMCases were very stable to pH under the tested conditions, mainly in the pH range between 3.0 and 7.0, which account for the most frequent conditions in the industry. Comparing with work of MISHRA and RAO (1984) all the enzymes presented good thermostability and xylanase of *Termoascus aurantiacus* was more thermostable than xylanase of *T. Reseei* QM 9414 (50ºC for one hour) and that *of P. funiculosum* (50ºC for 30 minutes). The enzymes of this strain of *T. aurantiacus* appeared to be as stable as other *T. aurantiacus* already reported. It is accept that conditions of growth and the nature of medium may stabilize the enzyme against thermal denaturation. The presence of protease induced by different composition of the medium may also affect the enzyme stability (GOMES et al, 1994; GOMES et al, 2000).

Crude β-glucosidase from *T. aurantiacus* were most active at pH 4.5, with an apparent optimum temperature at 75°C. Kalogeris et al., (2003) characterized a β-glucosidase produced by the fungus *T. aurantiacus*, and obtained the best performance of the enzyme in pH 4.5 and 80°C. The β-glucosidase produced by the fungus *T. aurantiacus* retained more than 90% of its original activity at pH values 4.5-8.0 after incubation for 24 h, and it maintained more than 85% of its activity after 1 h at 70°C (Table 5). These results are very similar to those obtained with purified enzymes by LEITE et al., (2007b). Whereas, β-glucosidase produced by the *T. aurantiacus* microorganism, in a non-purified extract, showed a higher stability to pH, when compared with to the same purified enzyme (LEITE et al., 2007b; LEITE et al, 2008).

Table 5. Effect of pH and temperature on activity of xylanse, CMCase and β-glucosidase from *T. aurantiacus* CBMAI-756 (DaSILVA et al, 2005; LEITE et al., 2008).

	Xylanase	CMCase	β-glucosidase
Optimum pH	5.0-5.5	5.0	4.5
Optimum temperature (°C)	75	75	75
Stability pH	3.5-8.0	3.0-8.0	4.5-8.0
Stability temperature (°C)	60	60	35-70

Effect of the temperature and pH in the activity and stability of crude enzymes from *Aureobasidium pullulans* ER-16 is shown in TABLE 6. Crude Xylanase from *A. pullulans* had a maximum activity at pH 5.0 and with temperature at 50°C and remained stable within a wide pH range. Xylanase remained stable after 24 h between pH 3.0–8.0. The optimum pH and temperature obtained for CMCase produced by this microorganism were 4.0–4.5 and 60°C, respectively. This CMCase remained stable within pH 3.5–7.5. Both enzymes had a considerable reduction in their activity after 1 h of incubation at temperatures more than 50°C. Thermal inactivation is commonly observed in xylanases and CMCases produced by mesophilic microorganisms (CARMONA et al 2005; LEITE et al 2007a).

Table 6. Effect of pH and temperature on activity of xylanse, CMCase and β-glucosidase from *Aureobasidium* pullulans ER-16 (LEITE et al., 2007a; LEITE et al., 2008).

	Xylanase	CMCase	β-glucosidase
Optimum pH	5.0	4.5	4.0
Optimum temperature (°C)	50	60	75
Stability pH	3.0-8.0	3.5-7.5	4.5-10.0
Stability temperature (°C)	50	50	75

Crude β-glucosidase from *Aureobasidium* pullulans ER-16 were most active at pH 4.0, with an apparent optimum temperature at 75°C. HAYASHI et al., (1993) obtained at pH 4.0 the best activity of β-glucosidase produced by the microorganism *Aureobasidium* sp. SAHA et al., (1994) obtained the best activity of β-glucosidase produced by *A. pullulans* at 75°C. The enzyme produced by the mesophilic microorganism *A. pullulans*, was found to be more stable (pH and temperature) than that produced by the thermophilic fungus *T. aurantiacus*. The former presented stability in a range of pH values 4.5–10.0, retaining 65% of activity at pH3.0, and it retained more than 90% of its activity after 1 h at 75°C (LEITE et al., 2008) (Table 6).

So, the fungus *Thermoascus aurantiacus* and the polymorphic fungus (yeast-like) *Aureobasidium pullulans* are good producers of β-glucosidases. It is believed that there is a correlation between the thermophilia of microorganisms and the thermostability of their enzymes (MAHESHWARI et al., 2000). Also, it is known that some mesophilic microorganisms, even psycrophiles, can also produce enzymes with high thermostability (ILLANES, 1999).

LEITE et al., (2007b) reported an ideal temperature at 80°C, for purified β-glucosidase from *A. pullulans* ER-16. As can be seen in Table 6, crude enzyme showed higher catalytic activity at 75°C. The physico-chemical characteristic differences between the crude and purified β-glucosidase, stimulated us to study the thermoinactivation of such enzymes in fermented extract, with a view to observe the denaturation profile of each protein at high temperatures.

The thermoinativation studies that β-glucosidase obtained from the mesophilic *A. pullulans* ER-16 is more stable than that obtained from the thermophilic *T. aurantiacus* CBMAI-756. The treatment of both β-glucosidases at 80°C resulted in a progressive inactivation of both enzymes, but the β-glucosidases obtained from mesophilic *A. pullulans* was much more thermostable than that obtained from thermophilic *T. aurantiacus* (Fig. 9). The half life time ($t_{(1/2)}$) of the enzyme from *A. pullulans* was 90 min against 30 min presented by the enzyme from *T. aurantiascus*. But still, the half life time was calculated following the equation: $t_{(1/2)} = \dfrac{0.693}{k_d}$

Where k_d is the thermoinativation constant (min^{-1}).

So, at 80°C, the $t_{(1/2)}$ for the enzyme of *T. aurantiacus* were 29.7 min for 80°C, 6.2 min for 82.5°C, 3.8 min for 85°C and 1.3 min for 87.5°C. For the β-glucosidase of *A. pullulans* the $t_{(1/2)}$ were 88.7 min for 80°C, 22.7 min for 82.5°C, 11 min for 85°C and 1.83 min for 87.5°C. These results at 80°C confirm the result obtained graphically in figure 9.

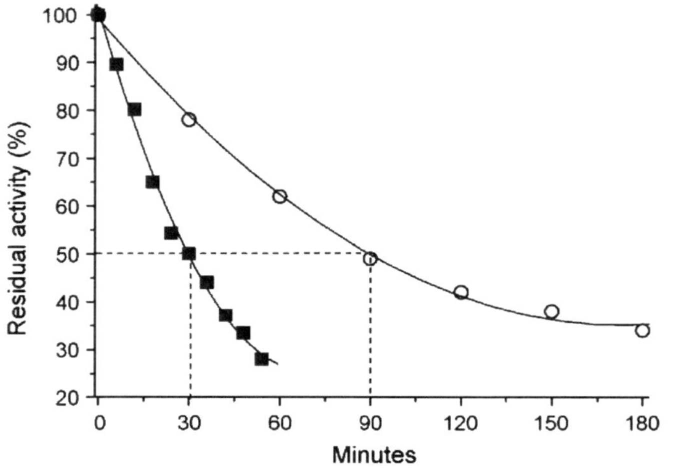

Figure 9. Time course of irreversible thermoinativation of β-glucosidases from *T. aurantiacus* (square) and *A. pullulans* (circle) at 80°C (LEITE et al., 2007b).

The thermoinactivation of both enzymes was measured at 80, 82.5, 85 and 87.5°C in acetate buffer 100 mM, pH 5.0. The linear regression of ln ra (residual activity) plotted against preincubation time shows a correlation coefficient (r) higher than 0.98 (Fig. 10). The slope of these plots were the k_d values used in the plot of ln k_d versus the reciprocal of the absolute temperature. Plots of ln ra *vs* time showed that thermal inactivation of β-glucosidase followed a first-order kinetic, as indicated by the linearity obtained by fitting the data (Fig. 10). This indicated irreversible inactivation by a monomolecular process or that inactivation by protein aggregation was not enough to cause lessen of linearity (CABOS et al, 2003). RASHID and SIDDIQUI (1998) also reported a first-order kinetic for β-glucosidase obtained from *A. niger*.

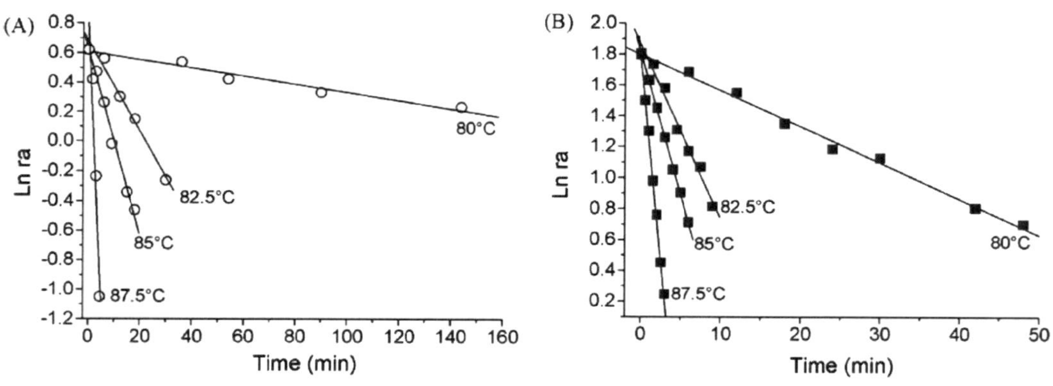

Figure 10. Linear regression of ln ra (residual activity) plotted against preincubation time at 80, 82.5, 85 and 87.5°C. (A) *A. pullulans* ER-16 and (B) *T. aurantiacus* CBMAI-756 (LEITE et al., 2007b).

Enzymes undergo partial unfolding when temperature rises above certain level, resulting in a change in native conformation and a loss of activity. The activity is fully reversible after short period of treatment at high temperature, but after longer periods only a small fraction of activity is regained, configuring a process of irreversible thermoinactivation (TOMAZIC; KLIBANOV, 1988). Residual activity was assayed immediately after cooling to 25°C and also after cooling the treated enzymes for a period of 24 h, since our calculated remaining activity was often the same, we can assume an irreversible process of thermoinactivation for both enzymes. This time of 24 h ensure that the thermal inactivation involves some irreversible mechanism; if there is refolding mechanism in the protein, after this time, the initial activity would be recouped for all assayed temperatures, and this definitely was not observed in this study.

Comparing the two enzymes, we can conclude that β-glucosidase obtained from this strain of the mesophilic *A. pullulans* is more resistant to thermal inactivation than β-glucosidase obtained from the thermophilic fungus *T. aurantiacus*. The comparison of thermal denaturation of purified β-glucosidase from a mesophilic *A. pullulans* and a thermophilic *T. aurantiacus* described in this work, clearly contributes to demonstrate that the complex understanding of enzymatic thermostability is not necessarily based only on its association with the thermophilia of the organism from which the enzyme was originated.

CONCLUSION

Nature represents an interminable source of cellulolytics and xylanolytics microorganisms and especially tropical countries such as Brazil, which presents a very diversified microbial flora, certainly shelters species of unknown microorganisms of optimum industrial interest. In this context, the exploration of new isolated strains is very important to provide a microorganism with properties that match with the conditions existing in the industrial environment (DaSILVA et al., 2005).

Because of this, our research group selected some microbial strains with industrial potential for cellulases and xylanases production. These strains were isolated from the environment in different Brazilian biomes such as the Amazon area and Cerrado area.

The application of microbial xylanases for biobleaching of pulp by the paper industry is intrinsically related to the absence of cellulases in the fermented medium, because the hydrolytic action of cellulases on cellulose fibers may decrease the quality of pulp, resulting in inferior paper. Hence, the presence of cellulolytic enzymes in the enzymatic extract produced by *A. pullulans, T. aurantiacus* and *N. spinosa* in SSF along with the low level of activity of xylanase in alkaline pH, impairs its use by the paper industry. On the other hand, the synergistic action of such enzymes benefits the attainment of fermentable sugars from agro-industrial residues that can be used to develop alternative fuels. This fact prompts us to continue with studies of these enzymes in order to contribute to the development of new sources of energy.

The production of cellulases and their application in the hydrolysis of lignocellulosic materials are technologies under development, for which certain tools and strategies can be applied in order to improve their productivity and economic costs. So, the concept of product engineering can be applied in production processes of cellulases in order to obtain enzyme preparations with ideal proportions between the various enzymes of the cellulolytic complex, particularly CMCase and β-glucosidase. The ideal option can be obtained by cocultivation of the overproducing strains of the major types of cellulase, or by separate production of cellulases and subsequent mixing of the extracts, at rates pre-optimized.

REFERENCES

BADHANA, A.K.; CHADHAA, B.S.; SONIAA, K.G.; SAINIA, H.S.; BHATB, M.K. *Enz. Microbial Technol.,* 35, 460–466, 2004.

BARBAGALLO, R.N.; SPAGNA, G.; PALMERI, R.; RESTUCCIA, C.; GIUDICI, P. Selection, characterization and comparison of β-glucosidase from mould and yeast employable for enological applications. *Enz. Microbial Technol.,* 35, 58-66, 2004.

BELANCIC, A.; GUNATA, Z.; VALLIER, M.J.; AGOSIN, E. β-glucosidase from the grape native yeast *Debaryomyces vanrijae*: purification, characterization, and its effect on monoterpene content of a muscat grape juice. *J. Agric. Food Chem.,* 51, 1453-1459, 2003.

BHAT, M. K.; BHAT, S.; Cellulose degrading enzymes and their potential industrial applications. *Biotechnol. Adv.*, 15, 583, 1997.

BHATIA, Y.; MISHRA, S.; BISARIA, V.S. Microbial β-glucosidases: cloning, properties, and applications. *Critical Reviews Biotechnol.,* 22, 375-407, 2002.

BIANCHI, V.L.D.; MORAES, I.O.; CAPALBO, D.M.F. Fermentação em estado sólido. *In: Biotecnologia industrial.* São Paulo, Edgard Blücher, v. 2, cap. 13, 247-276, 2001.

BRUINS, M.E.; JANSSEN, A.E.M.; BOOM, R.M. Thermozymes and their applications. *Appl Biochem Biotechnol.,* 90, 155–81, 2001.

CARDONA, C.A.; SÁNCHEZ, O.J. Fuel ethanol production: process design trends and integration opportunities. *Bioresour Technol.,* 98, 2415–2457, 2007.

CARMONA, E.C.; FIALHO, M.B.; BUCHGNANI, E.B.; COELHO, G.D.; BROCHETO-BRAGA, M.R.; JORGE, J.A. Production, purification and characterization of a minor form of xylanase from *Aspergillus versicolor*. *Process Biochem.*, 40, 359–364, 2005.

CARRÃO-PANIZZI, M.C.; BORDINGNON, J.R. Activity of β-glucosidase and levels of isoflavone glucosides in soybean cultivars affected by the environment. *Pesquisa Agropec. Brasileira,* 35, 873-878, 2000.

CASTRO, A.M.; PEREIRA Jr, N. Produção, propriedades e aplicação de celulases na hidrólise de resíduos agroindustriais. Quim. Nova, 33, 181-188, 2010.

COBOS, A.; ESTRADA, P. Effect of polyhydroxylic cosolvents on the thermostability and activity of xylanase from *Trichoderma reesei* QM 9414. *Enz. Microbial Technol.,* 33, 810–818, 2003.

COUTINHO, M.R. Obtenção de antocianinas presentes no repolho roxo (Brassica oleracea) 2002. Dissertação (Mestrado em Engenharia de Alimentos). Centro Tecnológico, Universidade Federal de Santa Catarina, Florianópolis.

DaSILVA, R.; GOMES, E.; FRANCO C.M.L. Pectinases, hemicelulases e celulases substrato, ação, produção e aplicação no processamento de alimentos. *Bol. Soc. Brasileira Ciênc. Tecnol. Alim.*, 31, 249-250, 1997.

DaSILVA, R.; Lago, E.S.; Merheb, C.W.; Macchione, M.M.; Park, Y.K.; Gomes, E. Production of xylanase and CMCase on solid state fermentation in different residues by *Thermoascus aurantiacus* miehe. Brazilian J. Microbiol., 36, 235-241, 2005.

DILLON, A.J.P. Celulases. *In: Enzimas como agentes biotecnológicos*. Ribeirão Preto, Legis Summa, cap.14, 243-269, 2004.

DURAND, H.; SOUCAILLE, P.; TIRABY, G. Comparative study of cellulases and hemicellulases from four fungi: mesophiles *Trichoderma reesei* and *Penicillium* sp. and themophiles *Thielavia terrestris* and *Sporotrichum cellulophilum*. *Enz. Microbial. Technol.*, 6, 175-180, 1984.

ESTEVES, E.A.; MONTEIRO, J.B.R. Efeitos benéficos das isoflavonas de soja em doenças crônicas. *Rev. Nutr. PUCCAMP*, 14, 43 - 52, 2001.

FILHO, E.X.F. Xilanases. In: Enzimas como agentes biotecnológicos. Ribeirão Preto, Legis Summa, cap. 8, p. 137-148, 2004.

GENOVESE, M.I. & LAJOLO, F.M. Determinação de isoflavonas em derivados de soja. *Ciênc.Tecnol. Alim.*, 21, 86-93, 2001.

GOMES, D.J.; GOMES, J.; STEINER, W. Production of highly thermostable xylanase by a wild strain of thermophilic fungus *Thermoascus aurantiacus* and partial characterization of the enzyme. *J. Biotechnol.,* 37, 11-22, 1994.

GOMES, I.; GOMES, J.; GOMES, D.J.; STEINER, W. Simultaneous production of high activities of thermostable endoglucanase and β-glucosidase by the wild thermophilic fungus *Thermoascus aurantiacus*. *Appl. Microbiol. Biotechnol.*, 53, 461-468, 2000.

GRAMINHA, E.B.N.; GONÇALVES, A.Z.L.; PIROTA, R.D.P.B.; BALSALOBRE, M.A.A.; DaSILVA, R.; GOMES, E. Enzyme production by solid-state fermentation: Application to animal nutrition. *Anim. Feed Scienc. Technol.*, 144, 1–22, 2008.

GROMIHA, M.M. Important inter-residue contacts for enhancing the thermal stability of thermophilic proteins. *Biophys Chem.*, 91, 91–97, 2001.

GUEGUEN, Y.; CHEMARDIN, P.; PIEN, S.; ARNAUD, A.; GALZY, P. Enhancement of aromatic quality of Muscat wine by the use of immobilized β-glucosidase. *J. Biotechnol.*, 55, 151-156, 1997.

HAQUE, M.A.; SHAMS-UD-DIN, M.; HAQUE, A. The effect of aqueous extracted wheat bran on the baking quality of biscuit. *Int J Food Sci Technol.*, 37, 453–462, 2002.

HEMINGWAY K.M.; ALSTON, M.J.; CHAPPELL, C.G.; TAYLOR, A.J. Carbohydrate-flavour conjugates in wine. *Carbohyd. Polym.*, 38, 283–286, 1999.

HERNÁNDEZ, L.F.; ESPINOSA, J.C.; FERNÁNDEZ-GONZÁLES, M.; BRIONES, A. β-glucosidase activity in *Saccharomyces cerevisiae* wine strain. *Int. J. Food Microbiol.*, 80, 171-176, 2003.

IEMBO, T; DaSILVA, R; PAGNOCCA, F.C; GOMES, E. Production, characterization and properties of β-glucosidase and β-xylosidase from a strain of *Aureobasidium* sp. *Appl Biochem Microbiol.*, 38, 549–52, 2002.

ILLANES, A. Stability of biocatalysts. *Electr J Biotechnol.*, 2, 1–9, 1999.

KALOGERIS, E.; INIOTAKI, F., TOPAKAS, E.; CHRISKOPOULOS, P.; KEKOS, D.; MACRIS, B.J.; Performance of intermittent agitation rotating drum type bioreactor for solid-state fermentation of wheat straw. *Bioresour Technol.*, 86, 207–213, 2003.

KANG, S.W.; PARK, Y.S.; LEE, J.S.; HONG, S.I.; KIM, S.W. Production of cellulases and hemicellulases by *Aspergillus niger* KK2 from lignocellulosic biomass. *Bioresour Technol.,* 91,153–156, 2004.

KAWAMORI, M.; TAKAYAMA K.; TAKASAWA S. Production of cellulose by thermophilic fungus. *Thermoascus aurantiacus* A-131., *Agric. Biol. Chem.,* 51, 647-654, 1987.

LEITE, R.S.R., BOCCHINI, D.A., MARTINS, E.S., SILVA, D., GOMES, E., SILVA, R. Production of cellulolytic and hemicellulolytic enzymes from *Aureobasidium pulluans* on solid state fermentation. *Appl. Biochem. Biotechnol.,* 137, 281-288, 2007a.

LEITE, R.S.R., GOMES, E., DaSILVA, R. Characterization and comparison of thermostability of purified β-glucosidases from a mesophilic *Aureobasidium pullulans* and a thermophilic *Thermoascus aurantiacus*. *Process Biochem.,* 42, 1101-1106, 2007b.

LEITE, R.S.R., ALVES-PRADO, H.F.; CABRAL, H.; PAGNOCCA, F.C.; GOMES, E., DaSILVA, R. Production and characteristics comparison of crude β-glucosidases produced by microorganisms *Thermoascus aurantiacus* e *Aureobasidium pullulans* in agricultural wastes. Enz. Microbial Technol., 43, 391–395, 2008.

LIMA, A.F. Produção e caracterização de β-glicosidase vegetal e microbiana e sua aplicação para conversão de isoflavonas glicosiladas em isoflavonas agliconas 2003. Dissertação (Mestrado em Ciências de Alimentos). Faculdade de Engenharia de Alimentos, Universidade Estadual de Campinas, Campinas.

LOPES, R.M.; OLIVEIRA, T.T.; NAGEM, T.J.; SILVA, P.A. Flavonóides. *Rev. Biotecnol. Ciênc. Desenvolv.*, 17, 18-22, 2000.

LYND, L.R.; WEIMER, P.J.; van ZY, W.H.; PRETORIUS, I.S. Microbial Cellulose Utilization: Fundamentals and Biotechnology. *Microbiol. Molec. Biolog. Reviews,* Sept., 506–577, 2002.

MAHESHWARI, R.; BHARADWAJ, G.; BHAT, M.K. Thermophilic fungi: physiology and enzymes. *Microbiol. Mol. Biol.* 64, 461–88, 2000.

MAICAS, S.; MATEO J.J. Hydrolysis of terpenyl glycosides in grape juice and other fruit juices: a review. *Appl. Microbiol. Biotechnol.*, 67, 322-335, 2005.

MARTINS, E. Produção de enzimas pectinolíticas pelo fungo termofílico *Thermoascus aurantiacus* através de fermentação em estado sólido utilizando resíduos agro-industriais 2003. Dissertação (Mestrado em Ciências e Tecnologia de Alimentos). Instituto de Biociências, Letras e Ciências Exatas, Universidade Estadual Paulista, São José do Rio Preto.

MATSUDA, S.; NORIMOTO, F.; MATSUMOTO, Y.; OHBA, R.; TERAMOTO, Y.; OHTA, N.; VEDA, S. Solubilization of a novel isoflavone glycoside hydrolyzing β-glucosidase from *Lactobacillus casei* subsp. *rhamnosus*. *J. Ferment. Bioeng.*, 77, 439–441, 1994.

MISHRA, C.; RAO, M. Production of xylanolytic enzymes in association with the cellulytic activities of Penicillium funiculosum. *Enz. Microbiol. Technol.,* 7, 295-299, 1985.

OTIENO, D.O.; SHAH, N.P. Endogenous β-glucosidase and β-galactosidase activities from selected probiotic micro-organisms and their role in isoflavone biotransformation in soymilk. *J. Appl. Microbiol.*, 103, 910-917, 2007.

PALMA-FERNANDEZ, E.R.D.; GOMES, E.; DaSILVA, R. Purification and Characterization of two β-Glucosidases from thermophilic fungus *Thermoascus aurantiacus* Miehe. *Folia microbiol.,* 47, 685-690, 2002.

PANDEY, A. Recent developments in solid state fermentation. *Process Biochem.*, 27, 109-117, 1992.

PANDEY, A. Solid – state fermentation. *Biochem. Engineer. J.,* 13, 81-84, 2003.

PARK, Y.K.; AGUIAR, C.L.; ALENCAR, S.M.; SCAMPARINI, A.R.P. Biotransformação de isoflavonas de soja. *Biotecnol. Ciênc. Desenvolv*., 20, 12-14, 2001.

PARRY, N.J.; BEEVER, D.E.; OWEN, E.; VANDENBERGHE, I.; VAN BEEUMEN, J. Biochemical characterization and mechanism of action of a thermostable β-glucosidase purified from *Thermoascus aurantiacus*. *Biochem. J.*, 353, 117-127, 2001.

J. PÉREZ, J.; MUNÕZ-DORADO, J.; RUBIA, T.; MARTÍNEZ, J. Biodegradation and biological treatments of cellulose, hemicellulose and lignin: an overview. *Int. Microbiol.*, 5, 53–63, 2002.

PETERSON, G; BARNES, S. Genistein and biochanin A inhibit the growth of human prostate cancer cells but not epidermal factor receptor tyrosine autophosphorylation. *Prostate*. 22, 335–345, 1993.

RAFII, F.; DAVIS, C.; PARK, M.; HEINZE, T.M.; BEGER, R.D. Variations in metabolism of the soy isoflavonoid daidzein by human intestinal microfloras from different individuals. *Arch. Microbiol.,* 180, 11-16, 2003.

RASHID, M.H.; SIDDIQUI, K.S. Thermodynamic and kinetic study of stability of the native and chemically modified β-glucosidase from *Aspergillus niger*. *Process Biochem.* 33, 109–115, 1998.

RIDGWAY, T., TUCKER, G., AND WISEMAN, H. Novel bioconversions for the production of designer antioxidant and colourant flavonoides using polyphenol oxidases. *Biotechnol. Genet. Eng. Rev.,* 14, 165-190, 1997.

RIEDL, K.M.; ZHANG, Y.C.; SCHWARTZ, S.J.; VODOVOTZ, Y. Optimizing dough proofing conditions to enhance isoflavone aglycones in soy bread. *J. Agric. Food Chem.,* 53, 8253-8258, 2005.

ROMERO, M. D., AGUADO, J., GONZALES, L., and LADERO, M. Cellulase production by *Neurospora crassa* on wheat straw. *Enz. Microbial. Technol.,* 25, 244–250, 1999.

SAHA, B.C.; FREER, S.N.; BOTHAST, R.J. Production, purification and properties of thermostable β-glucosidases from a color variant strain of *Aureobasidium pullulans*. *Appl Environ Microbiol.,* 60, 3774–3780, 1994.

SAID, S.; PIETRO, R. Aplicação industrial das hidrolases. *In: Enzimas de interesse industrial e biotecnológico*, Teresópolis, Eventos, cap. 5, 53-77, 2002.

SPAGNA, G.; BARBAGALLO, R.N.; PALMERI, R.; RESTUCCIA, C.; GIUDICI, P. Properties of endogenous β-glucosidase of *Saccharomyces cerevisiae* strain isolated from Sicilian musts and wines. *Enz. Microbial Technol.,* 31, 1030-1035, 2002.

TOMAZIC, S.J; KLIBANOV, A.M. Mechanisms of irreversible thermal inactivation of *Bacillus* α-amylases. *J Biol Chem*., 263, 3086–3091, 1988.

TRAON-MASSON, M.P.; PELLERIN, P. Purification and characterization of two β-D-glucosidases an *Aspergillus niger* enzyme preparation: affinity and specificity toward glucosylated compounds characteristic of the processing of fruits. *Enz. Microbial Technol.,* 22, 374-382, 1998.

VILLENA, M.A.; IRANZO, J.F.U.; GUNDLLAPALLI, S.B.; OTERO, R.R.C.; PÉREZ, A.I.B. Characterization of an exocellular β-glucosidase from *Debaryomyces pseudopolymorphus*. *Enz. Microbial Technol.*, 39, 229-234, 2006.

VILLENA, M.A.; IRANZO, J.F.U.; PÉREZ, A.I.B. β-Glucosidase activity in wine yeasts: Application in enology. *Enz. Microbial Technol.,* 40, 420–425, 2007.

YOSHIOKA, H.; ANRAKU, S.I.; HAYASHIDA, S. Production and Purification of a Novel type of CMCase from *Humicola grisea* var. *thermoidea* YH-78. *Agric. Biol. Chem.,* 46, 75-82, 1982.

ZANOELO, F.F.; POLIZELI, M.L.T.M.; TERENZI, H.F.; JORGE, J.A. β-Glucosidase activity from thermophilic fungus *Scytalidium thermophilum* is stimulated by glucose and xylose. *FEMS Microbiol Lett.*, 240, 137–143, 2004.

ZHANG, Y.H.P.; LYND, L.R. Toward an aggregated understanding of enzymatic hydrolysis of cellulose: Noncomplexed cellulase systems. *Biotechnol. Bioeng.*, 88, 797, 2004.

In: Cellulase: Types and Action, Mechanism, and Uses
Editor: Adam E. Golan
ISBN: 978-1-61761-983-0

Chapter 7

CELLULASES USES OR APPLICATIONS

***Yehia A.-G.Mahmoud**[*1] **and Tarek M. Mohamed**[2]*

Botany Department[1] ,Biochmistry Division, Chemistry Department [2], Tanta University, Faculty of Science Tanta 31527, Egypt

ABSTRACT

Cellulases are key industrial enzymes used to breakdown agriculture biomass to fermentable sugars. Cellulase has been used on the market as an industrial enzyme preparation and used as a main component of various products, such as detergents, fiber treating agents, paper pulp, additives for feed, and digestants. Cellulase is also used for commercial food processing in coffee. It performs hydrolysis of cellulose during drying of beans. Due to increasing environmental concerns and constraints being imposed on textile industry, cellulase treatment of cotton fabrics is an environmentally friendly way of improving the property of the fabrics. Furthermore, Cellulases are being used also in textiles for removing excess dye from denim fabric in pre-faded blue jeans (biostoning), also in removing the microfibirle which stick out from cotton fabrics after several washing. Restoring the softness and color brightness of cotton fabric could be achieved by using the cellulases. Cellulases can be used as a supplement in animal feed to decrease the production of fecal waste by increasing the digestibility of the feed. Cellulases can also be used to increase the efficiency of alcoholic fermentations (e.g., in beer brewing) by converting undigestible biomass into fermentable sugars. Ethanol is an alcohol made by fermenting and distilling simple sugars. As a result, ethanol can be produced from any biological feedstock that contains appreciable amounts of sugar or materials that can be converted into sugar such as starch or cellulose. Biofuels are liquid fuels produced from agriculture biomass using cellulases and other different enzymes. Agriculture biomass is available on a renewable or recurring basis, including agricultural crops and trees, wood and wood wastes and residues, plants (including aquatic plants), grasses, residues, fibers, and animal wastes, municipal wastes, and other waste materials. Biofuels (Types of biofuels include ethanol, biodiesel, methanol, and reformulated gasoline components) are

[*] Corresponding author: E-mail: yehiam2001@yahoo.com Fax:20403350804, Al-Baha University, Faculty of Science, Biology department, Biotechnology Unit, Al-Baha, Saudi Arabia

primarily used as transportation fuels for cars, trucks, buses, airplanes, and trains. As a result, their principal competitors are gasoline and diesel fuel. Cellulase produced by the organisms isolated from Rumen Fluid of Cattle was used for biopolishing.

Cellulases are used also, in removing microbial slime in slime covered surfaces and maintaining a slime-free surface as in exposed cooling tower surfaces and in waste water treatment and paper making. This method comprises utilizing an enzyme blend in 2 to 100 parts per million (ppm) of cellulase, α-amylase and protease. Such enzyme blends have been found specifically to digest microbial slime and reduce microbial attachment and biofilm.

KeyWords: Cellulose, Cellulases, Fungi, Bacteria, fermentation; Industrial applications and wastes management

CELLULASES USES OR APPLICATIONS

Cellulose is earth's major biopolymer and is of tremendous economic importance around the globe. Cellulose is the major constituent of raw materials like cotton (over 94%) and wood (over 50%). Cellulose is a long-chain polymer polysaccharide carbohydrate, of beta-glucose. It forms the primary structural component of plants and is not digestible by humans. Cellulose is the principal constituent of the cell wall of plants. Cellulose from major land plants as forest trees and cotton is assembled from glucose, which is produced in the living plant cell from photosynthesis. In the oceans, however, unicellular plankton produces most cellulose or algae using the same type of carbon dioxide fixation found in photosynthesis of land plants. It is estimated that the amount of carbon assimilated by plants throughout the year is about 200 billion tones. Plants in the form of structural polysaccharides, which human beings cannot degrade, store most of this energy. Cellulose has great commercial importance. Hence the study of cellulose, the enzyme essential for cellulose hydrolysis is important. There are many on going research on various aspects of cellulose.

Cellulases are a group of enzymes catalyzing an enzymatic reaction system in which cellulose is decomposed into glucose, cellobiose or cello-oligosaccharides. Successful utilization of cellulosic materials as renewable carbon sources is dependent on the development of economically feasible process technologies for cellulase production, and for the enzymatic hydrolysis of cellulosic materials to low molecular weight products such as hexoses and pentoses.

A cellulosic enzyme system consists of three major components: endo-ß-glucanase (EC 3.2.1.4), exo-ß-glucanase (EC 3.2.1.91) and ß-glucosidase (EC 3.2.1.21). The mechanism of action of each enzyme is as following:

a) Endo-ß -glucanase, 1, 4-ß-D-glucan glucanohydrolase, CMCase, Cx: "random" scission of cellulose chains yielding glucose and cello-oligosaccharides.
b) Exo- ß -glucanase, 1, 4-ß - D-glucan cellobiohydrolase, Avicelase, C1: exo-attack on the non-reducing end of cellulase with cellobiose as the primary structure.
c) ß-glucosidase, cellobiase: hydrolysis of cellobiose to glucose.

Reese *et al.* (1950) proposed that exo-ß-glucanase causes a disruption in cellulose hydrogen bonding, followed by hydrolysis of the accessible cellulose with endo-ß-gucanase. According to their hypothesis, in a synergistic sequence of events, endo-ß-glucanase acts randomly on the cellulose chain, while exo-ß-glucanase acts on exposed chain ends by splitting off cellobiose or glucose. Cellobiose is subsequently hydrolyzed by ß -glucosidase to glucose. This hypothesis is however the opposite of that proposed earlier by some researchers, and indicates that three, rather that two enzymes are essential for the decomposition of cellulosic biomass (Fig.1).

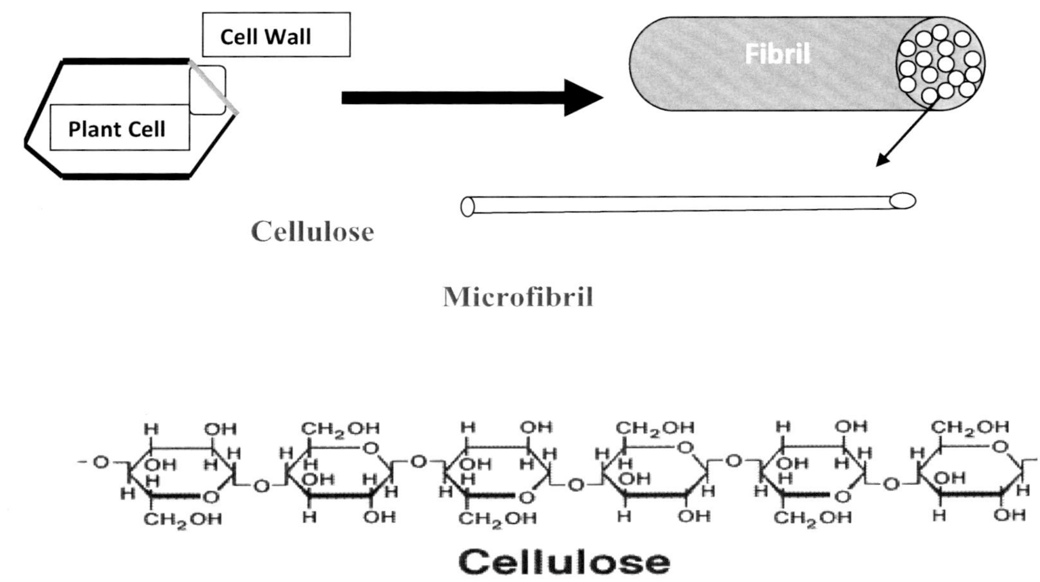

Figure 1. Plant cell wall fibrils.

Cellulases are key industrial enzymes used to breakdown biomass to fermentable sugars.Cellulase has great commercial importance. Cellulases now become in the third largest industrial enzyme worldwide, by dollar volume, because of their different uses in cotton processing, paper recycling, as detergent enzymes, in juice extraction, and as animal feed additives. Furthermore, celullases might become the largest volume industrial enzyme, if ethanol, butanol, or some other fermentation product of sugars, produced from biomass by enzymes, becomes a major transportation fuel. Currently, industrial cellulases are almost all produced from aerobic cellulolytic fungi, such as *Hypocrea jecorina* (*Trichoderma reesei*) or *Humicola insolens*. This is due to the ability of engineered strains of these organisms to produce extremely large amounts of crude cellulases might reach over 100 g per liter (Schulein 1998).

The basic idea behind the research is to enhance cellulase production from the microorganisms so that it could be used to finally degrade the huge amount of cellulosic biomass, which is widely available. Microorganisms bring about most of the cellulose degradation occurring in nature. They meet this challenge with the aid of a multi-enzyme system. They include fungi and bacteria, aerobes and anaerobes, mesophiles and thermophiles and occupy a variety of habitats. Aerobic bacteria produced numerous individual, extracellular enzymes with binding modules for different cellulose conformations. Anaerobic bacteria possess a unique extracellular multienzyme complex, called cellulosome. Binding to a non-catalytic structural protein (scaffoldin) stimulates activity of the single components towards the crystalline substrate.

Microorganisms Producing Cellulases

Cellulolytic microbes are primary carbohydrate degraders and are generally unable to use proteins or lipids as energy sources for growth (Lynd, et al.2002). Cellulolytic microbes notably the bacteria *Cellulomonas* and *Cyophaga* and most fungi can utilize a variety of carbohydrates in addition to cellulase (Poulsen and Petersen, 1988; Rajoka and Malik, 1997), which the anaerobic cellulolytic species have a restricted carbohydrate range, limited to cellulose and or its hydrolytic products (Rajoka and Malik, 1997; Ng and Zeikus, 1982 and Thurston, et al. 1993).

Table (1). Cellulolytic Fungi detected by the plate screening method and their relative activity as determined on CMC medium (Pečiulytė, 2007).

Fungal species	Ratio to total number of cellulolytic* (%)	Number of strains tested on CMC medium	(I_{CMC})
Acremonium strictum W. Gams	0.8	6	1,32 ± 0.08
Arthrobotrys oligospora Fresen.	0.2	2	0,42 ± 0,03
Aspergillus fumigatus Fresen	0.6	4	0.73 ± 0,04
Aspergillus niger var. *niger* Tiegh.	2.1	8	2.21 ± 0,97
Aspergillus repens (Corda) Sacc.	4.7	9	1,54 ± 0,54
Botyrotrichum piluliferum Sacc. & Marchal	0.9	2	0,97 ± 0,11
Cladosporium cladosporioides (Fresen.) G.A. de Vries	1.5	5	0,97 ± 0,07
Chaetomium globosum Kunze	0.5	2	0.88 ± 0.04
Geomyces pannorum (Link) Singler & J.W. Carmich. (= *Chrysosporium pannorum* (Link) S. Hughes)	1.3	6	1.29 ± 0.31
Epicoccum nigrum Link (= *E. purpurescens* Ehrenb.)	0.2	3	0.89 ± 0.45
Fusarium oxysporum Schltdt.	2.1	2	1.31 ± 0.27
Geotrichum candidum Link	0.5	1	0.83 ± 0.14
Gliocladium roseum Bainier (= *Clonostachys rosea* f. *rosea* (Link) Schroers, Samuels, Seifert & W. Gams	1.4	3	1.37 ± 0.25
Gliomastix murorum var. *murorum* (Corda) S. Hughes) (= *Acremonium murorum* (Corda) W. Gams)	3.6	8	2.07 ± 0.85
Lecanicillium lecanii (Zimm.) Zare & W. Gams (= *V. lecanii* (Zimm.) Viégas)	0.7	4	0.66 ± 0.03
Mucor racemosus Fresen.	0.1	1	0.28 ± 0.02
Myrothecium verrucaria (Alb. & Schwen.) Ditmar	1.1	2	1.04 ± 0.26
Paecilomyces lilacinus (Thom) Samson	1.9	3	1.62 ± 0.75
Penicillium brevicompactum Dierckx	16.2	11	1.31 ± 0.14
Penicillium fellutanum Biourge	3.5	4	0.54 ± 0.22
Penicillium funiculosum Thom	21.1	12	1.87 ± 0.13
Penicillium purpurogenum Stoll	5.9	5	1.45 ± 0.21
Penicillium simplicissimum (Oudem.) Thom	15.7	13	1.11 ± 0.04
Trichoderma harzianum Rifai	4.3	7	1.00 ± 0.12
Trichoderma longibrachiatum Rifai	1.6	4	0.98 ± 0.04
Stachybotrys chartarum (Ehrenb.) S. Hughes	7.5	6	1.85 ± 0.23

Celluloytic on the medium with the phosphoric acid swollen cellulose

The ability secrete large amounts of extracellular proteins is characteric of certain fungi and such strains are most suited for production of higher levels of extracellular cellulases. One of the most extensively studied fungi is Trichoderma reesei, which converts native as well as derived cellulose to glucose. Most commonly studied cellulolytic organisms include: fungal species-*Trichoderma*, *Humicola, Penicillium, Aspergillus*; Bacteria-Bacilli, *Pseudomonads*, *Cellulomonase*; and Actinomycetes-Streptomyces, *Actinomucor* and *Streptomyces.*

While several fungi can metabolize cellulose as an energy source, only few strains are capable of secreting a complex of cellulase enzymes which could have practical application in the enzymatic hydrolysis of cellulose (Table 1). Besides *T. Reesei*, other fungi like *Humicola, Penicillium* and *Aspergillus* have the ability to yield high levels of extracellular cellulases (Hayashida, et al. 1988; Chaabouni, et al.1995; Schulein 1997; van-Den Broeck, et al.2001 and Jorgensen, et al.2003). Anaerobic bacteria such as Cellulomonase, Cellovibrio and Cytophaga are capable of cellulose degradation in pure culture (Lynd, et al. 2002). However, the microbes commercially exploited for cellulase preparation are most limited to *T.reesei, H.insolens*, *A.niger*, *Thermomonospora fusca*, *Bacillus* sp and a few other organisms

The most complex and best investigated cellulosome is that of the thermophilic bacterium *Clostridium thermocellum* (Aubert et al., 1987). Cellulase is produced by the fermentation of a nonpathogenic, nontoxicogenic strain of fungi *Trichoderma viride*, which is capable of decomposing cellulosic polysaccharides into smaller fragments, primarily glucose. Cellulase is also derived from *Aspergillus niger (REF).*

Cellulases have a wide range of applications. The main potential applications are in food, animal feed, textiles, fuel and chemical industries (Coughlan, 1985; Mandels, 1985; Beguin and Aubert, 1994; Bhat and Baht, 1997; Ganga et al. 1999).

1. Cellulases Applications

1.1 In Food Industry

The concept of organic foods is getting popularity these days. Chemical additives have shown ill effects on human health. Food processing employing biochemicals can compensate the problem as these are easily denatured by heat treatment therefore, render the foods chemical free and safer. The produced carboxymethyl cellulase was used in bread making process that showed significant effects on farinographic and mixographic parameters. CMCase has a predominant role in bread industry due to its imperative role in pentosans hydrolysis. No doubt, pentosans are minor component of wheat flour, but play an important role in dough rheology and bread quality due to their high water binding capacity. Soluble pentasans are implicated in the dough elasticity and the hydrolysis of the insoluble pentosans promotes changes.

Other applications of cellulases in food processing are including the extraction of fruit juices and oil from seeds, in the clarification of fruit juices, in the removing of external soybean coat, in production of fermented soybean foods such as soy sauce and miso, in the isolation of proteins from soybean and coconut, and in the extraction of agar from seaweeds.

Also, cellulases used to increase the soaking efficiency and homogenous water absorption of cereals, used to efficiently isolate starch from corn and sweet potato, used in the in gelatinization of seaweeds to increase digestibility vegetables and soup mixtures. Furthermore, cellulases used to digest ball-milled lignocelluloses which can be used as food additive (Coughlan, 1985; Mandels, 1985; Beguin and Aubert, 1994; Bhat and Bhat, 1997).

Additionally, cellulases also are used, in improving the nutritive quality of fermented foods, in improving rehydrability of dried vegetables and soup mixtures, in the production of cello-oligosaccharides, glucose and other soluble sugar from cellulose wastes and used also in the removing of plant cell wall which will facilitate the release of flavors, enzymes polysaccharide and proteins (Mendels, 1985)

2. In Brewing and Wine Industries

Cellulases are used to hydrolyze β-1, 3 and β-1, 4 glucan which is present in low grade barley, also, it help in filtration of beer. Cellulases are using to increase the aroma in wines. The recombinant yeast is producing β-1, 3 and β-1, 4-gluconases which used in brewing industry (Beguin and Aubert, 1994; Ganga et al. 1999).

3. In Animal Feed Industry

Mammals normally do not have this enzyme but ruminants often carry bacteria which have cellulase and can allow them to utilize cellulose as a carbohydrate source. Cellulases are used as supplement in feed for ruminant and monogasteric animals to decrease the production of fecal waste by increasing the digestibility of the feed. Also, it is used in pretreating lignocellulosic material, dehulling of cereal grains, treating silage to improve the digestibility of ruminanta and monogasteric animals (Mandel, 1985).

Currently, there is a great deal of interest in using enzyme preparations containing high levels of cellulase and hemicellulase activities for improving the feed utilization, milk yield and body weight gain by ruminants. Nevertheless, the successful use of these enzymes in ruminant diet depends on: (1) their stability in the feed (during and after processing) and in the rumen; (2) the ability of enzyme components to hydrolyse plant cell wall polysaccharides; and (3) the ability of the animals to use the reaction products efficiently. Therefore, the enzyme preparations should be characterized by *in vitro* and *in vivo* experiments and should contain essential enzyme activities for different applications in order to guarantee success.

The forage diet of ruminants, which contains cellulose, hemicellulose, pectin and lignin, is more complex than the cereal-based diet of poultry. Enzyme preparations containing high levels of cellulase, hemicellulase and pectinase have been used to improve the nutritive quality of forages (Graham and Balnave, 1995; Kung et al., 199; Lewis et al., 1996). Nevertheless, the results with the addition of enzyme preparations containing cellulase, hemicellulase and pectinase to ruminant diet are somewhat inconsistent. Several studies have shown substantial improvements in feed digestibility and animal performance (Burroughs *et al.*, 1960; Rust *et al.*, 1965), while some researchers reported either negative effects or none at all (Perry *et al.*, 1966; Theurer et al., 1963). Furthermore, Beauchemin *et al.* (1995) reported that the addition of commercial enzyme preparations containing cellulase and xylanase to hay diet increased the live weight gain of cattle by as much as 35%. Similarly, a 5–25% increase in milk yield has been reported in the case of dairy cows fed with forage treated with commercial fibrolytic enzymes (Lewis et al., 1996; Stokes and Zheng, 1995). Thus, the overall success in improving the fiber digestion and ruminant performance may be limited. This could mainly be due to the presence of hydrophobic cuticle, lignin and its close association with cell wall polysaccharides and the nature of lignocellulose, which prevents the efficient utilization of fibre in the rumen. Hence, considerable basic and applied research effort, together with improved enzymes, will be needed to enhance fibre digestion by ruminants and thus, their performance.

Attempts have also been made to clone the cellulases gene in transgenic animals which could secrete the required cellulases into the gastrointestinal tract of the animal and help in the digestion of roughage efficiently (Ali *et al.*, 1995; Hall *et al.*, 1993; Beguin and Aubert, 1994). Initially the cellulase produced by the organisms isolated from Rumen Fluid of Cattle was used for its characterization and further the same was used to test its different applications (Muthukrishnan, 2007).

4. In Textile Industry, Leather and Laundry Industries

4.1. In Textile Industry

Celluloytic enzymes are currently being used as means of finishing cellulose fabrics such as rayon, linen and cotton. Using the knowledge of the mode of action of endoglucanases and cellobiohydrolases, products for industrial applications that have controlled and predictable activities have been developed. These products are being used to " stone wash" denim, and biofinish other textiles to prevent fuzzing and pilling, increase smoothness and softness, and increase color brightness as mentioned by many researchers.

Seed coat fragments remaining in the cotton after bleaching spoil the aesthetic effect of the fabric during dyeing and printing, because they differ in color and morphology from the cellulose. Actually, seed coat fragments are part of the outer layer of cotton seed formed from mature or immature seeds during mechanical processing. They are usually black or black brown and may or may not have fibers and linear attached (American Society for Testing and Materials, 1985). To eliminate seed coat fragments, more concentrated chemical solutions and longer steaming are needed than when other impurities removing from row cotton. As an additional disadvantage, more drastic methods also increase the hazard of cellulose degradation (Csiszar *et al.*, 1998).

Cellulases seed-coat treatments apply before the alkaline scouring has two effects; the first is that the enzyme partially degrades the lignocellulosic structure of seed-coat fragments, producing a measurable weight loss. Consequently it enhances the penetration of the alkaline scouring solution and increases the alkaline degradation of seed coat fragments of the untreated seed-coat fragments. The materials remaining after enzymatic treatment losses an additional weight during subsequent alkaline scouring. Thus, the total weight loss, based upon initial seed-coat fragment is 78-86%. Second, the optimized double-stage pretreatment significantly reduces the seed coat fragment contents of cotton fabric. Cellulases hydrolyze the tiny fibers that attached seed-coat fragments to fabrics, and the free-floating fragments can then be removed by filtration. Thus, cellulases pretreatment before alkaline scouring increases both the degradation of seed coat fragments and their removal from desized cotton fabrics (Csiszar *et al.* 1998).

Modern fabrics are reinforced with an adhesive called "the size" before weaving. This prevents breaking of the warp threads. Starch is the most common ingredient of the size, combined with lower quantities of other materials such as gelatin and carboxymethyl cellulose (cellulose that has been treated to make it water soluble). Harsher chemicals such as alkalis or acids have been traditionally used for desizing, but are now being replaced by enzymes. Apart from the environmental benefit of replacing harsh chemicals with biodegradable enzymes, have turned out to be less harsh on the main fabric as well. Most desizing preparations declare Alpha-amylase activity but will contain protease and cellulase as well.

In the modern textile industry enzymes are used increasingly in the finishing of fabrics and clothes. The main component of cotton and other natural fibres is cellulose. Whilst most of the fibres are arranged as long, straight chains some small fibers can protrude from the yarn or fabric. The correct application of a cellulase enzyme can remove these rough protuberances giving a smoother, glossier brighter colored fabric. This technique has become known as biopolishing and results in not only a softer fabric but also improved color brightness. The same process has recently been adapted and included in some laundry detergents.

Enzymes have been used in the textiles industry since the turn of the century to remove starchy and waxy residues from raw materials and to give fabric a uniform finish. Global sales of enzymes used in the textiles industry reached $164.2 million in 1998 and $182.7 million in 2002. Genencor is one of the major producers of industrial enzymes.

The finishing of denim jeans has also become a popular application for cellulases in the textile industry. Denim was stonewashed traditionally, with pumice stones to fade the surface of the garment. A small application of cellulase can replace many of the stones resulting in less damage to the garments and machinery. This technique has become known as Biostoning and can result in much greater fading without high abrasive damage both to the actual fabric and any other accessories (buttons, rivets) on the fabric. Stonewashing enzymes are usually available as either "acid" cellulases (optimum activity around pH 4.5) and "neutral" cellulases (optimum activity at just below pH 7.0) (Bertabet and Btra, (2000).

Adding pumice stones gives the additional effect of a faded or worn look. The pumice abrades the surface of the jeans like sandpaper, removing some dye particles from the surfaces of the yarn. Pumice has been used since the introduction of stone washed jeans in the early 1980s. However, stone washing with pumice has some severe drawbacks. The quality of the abrasion process is difficult to control: Too little will not give the desired look. Too much can damage the fabric, particularly at the hems and waistbands. The outcome of a load of jeans is never uniform, with a significant percentage always getting ruined by too much abrasion. The process is also non-selective. Everything in the washing machines gets abraded, including the metal buttons and rivets on the jeans as well as the drum of the washing machine. This substantially reduces the quality of the products and the life of the equipment, and increases production costs. Acid washing jeans avoided some of these problems, but came with added dangers, expenses, and pollution. Environmental regulations have put intense pressure on the textiles industry to generate less pollution. Treating the wastewater and disposing of the sludge (i.e. used pumice or neutralized acid) represents a growing portion of the production costs for a pair of jeans. So what are we to do? Will we have to give up our comfortable jeans? Will prices skyrocket? Will we have to choose between fashion and the environment?

Never fear! Biology is here! A technique known as "biostoning" was introduced in Europe in 1989 and then quickly adopted in the US the following year. Biostoning is by Waste, pollution, quality variability, and imperfections are all reduced, and unlike pumice or acid, which get used up during the wash, enzymes can be recycled. Biostoning relies on the action of enzymes to selectively modify the fabric surface.

In the early days, one problem with biostoning was "back staining." that happens when loosened dye particles redeposit onto the back surface of the fabric, causing discoloration. A reddening of the dyes sometimes occurred too. Maintaining the pH of the wash load between 6-8 has successfully controlled both problems. Biostoning can achieve the same effect as traditional stone washing, but without the damaging abrasion of the fabric and equipment (Betrabet and Btra, 2000).

A small dose of enzymes can replace several dozen pounds of pumice stones. So productivity can be increased by 30-50% because the room formerly taken up by the pumice stones in the washing machines can now be filled with more jeans. Now, there is no need for the time-consuming and expensive task of removing stone fragments from the jeans once the wash is done. Also, there is also no pumice dust to endanger employee health or gritty sediment to clog drains.

Enzymes have been used in the leather industry for many years and more recently have been introduced into modern textile industries. The main applications of enzymes in the leather industry are proteases which help in the dehairing of the animal hides and lipases are used for degreasing.

4.2. Use of Cellulase in Laundry

The cellulase preparations capable of modifying the structure of cellulose fibrils are added to laundry detergents to improve the color brightness, hand feel and dirt removal from cotton and cotton blend garments. Most cotton or cotton blend garments, during repeated washings, tend to become fluffy and dull. This is mainly due to the presence of partially detached microfibrils on the surface of garments that can be removed by cellulases in order to restore a smooth surface and original color to the garment. Also, the degradation of microfibrils by cellulase, softens the garment and removes dirt particles trapped in the microfibril network. This is currently accomplished by adding a commercial cellulase preparation from *H. insolens*, active under mild alkaline conditions (pH 8.5–9.0), and at temperatures over 50°C in washing powders (Uhlig, 1998). Although, the amount of cellulase added represents approximately 0.4% of the total detergent cost, it is considered rather expensive and hence, alternative cellulase preparations are required to attract the worldwide laundry market.

5. Agriculture-Based Biofuels

Biomass Research and Development defines biomass as "any organic matter that is available on a renewable or recurring basis, including agricultural crops and trees, wood and wood wastes and residues, plants (including aquatic plants), grasses, residues, fibers, and animal wastes, municipal wastes, and other waste materials." In recent years, environmental pollution has become a global problem. Enormous growth of industrial activities has given rise to problems such as global warming, desertification, and acid deposition. These global problems are rooted in the materially rich lifestyles, which are supported by abundant and wasteful use of fossil fuels in industrialized countries. Rapidly increasing industrial activities in China, India, and in other developing countries implicates that these practices will inevitably contribute to deterioration of the global environment and to destruction of the global ecosystem. Changes in our key industrial systems are required in order to minimize the impact of environmental pollution. The recycling of materials, and thus minimizing the generation of waste, is a basic concept which must be implemented in order to meet the new demands of development in both industrialized and developing countries.

Cellulosic biomass offers a possible solution. It is a complex mixture of carbohydrate polymers known as cellulose, hemi-cellulose, lignin, and a small amount of compounds known as extractives. Examples of cellulosic biomass include agricultural and forestry residues, municipal solid waste, herbaceous and woody plants, and underused standing forests. Cellulose is composed of glucose molecules bonded together in long chains that form a crystalline structure. Cellulose is a fibrous, tough, water-insoluble substance. Hemi-cellulose is not soluble in water. It is a mixture of polymers made up from xylose, mannose, galactose, or arabinose. Hemi-cellulose is much less stable than cellulose. Lignin, which is present along with cellulose in trees, is a complex aromatic polymer of phenylpropane building blocks. Lignin is resistant to biological degradation (Bernfeld, 1955)

Biofuels are liquid fuels produced from biomass. Types of biofuels include ethanol, biodiesel, methanol, and reformulated gasoline components. The biofuels are primarily used as transportation fuels for cars, trucks, buses, airplanes and trains. As a result, their principal competitors are gasoline and diesel fuel. Unlike fossil fuels, which have a fixed resource base that declines with use, biofuels are produced from renewable feedstocks. Furthermore, under most circumstances biofuels are more environmentally friendly (in terms of emissions of toxins, volatile organic compounds, and greenhouse gases) than petroleum products.

Supporters of biofuels emphasize that biofuel plants generate value-added economic activity that increases demand for local feedstocks, which raises commodity prices, farm incomes, and rural employment.

Ethanol, or ethyl alcohol, is an alcohol made by fermenting and distilling simple sugars. As a result, ethanol can be produced from any biological feedstock that contains appreciable amounts of sugar or materials that can be converted into sugar such as starch or cellulose. Sugar beets and sugar cane are examples of feedstocks that contain sugar. Corn contains starch that can relatively easily be converted into sugar. In the United States corn is the principal ingredient used in the production of ethanol; in Brazil (traditionally the world's largest ethanol producer), sugar cane is the primary feedstock. A significant percentage of trees and grasses are made up of cellulose which can also be converted to sugar, although with more difficulty than required to convert starch. In order to produce ethanol from lignocellulosic materials, the bundels of lignocelluloses should be open in order to access the polymer chains of cellulose and hemicelluloses by a process of so called pretreatment, then polymers should be hydrolyzed in order to achieve monomer sugar solutions, which could be fermented by microorganisms to ethanol solution (Mash) and finally ethanol could be purified from mash by distillation and dehydration in the following figure. By product recovery, utilities (steam and electricity generation and cooling water), wastewater treatment, and eventually enzyme production are the other units which are demanded in ethanol production from lignocellulosic materials. The processes of the cellulosic materials go through the following stages to produce the fuel (Fig.2):

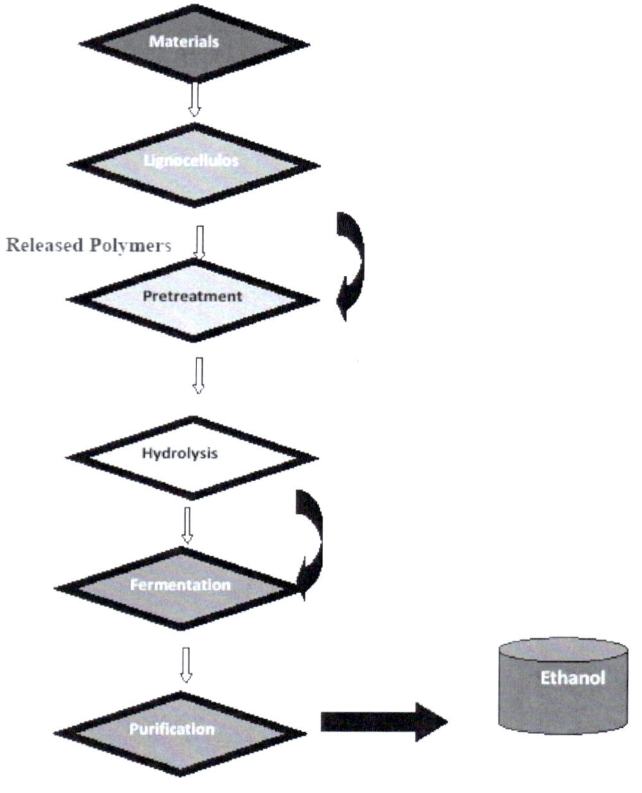

Figure 2. Ethanol production stages.

Most processes are designed to produce ethanol, as the technology for fermenting glucose to ethanol is very robust and high concentrations of ethanol can be achieved (15%). However, DuPont Company is trying to produce butanol from glucose the concentration is only about 2% because of its toxicity (David, 2009). There are two approaches being studied for producing liquid fuels from biomass with cellulases: in one, separate hydrolysis and fermentation, plant cell wall polymers are hydrolyzed by free enzymes in one step and the resulting sugars are fermented in the second step while in the other, called consolidated bioprocessing, the biomass will be converted into biofuel in a single step by either using an anaerobic bacterium that can hydrolyze plant cell walls and ferment the resulting sugars to ethanol or by simply combining the two SHF steps in a single vessel (Lynd et al. 2008).

Method of attacking and removing microbial slime in slime covered surfaces and maintaining a slime-free surface as in exposed cooling tower surfaces and in waste water treatment and paper making. This method comprises utilizing an enzyme blend in 2 to 100 parts per million (ppm) of cellulase, alpha-amylase and protease. Such enzyme blends have been found specifically to digest microbial slime and reduce microbial attachment and biofilm. A specific combination of polysaccharide degrading enzymes is a ratio of 2 parts cellulase to 1 alpha-amylase to 1 protease utilized in 2-100 parts per million. Broadly, the alpha-amylase must be at least 1 and the protease may vary from 0.5 to 1 part.

6. Cellulases and Hemicellulases in Pulp and Paper Biotechnology

Cellulases and hemicellulases have been used in the pulp and paper industry for different purposes. Commercial enzyme preparations contain various enzyme activities, where some may be vital, while others may be detrimental for a specific application. Therefore, enzyme mixtures or purified enzymes should be well characterized with respect to their substrate specificity and mode of action before using for a particular application in pulp and paper industry.

6.1. Bio-Mechanical Pulping

The mechanical pulping processes such as refining and grinding of the woody raw material lead to pulps with high content of fines, bulk and stiffness. Although these fibres are useful for producing different grades of papers, the main disadvantage of mechanical pulping is high-energy consumption. Bio-mechanical pulping using white-rot fungi resulted in substantial energy savings during refining, and improvements in hand-sheet strength properties (Akhtar, 1994; Leatham *et al.*, 1990). Unfortunately, these encouraging laboratory results have not yet been commercialized. Unrefined wood chips are generally less accessible to enzymatic modification; hence, the addition of an enzyme in mechanical pulping can be effective only after the initial refining.

The effects of enzymatic modification of coarse mechanical pulp using cellulase and hemicellulase from *Trichoderma* prior to secondary refining were studied (Pere *et al.*, 1996). Marginal modification of the above pulp, during secondary refining, led to an energy savings of 20% and 5% with cellobiohydrolase I and hemicellulase, respectively. Treatment with endoglucanase I from *Trichoderma*, slightly decreased the energy consumption at the expense of pulp quality, while no positive effect on energy consumption was observed with cellulase mixture. Energy consumption was reduced up to 30–40% with cellobiohydrolase I, when the refining was performed using low-intensity refiner. Use of cellobiohydrolase I also led to 10–15% energy savings during two-stage refining and resulted in increased tensile strength and high fibre qualities (Pere *et al.*, 1996).

6.2. Bio-Bleaching of Kraft Pulps

Use of hemicellulolytic enzymes was the first large-scale application of enzymes in the pulp and paper industry (Viikari *et al.*, 1986, 1987). This was based on the observation, that limited hydrolysis of hemicellulose in pulps by hemicellulases (mainly xylanases) increased the extractability of lignin from the kraft pulps and reduced the chlorine required in subsequent bleaching. Although the exact mechanism of action of xylanase in bio-bleaching is not known, it has been proposed that the xylanase either hydrolyzed the re-precipitated xylan partially or completely removed the xylan from the lignin-carbohydrate complexes. Both these processes were possible and allowed the enhanced leaching of entrapped lignin from the fibre cell wall and made the pulp more susceptible to the bleaching chemicals. The xylanase from *T. reesei* has been reported to act uniformly on all accessible surfaces of Kraft pulp and to be effective during bio-bleaching (Saake *et al.*, 1995; Suurnakki *et al.*, 1996a).

Compared to xylanase, mannanase has attracted minimal attention in bio-bleaching because of its limited action on most pulps. Also, the mechanism of mannanase-aided bleaching appears to differ from the xylanase-aided bleaching, since the distribution of glucomannan is different from xylan in pulps (Buchert *et al.*, 1992; Suurnakki *et al.*, 1996b, c). In case of mannanase-aided bleaching the composition and configuration of the outer surface of pulp fibres appear to be important (Suurnakki *et al.*, 1996c).

The role of xylanase in the de-lignifications of Kraft pulps has been extensively studied using two purified xylanases from T. reesei with different pI (5.5 and 9.0), pH optima and substrate specificities (Buchert et al., 1992; Tenkanen et al., 1992a, b). Interestingly, both xylanases performed almost in the same manner, in reducing kappa number and improving the brightness in subsequent chemical modifications (Buchert *et al.*, 1992). Purified *T. reesei* hemicellulases have also been used in bleach boosting of different types of pulps. The xylanase from T. reesei was most effective when used in conventionally cooked pulps, and the effect was more pronounced with pulps produced from northern pine than radiata pine (Suurnakki *et al.*, 1996a). Similar results have been reported with xylanases from other micro-organisms with respect to the origin of pulp and its production method (Allison et al., 1995; Nelson et al., 1995; Tolan, 1992).

It has been suggested that mannanase was most beneficial in pulp bleaching when used in combination with xylanase (Buchert et al., 1992; Suurnakki et al., 1996a), while the accessory enzymes such as b-xylosidase and a-L-arabinosidase played a minor role in xylanase aided bleaching of pulps (Kantelinen et al., 1993; Luonteri et al., 1996). Interestingly, the endoglucanase I from T. reesei has been shown to increase the bleach ability of pulps due to its xylanase activity (Buchert et al., 1994). Most of the commercial hemicellulase preparations currently used in bio-bleaching of different pulps originate from T. reesei.

6.3. Bio-Modification of Fibres

Cellulase and hemicellulase mixtures have been used for the modification of fiber properties with the aim of improving drainage, beatability and runability of the paper mills (Noe *et al.*, 1986; Pommier *et al.*, 1989, 1990). In these applications, the enzymatic treatment was performed either before or after beating of the pulps. The aim of cellulase and hemicellulase treatment prior to the refining process is either to improve the beatability response or to modify the fiber properties. The addition of cellulase and hemicellulase after beating is to improve drainage properties of pulps, which determine the speed of paper mills. A commercial cellulase/hemicellulase preparation, named Pergalase-A40, from *Trichoderma* has been used by many paper mills around the world for the production of release papers and wood-containing printing papers (Freiermuth *et al.*, 1994; Pommier *et al.*, 1990).

In the late 1980s, the possibility of improving the drainage rates of recycled fibers by cellulase, especially by endoglucanase was identified (Pommier *et al.*, 1989, 1990). This was subsequently confirmed by Kamaya (1996) using purified endoglucanases from *Trichoderma*. Both endoglucanases I and II from *Trichoderma* were equally effective in decreasing the Schoper-Riegler (SR) value of recycled soft wood craft pulp and indicated improved drainage, whilst cellobiohydrolase had no effect. Xylanase and mannanase treatment resulted in only a marginal improvement of the SR value. However, depending on the origin of the pulp, the efficiency of different enzymes might vary. For example, Pere *et al.* (1996) demonstrated the need of simultaneous solubilization of xylan and cellulose for the drainage improvement of reed cannary grass kraft pulp. By using endoglucanase I from *T. reesei*, which hydrolyzed both cellulose and xylan, Pere *et al.* (1996) showed a drainage improvement by 30%, while endoglucanase II from the same fungus, which was specific for cellulose, showed only a limited effect on the drainage property.

A detailed understanding of the action of different cellulase and hemicellulase components on different types of pulps is vital for the development of enzymatic modification of fibers. Mansfield et al. (1996) studied the action of a commercial cellulase preparation on different fractions of Douglas fir kraft pulp. They observed that the cellulase treatment decreased the defibrillation, which reduced the fibre coarseness. Also, with increasing dose of cellulase, the strength properties of fibre reduced. Pere *et al.* (1995) and Rahkamo *et al.* (1996) investigated the effect of major cellulase components from *T. reesei* on the fiber properties of unbleached soft wood kraft and dissolving pulps. They found that the cellobiohydrolases had moderate effect on fibre viscosity, while endoglucanases, especially endoglucanase II, dramatically decreased the pulp viscosity even at a low concentration. Nevertheless, cellobiohydrolase I treatment showed no effect on the hand-sheet properties even after PFI-refining, and suggested that this enzyme did not cause any structural damage to the fibres. On the other hand, endoglucanase II treatment damaged the strength properties, and indicated that this enzyme attacked cellulose fibres at sites where even low levels of hydrolysis resulted in large decrease in viscosity and led to a dramatic deterioration in the tensile index (Pere et al., 1996). Moreover, Oksanen et al. (1997) and Kamaya (1996) studied the effect of purified cellulase and hemicellulase components from *Trichoderma* on the beatability of pulp and technical properties of paper from bleached kraft pulps. Treatment of pulp with either cellobiohydrolase I or II had no effect on the development of pulp properties, whereas endoglucanase, especially endoglucanase II, improved the pulp beat ability, sheet density and other properties of the paper. Xylanase and mannanase, however, did not modify the pulp properties significantly, when less than 10% of the respective hemicellulose was hydrolyzed (Oksanen et al., 1997).

6.4. Bio-De-Inking

The application of enzymes in de-inking has been intensively studied in both laboratory and pilot scales, but the technique has not yet been commercialized (Buchert *et al.*, 1998).

The two principal approaches in using enzymes for de-inking include the (1) hydrolysis of soy-based ink carriers by lipase, and the release of ink from fibre surfaces by cellulases, xylanases and pectinases. Most applications proposed so far use cellulases and hemicellulases for the release of ink from the fibre surface by partial hydrolysis of carbohydrate molecules (Jeffries *et al.*, 1994; Prasad *et al.*, 1992, 1993). The main advantage of enzymatic deinking is the avoidance of the use of alkali. De-inking, using enzymes at acidic pH, also prevents the alkaline yellowing, simplifies the de-inking process, changes the ink particle size distribution and reduces the environmental pollution. In addition, the enzymatic de-inking improves the fibre brightness, strength properties, pulp freeness and cleanliness as well as reduces fine particles in the pulp. Xylanase treatment has been reported to increase the strength properties, while cellulase treatment improved the brightness and freeness of the pulp (Prasad et al., 1993). In fact, the enzymatic de-inking has a great potential both from commercial and environmental standpoints and expected to be commercialized in the near future.

6.5. Bio-Improvement of Drainage Properties and the Performance of Paper Mills

During mechanical pulping, various wood components such as pitch, lignin and hemicellulose are dissolved and released into the drainage. These components are collectively called 'dissolved and colloidal substances.' During peroxide bleaching of mechanical pulps, other wood components including pectin are also released. All these components often cause severe problems in paper mills including pitch depositions, specks in the paper and decreased de-watering. Enzymes, especially carbohydrases, which act on the above-mentioned colloidal substances, are expected to improve the overall performance of paper mills. Using a commercial enzyme preparation from *Trichoderma*, Kantelinen et al. (1995) demonstrated a remarkable decrease in the turbidity of thermo-mechanical pulping filtrates.

Also, the enzymatic treatment destabilized the lipophilic extractives in the filtrates and facilitated their attachment to thermo-mechanical pulping fibers. In addition, the same authors showed that the purified endoglucanase I from T. reesei was useful for disturbing the steric stability of colloidal pitch, while the xylanase from the same fungus was effective only at high concentrations.

6.6. Bio-Characterization of Pulp Fibers

Hydrolases acting on pulp fibres is useful tools for the characterization of fibers. Purified xylanase and mannanase from *T. reesei* have been successfully used for selective solubilization of xylan and glucomannan from different pulps (Buchert *et al.*, 1996a). In addition, either purified cellulase components or mixtures of cellulase and hemicellulase components have been used for partial or complete solubilization of pulp fibres and subsequent characterization of hydrolysis products by either NMR or HPLC (Buchert *et al.*, 1995; Teleman *et al.*, 1995; Tenkanen *et al.*, 1995). Selective enzymatic solubilization of xylan or glucomannan facilitates determination of the influence of the respective hemicellulosic components on fiber properties such as pore size distribution (Suurnakki *et al.*, 1997), location of lignin (Buchert *et al.*, 1996b), brightness reversion (Buchert *et al.*, 1997) and hornification (Oksanen *et al.*, 1997). Enzymatic solubilization of pulp carbohydrates under mild and non-destructive conditions is beneficial, especially in the analysis of acid-labile pulp components. The suitability of this approach has been verified in the structural analysis of Kraft xylan using *T. reesei* xylanase, which led to the identification of hexenuronic acid in Kraft pulps (Buchert *et al.*, 1995; Teleman *et al.*, 1995).

7. Waste Water Treatment

Improperly dispose of fat, oil and grease to drain, sewer lines can become clogged causing sewage to backup into basements of homes and businesses. Waste Digester is a bacterial/enzyme concentrate containing approximately five billion viable organisms per gram (155 billion/ounce) composed of aerobic, facultative and anaerobic bacteria from strains adapted for maximum efficiency within the conditions found in industrial and municipal waste treatment systems. Enzymes are added to Waste Digester blends. Waste Digester also works to reduce waste and odor in residential septic tanks and lines. Typical treatment for the powder is one cup every other week regarding a typical 1000 gallon tank.

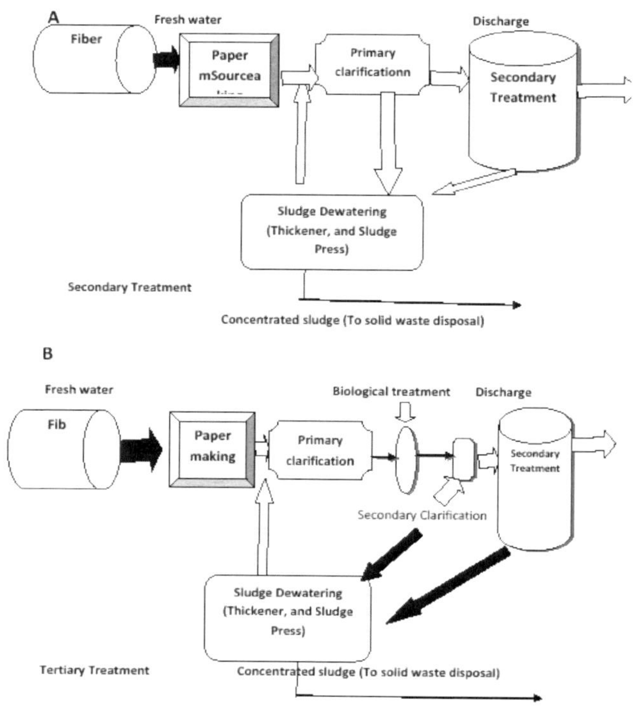

Figure 3. Waste Water Treatment in Pulp Industry.

Waste Digester, with its concentration of bacteria and enzymes, is more effective and efficient at digesting various toxicants in waste treatment systems, as opposed to the natural or existing strains. The biochemical capacity is greatly improved by the addition of enzymes, in every type of application, including lift stations, collection systems, scum pits and trickling filters. The select bacteria produce higher concentrations of enzymes and establish rapid flocculation with accompanying digestion rates, thus increasing efficiency of the waste treatment system. Higher concentration, varying compositions and rapid velocity of biological degradation are gained by using Waste Digester. It helps reduce B.O.D. (Biological Oxygen Demand) and suspended solids. The reduction of sludge volume that accompanies digestion is due to the liquefaction and gasification of organic matter and the destruction of water binding matter. The more efficient sludge digestion, with its decrease in sludge volume, in effect gives added capacity without capital expense.

Waste Digester can also be used on a regular basis for residential, rural, septic, leach fields to keep them moving properly!! Replacing a septic field is very expensive! People in rural areas should look into this application!! One of the main reasons for septic field failure; the clogging of the pores of the soil, preventing absorption. Grease and oil are the big culprits. Septic tanks hold everything from the house that goes down the drains. However, the septic field gets the overflow effluent. It is the liquid oils and grease that escape from the septic tank, that clog the septic field.

Concluding Remarks and Future Prospects

The progress in biotechnology of cellulases and related enzymes is truly remarkable and attracting worldwide attention. Currently, cellulases, hemicellulases and pectinases are widely used in food; brewing and wine, animal feed, textile and laundry, paper and pulp industries as well as in research and development. Some of these applications prefer one or two selected components of cellulase, hemicellulase or pectinase, while others require mixtures of cellulases, hemicellulases and pectinases for the biochemistry, genetics and protein, as well as on the structure–function relationships of cellulases including cellulosomes and related enzymes from bacteria and fungi, has led to speculation and anticipation of their enormous commercial potential in biotechnology and research.

In fact, the potential use of *C. thermocellum* cellulosome and its subunits in research, medicine and biotechnology has been elegantly described by Bayer and co-workers (Bayer *et al.*, 1994). Also, the non-catalytic domains of cellulase system, such as the CBD, and the promoters of cellobiohydrolase I from *T. reesei* and related enzymes will undoubtedly play a key role in future biotechnology and research. Hence, to meet the growing demand for cellulases and related enzymes and to realise their full potential in biotechnology and research, continued multidisciplinary research on basic and applied aspects is vital. These developments together with improved scientific knowledge are expected to pave the way for a remarkable success in the biotechnology of cellulases and related enzymes in the 21st century.

References

Akhtar, M. (1994).*Biochemical pulping of spen wood chips with three strains of Ceriporiopsis subvermispors*, Holzforschung, 48, 199-202.

Ali S, Hall J, Soole KL, Fontes CMCA, Hazlewood GP, Hirst BH, Gilbert HJ.(1995). *Targeted expression of microbial cellulases in transgenic animals*. In: Petersen SB, Svensson B, Pedersen S, editors. Carbohydrate Bioengineering. Progress in Biotechnology, Vol. 10. Amsterdam: Elsevier, pp. 279–93.

Allison RW, Clark TA, Ellis MJ. (1995*). Process effects on the response of softwood Kraft pulp to enzyme assisted bleaching*. Appita, 48, 201–6.

Aubert, J.P., Beguin, P., Millet, J. (1987). *Biochemistry and Genetics of Cellulose Degradation, Fungal and Bacterial Enzyme Systems and their Manipulation.* FEMS Symposium, 43, Academic Press, New York.pp: 11-30.

Aubert, J.P., Beguin, P., Millet, J. (1987). *Biochemistry and Genetics of Cellulose Degradation, Fungal and Bacterial Enzyme Systems and their Manipulation*. FEMS Symposium, 43, Academic Press, New York.pp, 11-30.

Bayer EA, Morag E, Lamed R. (1994). *The Cellulosome—a treasure-trove for biotechnology*. Trends Biotechnol.12, 379–86.

Beauchemin KA, Rode LM, Sewalt VJH. (1998), *Fibrolytic enzymes increase fiber digestibility and growth rate of steers fed dry forages*. Can J Anim Sci 95(75), 641–4.

Beguin P, Aubert JP. (1994). *The biological degradation of cellulose*. FEMS Microbiol Rev., 13, 25–58.

Bernfeld, P. (1955). *Amylases, alpha and beta, Methods*. Enzymol., 1, 149-58

Betrabet, S. M. BTRA *{The Bombay Textile Research Association} Final Technical Report. (2000). application of biotechnology in wet processing of textiles with special reference to impart speciality finishes and amylases for rapid desizing under Indian mill conditions*. 5, 24-26.

Betrabet, S. M. BTRA *{The Bombay Textile Research Association} Final Technical Report. (2000). application of biotechnology in wet processing of textiles with special reference to impart speciality finishes and amylases for rapid desizing under Indian mill conditions*. 5, 24-26.

Bhat, M.K. & Bhat, S. (1997): *Cellulose degrading enzymes and their potential industrial applications*. Biotechnol.Adv., 15, 583-620.

Buchert J, Bergnor E, Lindbland G, Viikari L, Ek M. (1997).*Significance of xylan and glucomannan in the brightness reversion of kraft pulps*. Tappi J., 80, 165–75.

Buchert J, Carlsson G, Viikari L, Strom G. (1996a).*Surface characterization of unbleached Kraft pulps by enzymatic peeling and ESCA*. Holzforschung, 50, 69–74.

Buchert J, Oksanen T, Pere J, Siika-aho M, Suurnakki A, Viikari L. *Applications of Trichoderma reesei enzymes in the pulp and paper industry*. In: Harman GF, Buchert J, Ranua M, Kantelinen A, Viikari L. (1992). The role of two Trichoderma reesei xylanases in the bleaching of pine Kraft pulp. Appl Microbiol Biotechnol., 37, 825–9.

Buchert J, Ranua M, Siika-Aho M, Pere J, Viikari L. (1994).*Trichoderma reesei cellulases in the bleaching of kraft pulp*. Appl Microbiol Bitechnol.,40, 941–5.

Buchert J, Suurnakki A, Tenkanen M, Viikari L. (1996b). *Enzymatic characterisation of pulps.* In: Jeffries TW, Viikari L, editors. Enzymes for pulp and paper processing. ACS Symp Ser, Vol. 655, pp. 38–43.

Buchert J, Teleman A, Harjunpaa V, Tenkanen M, Viikari L, Vuorinen T. (1995).*Effect of cooking and bleaching on the structure of xylan in conventional pine kraft pulp*. Tappi J., 78, 125–30.

Buchert, J., Suurnakki, A., Tenkanen, M. & Viikari, L. (1996). *Enzymatic characterization of pulps in, Enzymes for pulp and paper processing,* edited by T.W.Jeffries, L. Viikari, Acs Sympser, 655, 38-43.

Burroughs W, Woods W, Ewing SA, Greig J, Theurer B. (1960). *Enzyme additions to fattening cattle rations.* J Anim Sci, 19, 458–64.

Chaabouni, S.E., Belguith, H., Hassairi, I., M.Rad, K. & Euouz, R. (1995). *Optimization of cellulose production by Penicillium occitanis,* Appl. Microbiol.Biotechnol.43, 267-269.

Coughlan, M. P., K. Hon-Nami, H. Hon-Nami, L. Ljungdahl, J. J. Paulin, Csisza´r E, Szaka´cs G, Ruszna´k I. (1998). *Combining traditional cotton scouring with cellulase enzymatic treatment.* Textile Res J 68(3), 163–167.

Freiermuth B, Garrett M, Jokinen O. (1994*).The use of enzymes in the production of release papers*. Paper Technol., 25, 21–3.

Graham, H. & Balnave, D. (1995). *Dietary enzymes for increasing energy availability in: Biotechnology in Animals feeds and Animal feedings*, edited by R.J.Wallace & A. Chesson (Weinheim, Germany, VHC), 296-309.

Hall J, Simi A, Surani MA, Hazlewood GP, Clark AJ, Simons JP, Hirst BH, Gilbert HJ. (1993). *Manipulation of the repertoire of digestive enzymes secreted into the gastrointestinal tract of transgenic mice.* Bio/Technol. 11, 376–9.

Hayashida, S., Otta, K.& Mo, K. (1988), *Cellulases of Humicola insolens and Humicola grisea, in Methods in Enzymology*, Vol.160, edited by W.A.Wood & J.N. Abelson (Academic Press, New York), 323-332.

Kamaya Y. (1996). *Role of endoglucanase in enzymatic modification of bleached Kraft pulp.* J Ferm Bioeng.82, 549–53.

Kantelinen A, Jokinen O, Sarkki M-L, Pettersson C, Sundberg K, Eckerman C, Ekman R, Holmbom B. *Effects of enzymes on the stability of colloidal pitch.* In: Proc. 8th Int Symp Wood and Pulping Chemistry, Vol. 1. (1995). pp. 605–12.

Kantelinen A, Rantanen T, Buchert J, Viikari L. (1993). *Enzymatic solubilization of fibre-bound and isolated birch xylans.* J Biotechnol 28, 219–28.

Kubicek CP, editors. *Trichoderma & Gliocladium–Enzymes, biological control and commercial applications*, Vol. 2. pp. 343–63.

Kung, L.Jr, Kreck, E.M., Tung, R.S., Hession, A.O., Sheperd, A.C., Cohen, M.A., Kusters-van Someren MA, Beldman G, editors. *Xylans and xylanases. Progress in Biotechnol.* Vol. 7. Amsterdam: Elsevier, 1992a. pp. 547–50.

Leatham G, Myers G, Wegner T. (1990).*Biochemical pulping of aspen chips: energy savings resulting from different fungal treatments.* Tappi J., 73, 197–200.

Lewis, G.E., Hunt, C.W., Sanchez, W.K., Treacher, R., Pritchard, G.T.& Feng, P. (1996), *Effect of directed-fed of fibrolytic enzymes on the digestive characteristics of a forage-based diet fed to beef steers,* J.Animal Sci., 74, 3020-3028.

Luonteri E, Tenkanen M, Siika-Aho M, Buchert J, Viikari L. *a-Arabinosidases of Aspergillus terreus and their potentials in pulp and paper applications.* In: Lynd, L. R., Weimer, P.

J., van Zyl, W. H. and Pretorius, I. S. (2002*)*, Microbial Cellulose Utilization: Fundamentals and Biotechnology, Microbiology and Molecular Biology Reviews, 66 (3), 506–577

Mandels, M. 1985. *Applications of cellulases*. Biochem. Soc. Trans. 13, 414–419.

Mansfield SD, Wong KKY, De Jong E, Saddler JN. (1996). *Modification of douglas fir mechanical and kraft pulps by enzyme treatment.* Tappi J 79, 125–32.

Muthukrishnan R. (2007**)**. *Characterization of Cellulase from Organisms Isolated From Rumen Fluid,* Vol. 5 Issue 3, Pharmainfo.net

Nelson SL, Wong KKY, Saddler JN, Beatson RP. (1995). *The use of xylanase for peroxide bleaching of kraft pulps derived from different wood species*. Pulp Paper Canada, 96, T258–61

Ng, T. K. &J. G. Zeikus. 1982. *Differential metabolism of cellobiose and Of extracellular enzymes. Appl. Microbiol.* Biotechnol. 29, 480–484.

Noe P, Chevalier J, Mora F, Comtat J. (1986).*Action of enzymes in chemical pulp fibres. Part II: enzymatic beating.* J Wood Chem Technol., 6,167–84.

Oksanen T, Pere J, Buchert J, Viikari L. (1997). *The effect of T. reesei cellulases and hemicellulases on the paper technical properties of never-dried bleached Kraft pulp.* Cellulose 4, 329–39.

Pečiulytė D. (2007). *Isolation of cellulolytic fungi from waste paper gradual recycling materials*, EKOLOGIJA. 53. (4), 11–18

Pere, J., Paavilainen, L., Siika-Aho, M., Cheng, Z. And Viikarid, L. (1996). *Potential use of enzymes in drainage control of nonwood pulps, in Proc.3rd Int.* Vol.3 (Beijing, China, 421-430.

Pere, J., Siika-Aho, M. & Buchert, J. And Viikari, L. (1995). *Effects of purified T.Reesei cellulases on the fibre properties off Kraft pulp,* Tappi J., 78, 71-78.

Perry TW, Purkhiser ED. & Beeson WM. (1966).*Effects of supplemental enzymes on nitrogen balance, digestibility of energy and nutrients and on growth and feed efficiency of cattle.* J Anim Sci., 25, 760–4.

Pommier JC, Fuentes JL, Goma G. (1989).*Using enzymes to improve the product quality in the recycled paper industry. Part 1: the basic laboratory work.* Tappi J 72, 187–91.

Pommier JC, Goma G, Fuentes JL, Rousset C, Jokinen O. *Using enzymes to improve the process and the product quality in the recycled paper industry. Part 2: industrial applications.* Tappi J 1990; 73:197–202.

Poulsen, O. M. & L. W. Petersen. (1988). *Growth of Cellulomonas sp.ATCC 21399 on different polysaccharides as sole carbon source induction of extracellular enzymes.*Appl.Microbiol.Biotechnol.29, 480-484.

Prasad DY, Heitmann JA, Joyce TW. (1993). *Enzymatic de-inking of colored offset newsprint.* Nord Pulp Pap Res J., 8, 284.

Prasad, D.Y., Heitmann, J.A. & Joyce, T.W. (1992). *Enzyme de-inking of black and white letterpress printed newsprint waste*, Prog Paper Recycle, 1, 21-30.

Rahkamo L, Siika-Aho M, Vehvilainen M, Dolk M, Viikari L, Nou-Siainen P, Buchert J. (1996).*Modification of hardwood dissolving pulp with T. reesei cellulases*. Cellulose, 3, 153–63.

Rajoka, M. I., and K. A. Malik. (1997). *Cellulase production by Cellulomonas, cultured in media containing different cellulosic substrates*. Biores. Technol. 59, 21–27

Reese, E. T., R. G. H. Sui, and H. S. Levinson. (1950). *The biological degradation of soluble cellulose derivatives and its relationship to the mechanism of cellulose hydrolysis*. J. Bacteriol. 59,485–497.

Rust JW, Jacobsen NL, McGilliard AD, Hotchkiss DK. (1965). *Supplementation of dairy calf diets with enzymes. 1. Effect on nutrient utilization and on the composition of rumen fluid.* J Anim Sci .24, 156–60.

Saake B, Clark T, Puls J. *Investigations on the reaction mechanism of xylanases and mannanases on sprucewood chemical pulps*. Holzforschung 1995; 49:60–8.

Schu¨lein, M. (1997). *Enzymatic properties of cellulases from Humicola insolens*. J. Biotechnol. 57, 71–81.

Schu¨lein, M. (1998). *Kinetics of fungal cellulases*. Biochem. Soc. Trans. 26,164–167.

Srebotnik E, Messner K, editors. *Biotechnology in the pulp and paper industry*. Vienna: Facultas-Universitatsverlag, 1996. pp. 119–22.

Stokes MR, Zheng S. *The use of carbohydrase enzymes as feed additives for early lactation cows.* In: 23rd Biennial Conf. Rumen Function. Chicago, II, 1995. p. 35.

Suurnakki A, Clark T, Allison R, Viikari L, Buchert J. (1996a). *Xylanase- and mannanase-aided ECF and TCF bleaching*. Tappi J.79, 111.

Suurnakki A, Heijnesson A, Buchert J, Tenkanen M, Viikari L, Westermark U.(1996b). *Effect of pulp surfaces on enzyme-aided bleaching of Kraft pulps*. J Pulp Paper Sci., 22:J91–6.

Suurnäkki A., Niku-Paavola M-L., Buchert J., Viikari L. (2004). *Enzymes in pulp and paper processing.* In: AehleW. (ed.). Enzymes in Industry. Weinheim, Wiley-VCH. P.232–244, 437–439.

Swain, H.E. & Leedle, J.A.Z. (1997). *Effects of alive yeast culture and enzymes on in vitro ruminal fermentation and milk production of dairy cows*, J.Dairy Sci., 80, 2045-2051.

Teleman A, Harjunpaa V, Tenkanen M, Buchert J, Hausalo T, Drakenberg T, Vuorinen T. (1995). *Characterization of 4-deoxy-b-L-threo-hex-4-enopyranosyluronic acid attached to xylan in pine kraft pulp and pulping liquor by 1H and 13C NMR spectroscopy*. Carbohydr Res., 272, 55–71.

Tenkanen M, Buchert J, Puls J, Poutanen K, Viikari L. *Two main xylanases of Trichoderma reesei and their use in pulp processing.* In: Visser J, Voragen AGJ, Tenkanen M, Hausalo T, Siika-Aho M, Buchert J, Viikari L. Use of enzymes in combination with anion exchange chromatography in the analysis of carbohydrate composition of kraft pulps. In: Proc. 8th Int. Symp. Wood and Pulping Chemistry, Vol. III. 1995. pp. 189–94.

Tenkanen M, Puls J, Poutanen K. (1992b). *Two major xylanases of Trichoderma reesei.* Enzyme Microb Technol.14, 566–74.

heurer B, Woods W, Burroughs W. (1963). *Influence of enzyme supplements in lamb fattening rations*. J Anim Sci. 22, 150–4.

Thurston, B., K. A. Dawson, and H. J. Strobel. (1993). *Cellobiose versus glucose utilization by ruminal bacteria Ruminococcus albus.* Appl. Environ. Microbiol. 59, 2631–2637

Tolan JS. (1992).*The use of enzymes to enhance pulp bleaching*. In: Proceedings of Tappi Pulping Conference, Boston, MA, Book 1, pp. 13–17.

Uhlig, H. (1998). *Industrial enzymes and their applications in industry*. (John Wiley & Sons Inc.) New York, w. (ed.). Weinheim, Wiley-Vch. p. 435

Viikari L, Ranua M, Kantelinen A, Linko M, Sundquist J. (1987). *Application of enzymes in bleaching.* In: Proceedings of 4th International Symposium on Wood and Pulping Chemistry, Paris, Vol. 1, pp. 151–4.

Viikari L, Ranua M, Kantelinen A, Sundquist J, Linko M. (1986). *Bleaching with enzymes.* In: Proceedings of 3rd International Conference on Biotechnology in the Pulp and Paper Industry, STFI, Stockholm. pp. 67–9.

In: Cellulase: Types and Action, Mechanism, and Uses
Editor: Adam E. Golan
ISBN: 978-1-61761-983-0

Chapter 8

LIMITATION OF THE DEVELOPMENT ON CELLULOSE HYDROLYSIS BY CELLULASE ASSAY AND SEARCH FOR THE TRUE CELLULASE DEGRADING CRYSTALLINE CELLULOSE

Wenzhu Tang, Xiaoyi Chen, Hui Zhang, Fang Chen and Xianzhen Li*

Department of Bio & Food Engineering, Dalian College of Light Industry, Dalian 116034, PR China

1. ABSTRACT

Cellulose is the most abundant component of plant biomass found in nature that is almost exclusively in plant cell walls, whereas it cannot be effectively converted into the usable sugars due to the lower cellulase activity. Although various classes of cellulase have been isolated and the synergism between them has been studied in detail, the thorough degradation of natural cellulose cannot be observed in the depolymerization by cellulase system, presumably it is due to the cellulase assay methods and substrates used for determining cellulases. Assay using cellulose as substrate is useful for assessing the potential enzyme system but it cannot be used for searching novel individual cellulase because the total activity is determined with such substrates. Considering that the natural cellulose can be degraded by living microbes whereas cannot by the secreted cellulases, we can conclude that there are the true cellulases degrading natural cellulases not to be isolated yet. It is obvious that the substrate is the vital determinant for cellulase assay and the key problem for seeking the true cellulase is how to obtain single cellulose chains. It is believed that the thoughtful substrate for cellulase assay should be amorphous cellulose molecule in the form of single chains.

* Corresponding Author: Xianzhen Li, Department of Bio & Food Engineering, Dalian College of Light Industry, Dalian 116034, PR China, Tel: 86 (411) 86314195, Fax: 86 (411) 86323646, Email: xianzhen@dlpu.edu.cn, xianzhen@mail.com

2. Introduction

Cellulose is the most abundant natural resource being a recalcitrant substrate for enzymatic hydrolysis, and large amount of cellulose also is available as municipal and industrial waste contributing to pollution problem. Thus the much faster enzymatic degradation is vital to use the cellulosic biomass as a renewable source of energy via breakdown to sugars (Berner, 2003; Cox et al., 2000).

Cellulose is a homopolysaccharide comprised of glucose units linked by β-1,4-glycosidic bonds, cellobiose of which is the smallest repeating unit and can ultimately be converted into glucose (Sánchez, 2009; Zhang and Lynd, 2004). Although its chemical composition is very simple, the efficient degradation of cellulose requires the joint action of several types of cellulases. To date, the mechanism for the enzymatic hydrolysis has been explained in term of the sequential action, in which at least three cellulase types are needed to cooperate to digest efficiently crystalline cellulose, including endoglucanase (EC 3.2.1.4) cutting cellulose chain randomly, cellobiohydrolase (CBH, EC 3.2.1.91) cleaving cellobiose from the cellulose chain ends, and β-glucosidase (EC 3.2.1.21) hydrolyzing cellobiose or soluble cellodextrin into glucose. It was suggested that endo-cellulase act randomly within the amorphous region of cellulose to expose crystalline terminals, which was susceptible to attack by exo-cellulase. Therefore, most emphasis has been placed on the synergistic effect between cellulases, whereas the efficient hydrolysis has not been found in such study on the cooperative hydrolysis of crystalline cellulose. Moreover, some microbes are able to grow on cellulose leading to the thorough degradation of cellulose but the cellulase activity cannot be detected. In fact, the increasing evidences suggest that cellulose hydrolysis cannot be finished only by three known types of cellulases, presumably some specific enzymes were included in cellulose degradation. However such novel enzymes cannot be obtained yet, because all the substrates using for cellulase assay were cellulose derivative like carboxymethyl cellulose for endo-cellulase or native cellulose like filter paper for overall cellulase activity, rather than the true substrate of cellulose molecule. Considering that cellulase action is impacted by substrate property, the methodology of cellulase assay and its substrate for enzyme test has been realized to be very important for screening the potential of specific cellulase enzymes. In this chapter, we will focus on the cellulase assay in studying cellulose degradation, including substrates for cellulase assay and the role of cellulose property in determining cellulase component, the impact of model substrates on synergy among cellulases, and strategy seeking the novel true cellulases.

3. Cellulase System and Cellulolytic Mechanism

Microfibril is the structural backbone of cellulose forming amorphous and crystalline region, its degradation requires multiple enzymes acting in concert to accommodate this structural heterogeneity (Lynd et al., 2002). Both aerobic and anaerobic microbes have been found to be capable of degrading cellulose, but their secreted cellulases are different. The aerobic fungi or bacteria produce the individual cellulases extracellularly (Okada et al., 1998) and the anaerobic fungi or bacteria produce large cellulosome complexes organized on the cell surface (Nicholson et al., 2005; Doi and Tamaru, 2001). No matter what the cellulase class is, three classes of enzymes must have been involved in cellulose biodegradation (Lynd et al., 2002; Zhang and Lynd, 2004). Endoglucanase fragments cellulose by hydrolyzing internal β-1,4-glucosidic bonds randomly at exposed position to generate new chain terminals. Endoglucanase can be produced by many bacteria, fungi and plant with different catalytic modules (Zhou et al., 2008). Fungal endoglucanases generally have a catalytic module with or without a carbohydrate-binding module (CBM), while bacterial endoglucanases possess multiple catalytic modules, CBMs and other domains. Exoglucanase or cellobiohydrolase processively cleaves cellulose chain from the chain end to release cellobiose or glucose. Cellobiohydrolase is able to degrade the crystalline part of cellulose. In the culture media of typical cellulolytic fungi, cellobiohydrolases are the most abundant among the secreted protein (Zhou et al., 2008), although some cellulolytic bacteria may not produce cellobiohydrolase (Xie et al., 2007). Fungal cellobiohydrolases possess a catalytic module with or without CBM, while bacterial cellobiohydrolases may possess more than one catalytic module, more than one CBM of different families, and other functionally known or unknown domains. β-Glucosidase hydrolyzes cellobiose to glucose for eliminating cellobiose inhibition. Different β-glucosidases are produced by various bacteria, fungi and plant with different catalytic modules (Eyzaguirre et al., 2005; Hrmova et al., 2002). The β-glucosidase produced by aerobic fungi is the extracellular protein but that produced by anaerobic bacteria usually is kept in cytoplasm.

Endoglucanase also is referred to as carboxymethylcellulases because its enzyme activity is usually determined using the carboxymethylcellulose (CMC) (Zhou et al., 2004). According to the classical hydrolytic mode, endoglucanase initiates cellulose depolymerization by attacking the amorphous region of the cellulose, providing more free chain ends for cellobiohydrolase, and cellobiohydrolase creates a substrate-binding tunnel with their extended loops surrounding cellulose (Dashtban et al., 2009). The enzymatic depolymerization by endoglucanase and exoglucanase is thought to be the rate-limiting step for the whole cellulose hydrolytic process (Zhang and Lynd, 2004).

Considerable progress has been made in elucidating the cellulolytic mechanism with soluble oligosaccharide as substrate, but the exact biochemical mechanisms involved in natural cellulose depolymerization still remain unknown although largely inaccessible nature of the cellulosic substrate has also been much discussed. The disruption of the highly ordered regions in cellulose structure can facilitate the exposure of inaccessible cellulose chains, thereby enhancing enzyme access to cellulose and promoting the hydrolytic attack of cellulases. However, there is not any efficient pretreatment process so far can be obtained to increase the accessibility of cellulose. In studying the morphological and structural changes of cotton fiber, it was found that the treatment with CBM could promote the non-hydrolytic disruption of crystalline cellulose by weakening and splitting the hydrogen bonds, thereby loosening cellulose chains (Wang et al., 2008). Anaerobic *Clostridium thermocellum* produces an insoluble substance when growing on cellulose, which has a strong affinity for crystalline cellulose and is the part of cellulolytic system required for the efficient enzymatic degradation of cellulose (Kopecny and Hodrova, 1997). Expansins are produced in plant tissue, which are known for their unique loosening and weakening effect on the cellulosic network of plant cell wall (Cosgrove, 2005). Although their mechanism attacking cellulose has not been understanded, the most of these swelling or delaminating agents contain a carbohydrate-binding surface, which may play an important role in the non-hydrolytic amorphogenesis (Arantes and Saddler, 2010).

In our early search of soil samples for cellulolytic bacteria, a *Streptomycete* strain was obtained (Li, 1997). This strain completely degraded filter paper, α-cellulose, and Avicel in 2 or 3 days. It is interested that almost no reducing sugar can be detected during incubation in medium containing cellulose as the sole carbon source. A low-weight-mass enzyme with fragmentation activity was purified, which is responsible for fragmentation of filter paper without release of soluble sugar and decomposition of filter paper fibre chain (Li and Gao, 1997). Non-enzymatic systems have been proposed to explain the initial opening of the lignocellulose, following which cellulase could diffuse and cause further degradation. Formation of short fibre probably occurred either by disturbances in the intra- and inter-chain hydrogen bonds or by non-hydrolytic fibre disruption (Li and Gao, 1997).

The *Clostridium thermocellum* is generally described to be specialized in the degradation of crystalline cellulose and indeed cannot readily utilize carbohydrates other than cellodextrin (Demain et al., 2005). It produces cellulosome being the most effective cellulolytic complex known so far. About 30 genes involved in cellulosome formation have hitherto been isolated by screening of genomic libraries (Zverlov et al., 2005). The cellulolytic bacterium *Clostridium cellulolyticum* also produces cellulosome, which is efficient for degradation of crystalline cellulose (Belaich et al., 2002). The catalytic subunits of the cellulosome are bound to a non-catalytic scaffolding protein called CipC, which contains a cellulose binding domain, two X2 domains of unknown function, and eight reiterated domains called cohesins. The binding of each cellulosomal enzyme to the scaffoldin is accomplished by means of a dockerin domain, generally located at the C-termini of the enzymes, which interacts with one of the eight cohesins of CipC.

Cellulases produced by *Saccharophagus degradans* consist of two cellodextrinases, three cellobiases, a cellodextrin phosphorylase, and a cellobiose phosphorylase (Taylor et al., 2006). Most of these enzymes exhibit modular architecture, and some contain novel combination of catalytic and substrate binding modules. In its cellulose degradation the endoglucanases hydrolyze individual cellulose chains to cellodextrins, which are cleaved to glucose and oligocellodextrins by cellodextrinases. Cellobiose is hydrolyzed to glucose at the cell surface by cellobiase and then imported to the cytoplasm. Alternatively, short cellodextrins and cellobiose is possibly imported into cytoplasm for further metabolism by the cellobiases or phosphorylases.

4. Synergism of Cellulose Degradation

It has been known that microbial cellulolysis occurs both in aerobic and anaerobic biotopes and different strategies have been developed to degrade cellulosic substrates. In general, aerobic microbes secrete lots of free cellulases degrading cellulose synergistically, and anaerobic microbes produce large cellulosome complex depolymerizing both crystalline and amorphous cellulose (Minardon et al., 2007). The known synergistic action includes endo-exo synergy between endoglucanase and exoglucanase, exo-exo synergy between exoglucanase catalyzing from the reducing and non-reducing chain ends of cellulose, synergy between exoglucanase and β-glucosidase, and intramolecular synergy between catalytic domains and CBMs (Ramirez-Ramirez et al., 2008; Karboune et al., 2008).

It is believed that a minimal cellulase mixture for cellulose degradation needs four complementary enzymes (CBH-I, CBH-II, endoglucanase, and β-Glucosidase), whereas cellulolytic organisms generally produce more than four cellulases under culture condition (Martinez et al., 2008; Lykidis et al., 2007; Xie et al., 2007).

The most widely accepted synergism is a sequential action, it is presumed that endoglucanase initiates attack on cellulose to form new chain ends, which then serve as attack points for processive hydrolysis by cellobiohydrolase. Endoglucanase and cellobiohydrolase hydrolyze insoluble cellulose on the cellulose surface to soluble cellodextrin, followed by β-glucosidase-mediated hydrolysis of cellobiose to glucose (Lynd et al., 2002; Zhang and Lynd, 2004). The primary depolymerization reaction from long solid cellulose chain to short soluble cellodextrin is the rate-limiting step for the overall hydrolysis process (Lynd et al., 2002; Zhang and Lynd, 2004). The cellobiohydrolase Cel48C acts synergistically with *Paenibacillus* endoglucanase Cel9B or *Thermobifida fusca* cellulases Cel6A and Cel6B on microcrystalline cellulose or filter paper (Sánchez et al., 2004). Maximum synergism rate was obtained with a mixture of *Paenibacillus* cellulases Cel9B and Cel48C, accompanied by *Thermobifida* exocellulase Cel6B. It appears that two cellulases give synergism only if they attack insoluble cellulose at different sites and if they create new sites for each other as a result of their activity (Wilson, 2004). It is interest that pretreatment of crystalline cellulose with an endocellulase can make a better substrate for exocellulase, but pretreatment with an exocellulase cannot increase the activity of endocellulases on the pretreated cellulose.

Cellulolytic microbes could produce various cellobiohydrolases showing different substrate specificities. Cellobiohydrolyases Cel7 and Cel6 is synergistic in hydrolyzing cellulose, a phenomenon attributed to their specificities at opposing cellulose chain ends, which is similar to synergistic degradation in cellobiohydrolases Cel5, 9, 48, and 74 (Yaoi et al., 2007). Several explanations have been proposed to account for the cooperation between two exocellulases, one of which was presumed that two types of non-reducing ends occur in cellulose on which each CBH specifically acts, resulting in the increased activity when both enzymes are present (Wood, 1991). This finding has led to the proposal that the differences in the chain end preference and in the action directionality of the two CBH were responsible for the "exo-exo" synergy (Barr et al., 1996). The synergism also occurs in cellobiohydrolases with different specificities produced in one strain (Berger et al., 2007; Karboune et al., 2008).

The accessibility of cellulose to the reactive sites on the surface of native cellulose is a rate-limiting factor in the enzymatic hydrolysis (Sánchez et al., 2004). The adsorption capacity of the individual cellulase is reflected by their modular structure, most of which consist of at least a catalytic domain and a CBM. Different endoglucanases have different preference for cellulose regions to bind or the site of glycosidic bond to cleave due to the difference in CBM specificity, multiplicity, linkage to the catalytic module. There is no significant synergism on the purified homogeneous cellulose among different endoglucanases, whereas the endoglucanases are beneficial to degradation on polymorphous and heterogeneous cellulose (Wilson, 2008). It is supposed that endoglucanase Cel12 without CBM acts first on amorphous in the outer part of cellulose, and endoglucanase with CBM then acts on crystalline in the inner part of cellulose (Rabinovich et al., 2002). Brown rot doesn't produce cellobiohydrolyase but its processive endocellulases may act on crystalline cellulose and cooperate with other endoglucanases to constitute a complete cellulolytic system (Cohen et al., 2005).

Some non-hydrolytic proteins such as plant expansins are known to be capable of loosening or disrupting the packaging of the plant cell wall (Cosgrove, 2000). Recently expansin-like protein genes have also been cloned from some microbes and their recombinant proteins show cell wall disrupting activity (Kim et al., 2009; Kerff et al., 2008). Expansin-like modules usually occur in the swollenin consisted of an expansin-like domain linked to a CBM and other modules often found in cellulases (Saloheimo et al., 2002). It can disperse cotton fiber or weaken filter paper without releasing detectable reducing sugar but does not show any significant cellulose-hydrolytic activity, indicating a non-catalytic nature of protein (Yao et al., 2008). Such disruption activity is believed to confer a synergistic effect to the enzymatic hydrolysis by enabling it to be more accessible to the enzyme (Cosgrove, 2000). It has been confirmed that expansin-like proteins show a synergistic effect with *Trichoderma* endoglucanase (Cosgrove, 2007). Recently, an expansin-like module has been found in a putative plant endoglucanase, suggesting possible interaction with endoglucanase activity (Bouzarelou et al., 2008). Another expansin purified from corn stover also was found to have a synergistic effect with cellulase when hydrolyzing filter paper (Han and Chen, 2007). Kim et al. (2009) found an expansin-like protein BsEXLX1 from *Bacillus subtilis* and expressed in *Escherichia coli*. The recombinant protein exhibited cellulose-weakening activities when used filter paper as substrate, and displayed a significant synergism with cellulase. The synergistic activity was 5.7-fold greater than the individual cellulase activity, however, such synergistic effect was observed only if a low dosage of cellulase was used.

The synergy is related to the crystalline degree of the substrate, and acid hydrolysis can increase the crystallinity of bacterial cellulose leading to lower synergy between Cel7B and Cel7A, because the action of the endoglucanase Cel7B becomes less important (Samejima et al., 1998). Most native CBMs are part of various biomass-active enzymes, but some exist as either independent binding proteins or apparently non-catalytic proteins (Moser et al., 2008; Shoseyov et al., 2006). The synergism of CBM with the catalytic domain was reported on the cotton fiber but not on bacterial microcrystalline cellulose (Din et al., 1994). The key role of CBM is postulated to anchor enzyme to targeted carbohydrate substrate or direct enzyme to specific region in complex biomass substance (Blake et al., 2006).

The cellulosome is believed to allow concerted enzyme activity in close proximity to the bacterial cell, enabling optimum synergism between the cellulases presented on the cellulosome. The synergistic action has been illustrated in cellulosome resulting in more efficient degradation of natural cellulose (Schwarz, 2001). Such synergism generally exceeds the potential of non-cellulosomal degrading systems due to the structural organization, concentrating the hydrolytic enzymes to specific sites to the substrate, the large number of different types of hydrolytic enzymes, and cooperative action with non-cellulosomal enzymes (Murashima et al., 2002). The presence of abundant cellulosomal enzymes results in a large variety of enzyme combination in cellulosome and a functional variety of cellulosomes capable of attacking different types of cellulose. Synergy in sequential or non-sequential mode was observed between three cellulases: endoglucanases EngE and EngH, and exoglucanase ExgS (Murashima et al., 2002). Optimal synergy for these three enzymes was observed when a proper ratio of the enzymes was present.

The synergism was also found to occur between specific cellulosomal enzymes and non-cellulosomal enzymes. It is exemplified by the action of cellulosomal hemicellulase XynA and non-cellulosomal hemicellulases ArfA and BgaA (Kosugi et al., 2002). When corn cell wall was used as the substrate, synergy between XynA and ArfA or XynA and BgaA was observed, and the greatest degree of synergy was observed when XynA, ArfA and BgaA were used together. In this case, simultaneous addition of the enzymes, but not sequential addition, showed the highest degree of synergy (Murashim et al., 2003). Thus the action of non-cellulosomal ArfA and BgaA also enhances the activity of XynA probably by allowing better access of xylanase to the substrate. When cellulosomes were used simultaneously with non-cellulosomal β-glucosidase BglA, cellulose was degraded to glucose more rapidly than with cellulosomes alone (Kosugi et al., 2006). BglA was found to degrade the cellooligosaccharides produced by cellulosomes to glucose more rapidly than it was able to degrade cellobiose.

Although the synergy among the individual purified cellulases has been confirmed in cellulose degradation, the thorough depolymerization of natural cellulose in such synergic action cannot be observed so far. However, the cellulosome could digest cellulose completely using the similar synergic action. It is presumed that some unknown cellulases occurred in the cellulosome, which probably are the true cellulases degrading cellulose molecules. Being different from cellulosome, the individual cellulases were purified and detected with the classical assay methods leading to the lose of true cellulases, therefore such artificial synergy between the purified cellulases is not able to finishing cellulose degradation.

5. Cellulase Assay and Substrate

Suitable assays are important for the research and development of cellulases from preliminary screening of candidates. The basic approaches for cellulase assay can be classed as the individual cellulase activity and the total cellulase activity (Zhou et al., 2004). The cellulase determination was usually performed on Avicel, acid swollen cellulose, carboxymethylcellulose, bacterial microcrystalline cellulose, filter paper, or cellodextrins. The enzymatic activities generally were detected by measuring the amount of reducing sugars released.

Like α-amylase decreasing the degree of polymerization of starch, endoglucanase was defended as cutting off inner glycosidic bond randomly. Therefore endoglucanase activities are often measured on a soluble cellulose derivative, such as CMC, based on the reduction in substrate viscosity significantly with little hydrolysis (Zhang and Lynd, 2004). Endoglucanase activities can also be measured based on an increase in reducing ends determined by a reducing sugar assay. It is strongly recommended that the non-purified endoglucanase activities are not measured by reducing end methods because exoglucanases also increase the number of reducing ends. Sometimes soluble oligosaccharides and their chromophore-substituted substrates are used to determine endoglucanase activities based on the release of chromophores or the formation of shorter oligosaccharide (Zverlov et al., 2005).

Endoglucanase activities can also be detected on agar plate by staining residual polysaccharide with dyes considering that the dyes are adsorbed only by long chains of polysaccharides (Murashima et al., 2002; Toyama et al., 2008). This is a semi-quantitative method, and is well suit to monitor large numbers of samples. Its precision is limited because of the relationship between the cleared zone diameters and the logarithm of enzyme activities. Moreover, the processive action of exoglucanase is blocked by carboxymethyl substitutions, which prohibits cellulose chain from shortening (Demain et al., 2005). Quartz crystal microbalance, surface plasmon resonance and ellipsometry can be used to measure cellulase action on cellulose film or nanofibrils (Ahola et al., 2008; Eriksson et al., 2005). However, these methods are more suited for basic enzymology study, but not high throughput screening or evaluation.

Considering that exoglucanases show relatively high activity on Avicel but little activity on soluble CMC, exoglucanase activity usually is measured using Avicel as substrate. Unfortunately, the hydrolysis of crystalline cellulose requires synergistic reaction of both endoglucanase and exoglucanase. The amorphous cellulose and soluble cellodextrins are substrates for purified exoglucanases, therefore, there are no substrates specific for exoglucanases within the cellulase mixtures (Wood and Bhat, 1988).

A selective assay was reported for exoglucanases in the presence of endoglucanases and β-glucosidases with p-nitrophenyl-β-D-cellobioside as substrate (Deshpande et al., 1984). Because the exoglucanase activities must be quantified in the assay procedure in the presence of added purified endoglucanases, a little endoglucanase should be added. However, this technique also has its own limitation (Deshpande et al., 1984): (1) CBH II activity cannot be measured using p-nitrophenyl-β-D-cellobioside, (2) the specific activity of the available purified endoglucanases may not be representative of all existing endoglucanases in the mixture, and (3) the product ratio from endoglucanase actions may be influenced by the presence of exoglucanases.

The β-D-glucosidases can hydrolyze cellobiose and other cellodextrins to produce glucose in the aqueous phase. The hydrolytic rates decrease markedly as the substrate DPs increase (Zhang and Lynd, 2004). Cellobiose should be the better substrate for measuring β-D-glucosidase activities because it can not be hydrolyzed by both endoglucanase and exoglucanase. If the longer cellodextrin was used, it will be hydrolyzed by endoglucanases or exoglucanases (Zhang and Lynd, 2004). The β-D-glucosidases are very amenable to a wide range of simple sensitive assay methods, based on colored or fluorescent products released from β-D-1,4-glucopyranoside derivatives (Setlow et al., 2004).

The most common total cellulase activity assay is the filter paper assay (FPA) using Whatman No. 1 filter paper as substrate, which was established and published by the International Union of Pure and Applied Chemistry (IUPAC) (Ghose, 1987). Such assay requires more than 3.6% hydrolysis of the filter paper to ensure both amorphous and crystalline fractions being hydrolyzed during determination, therefore, a serial dilution of enzyme is required to get the fixed degree of hydrolysis (Coward-Kelly et al., 2003; Decker et al., 2003). The result reliability is related to β-D-glucosidase level present in cellulase mixture, because the produced reducing value is strongly influenced by the reducing end ratio of glucose, cellobiose, and longer cellodextrins (Kongruang et al., 2004; Zhang and Lynd, 2005). In addition, the cutting method of filter paper also influences enzyme activity, because the different paper cutting methods could lead to different accessible reducing ends of the substrate (Zhang and Lynd, 2005).

A new method to determine cellulase activity recently was developed using a quartz crystal microbalance (QCM) technique (Hu et al., 2009). In this method QCM frequency change is used to measure the solution viscosity and density changes in the solution incubating the cellulose substrate after enzymatic hydrolysis. The results are then used to quantify the enzyme activity. It is shown that the QCM technique provides results closer to those obtained by measuring the actual reducing sugars. The elimination of the use of color development in the standard redox methods makes the QCM platform easier to implement. It also allows more flexibility in terms of the nature of the substrate.

Actually, the purified cellulose used for studying cellulase varies considerably in fine structural features, and the choice of substrate for such study undoubtedly affects the result obtained. Holocelluloses usually are produced by delignification of wood, which contain substantial amount of hemicelluloses and often have a low bulk density suggestive of some swelling of cellulose fibers. Microcrystalline celluloses are nearly pure cellulose, and both hemicelluloses and the more extensive amorphous regions have been removed by the diluted acid treatment. Commercial microcrystalline celluloses differ primarily in particle size distribution, which has significant implications for the rate of hydrolysis and utilization. Bacterial cellulose is highly crystalline but differs from plant cellulose in the arrangement of glucosyl units within the unit cells of the crystallites (Weimer et al., 2000). Studies with pure celluloses indicate that amorphous celluloses are degraded 5 to 10 times more rapidly than are highly crystalline celluloses by both fungal enzymes (Gama et al., 1994).

The bacterial cellulose has been used as a substrate of choice for cellulase studies because it has several advantages in comparison to that of plant origin (Gilkes et al., 1992). It has a more homogeneous structure, higher crystallinity and is available in a never-dried form. Moderate hydrolysis of bacterial cellulose with HCl yields a product with a reduced degree of polymerization and higher crystallinity, bacterial microcrystalline cellulose. This highly crystalline cellulose with a simple morphology has been used as a model substrate to study the mechanism of crystal erosion (Chanzy and Henrissat, 1985).

The variable structural complexity of pure cellulose and the difficulty of working with insoluble substrates have led to the wide use of the highly soluble cellulose ether CMC. However, the use of CMC as an enzymatic substrate has weakened the meaning of the term "cellulolytic," because many organisms that cannot degrade cellulose can hydrolyze CMC via mixed β-glucan enzymes (Fields et al., 1998). Because of the substituted nature of the hydrolytic products, relatively few microbes can use CMC as a growth substrate. Utilization of cellulosic biomass is more complex than is that of pure cellulose, because of not only the former complex composition but also the diverse architecture of plant cells themselves.

It is obvious that many widely used substrates are reactive toward more than one class of enzymes. Therefore, the minute amount of one enzyme presented as impurity or background can significantly impact an assay on a targeted enzyme due to the synergism among cellulase components, when both are active on a selected substrate, leading to a severe overestimation of the targeted enzyme's activity.

6. Potential Strategy for Seeking True Cellulase

Like starch hydrolysis catalyzed by α-amylase and glucoamylase, cellulose degradation requires the catalysis of endo-, exo-glucanase and glucosidase. However differing from starch, natural cellulose occurs in elementary fibril with crystalline and amorphous region. Moreover cellulose is insoluble substrate even after pretreatment.

To quantify cellulase activity, enzyme assay has been carried out mostly on pure cellulose or cellulose derivatives (Zhou et al., 2004). CMC is used as the substrate when quantifying endoglucanase activity (Zhou et al., 2004). Cotton and cellodextrin can be used as the substrate for measuring exoglucanase activity (Bisaria and Ghose, 1981). Filter paper is normally used as the substrate to determine saccharifying cellulase activity (Nordmark et al., 2007). Cellulose-azure is usually used as a substrate for the total cellulase activity (Lai et al., 2006). Considerable research is aimed at obtaining novel cellulase systems for application in textile, food, and biomass conversion industries. Assay using cellulose as substrate is useful for assessing the potential enzyme system because its hydrolytic rate is presumably dependent on the well-documented synergism (Boisset et al., 2001; Murashima et al., 2002). However, it cannot be used to looking for novel individual cellulase due to the total cellulase activity determined with such substrates.

Endo-β-1,4-glucanase randomly cleaves internal β-1,4-glycosidic bonds solubilizing cellulose and releasing cellodextrin with different degrees of polymerization (Tokuda et al., 1997; Wantanabe et al., 2002). Its standard substrate should be amorphous cellulose but its activity always is determined using soluble cellulose derivative CMC as substrate, in which the nature of cellulose has been modified by substitute of carboxymethyl group. Cellobiohydrolase attacks the non-reducing end of crystalline cellulose chain to cleave off cellobiose but its activity was assayed with soluble cellodextrin (Lynd et al., 2002). It was found that the catalytic mode is confused with their substrate specificities, in which endoglucanases show a high level of activity on soluble cellulose derivative and very low level on microcrystalline cellulose, while exoglucanases show relatively high level of activity on microcrystalline cellulose (Lynd et al., 2002).

The fungus *Trichoderma reesei* is considered to be the most efficient cellulase producer, and has a long history in the production of hydrolytic enzyme. However, these cellulases determined by traditional methods usually show little activity on the crystalline cellulose although it can generate all three types of cellulase (Wang et al., 2004). Being different from fungal cellulases, anaerobic bacterium *Clostridium thermocellum* produces cellulosome having high activity against crystalline cellulose. The true cellulase activity presumably occurs in such cellulosome because of its ability to completely solubilize crystalline forms of cellulose (Zhang and Lynd, 2005). This suggests that some unknown cellulases cannot be detected yet with those substrates. Therefore, both building up a new assay and generating novel substrates are very important for discovering novel true cellulase.

The hydrolytic rate of starch is about 100-fold faster than hydrolysis rates for cellulose under condition using crystalline substrate (Klyosov, 1988). Actually the large difference in the hydrolytic rate of cellulose and starch is due primarily to the substrate characteristics rather than to β-linked glucosidic bonds being more difficult to hydrolyze than α-linked glucosidic bonds (Zhang and Lynd, 2004). Several rate-limiting determinants have been proposed, including crystallinity, degree of polymerization, particle size, and accessible surface area (Fan et al., 1980). Crystallinity is widely regarded as a major feature influencing cellulose hydrolysis (Wyman, 2007). Natural cellulose never occurs as a single chain, but exists as coupling of many adjacent parallel. Such parallel array results in a crystalline structure with low accessibility (Notley et al., 2004). The hydrophobic cellulose sheet makes crystalline cellulose resistant to acid hydrolysis (Matthews et al., 2006). The strong interchain hydrogen-bonding network makes crystalline cellulose resistant to enzymatic hydrolysis (Nishiyama et al., 2002). Additionally, access to the crystalline cellulose cores also is restricted by a coating of amorphous cellulose and hemicellulose (Ding and Himmel, 2006).

Pretreatment aimed at converting crystalline cellulose to amorphous substrate is very difficult so far.

Since cellulose hydrolysis is a surface phenomenon, available surface area is a potential rate-limiting feature, although there remains some debate about what constitutes the "available" surface area (Mansfield et al., 1999). Cellulose in the plant cell wall is not readily available to enzymatic hydrolysis due to the low accessibility of crystalline cellulose preventing cellulases from efficient degradation (Zhang et al., 2007). Although cellulose could be slowly eroded by surface shaving, the cellulose chain in the crystalline region is difficult to be delaminated or loosened, thereby making the individual cellulose molecules more accessible and available for interaction with cellulase. The generation of short fiber has been found in fiber swelling of cellulose before any detectable reducing sugar is released during enzymatic hydrolysis (Coughlan, 1985).

Cellulase must overcome several obstacles for successful catalysis, specifically they must access the insoluble substrate, disrupt the packaging of the highly ordered polymer and direct a single polymer chain into the active site cleft of the enzyme (Kim et al., 2009). As a result, the specific activity of cellulase should be increased by 40- to 100-fold to show an equivalent degrading efficiency comparing with amylase (Merino and Cherry, 2007). Otherwise, it will be necessary to look for another true cellulase degrading natural cellulose. It presumed that three properties of cellulose and starch influence hydrolysis rate except any difference in the intrinsic reactivity between β-linked and α-linked glucans: the fraction of bonds accessible for insoluble substrate, the availability of chain ends for insoluble substrates, and the solubility of hydrolysis products. Therefore, the substrate for cellulase assay is the vital factor limiting to look for true cellulase. It is obvious that the thoughtful substrate for cellulase assay should be amorphous cellulose molecule in the form of single chains. Therefore a key problem for seeking true cellulase is how to obtain single cellulose chains.

Xanthan is an extracellular heteropolysaccharide produced by *Xanthanmonas campestris* pv. *campestris*, and is composed of a cellulosic backbone with linear trisaccharide side chains, consisting of a mannosylglucuronyl-mannose sequence attached α-1,3 to alternating glucosyl residues (García-Ochoa et al. 2000). The internal and terminal mannosyl residues of the side chain are mostly acetylated and pyruvylated respectively (Sandford et al., 1997). Based on the chemical structure of xanthan, the remained main chains after removing all trisaccharide side chains just are cellulose molecule. Such amorphous cellulose should be the best substrate used for assaying endocellulase. It was reported that α-manosidase can cleave α-1,3-linkage between backbone glucose and mannose in side chain. Recently, we isolated a xanthan-degrading bacterium *Microbacterium* sp. strain XT11, which produced extracellular xanthan-degrading enzymes degrading xanthan to its constituted unit (Qian et al., 2007). A novel α-manosidase was isolated from the culture of *Microbacterium* sp. XT11 growing on xanthan, which can produce insoluble cellulse chains (data not shown) for screening the true cellulases.

REFERENCES

Ahola S, Turon X, Osterberg M, Laine J, Rojas OJ. *Enzymatic hydrolysis of native cellulose nanofibrils and other cellulose model films: effect of surface structure*. Langmuir. 2008, 24:11592-11599.

Arantes V, Saddler JN. *Access to cellulose limits the efficiency of enzymatic hydrolysis: the role of amorphogenesis.* Arantes Saddler Biotechnol Biofuels. 2010, 3: 4-12.

Barr BK, Hsieh Y-L, Ganem B, Wilson DB. *Identification of two functionally different classes of exocellulases.* Biochemistry. 1996, 35:586-592.

Belaich A, Parsiegla G, Gal L, Villard C, Haser R, Belaich JP. *Cel9M, a new family 9 cellulase of the Clostridium cellulolyticum cellulosome.* J Bacteriol. 2002, 184:1378-1384.

Berger E, Zhang D, Zverlov VV, Schwarz WH. *Two noncellulosomal cellulases of Clostridium thermocellum, Cel9I and Cel48Y, hydrolyse crystalline cellulose synergistically.* FEMS Microbiol Lett. 2007, 268:194-201.

Berner RA. *The long-term carbon cycle, fossil fuels and atmospheric composition.* Nature. 2003, 426:323-326.

Bisaria VS, Ghose TK. *Biodegradation of cellulosic materials: substrates, microrganisms, enzyme and products.* Enzyme Microbiol Technol. 1981, 3:90-104.

Blake AW, McCartney L, Flint JE, Bolam DN, Boraston AB, Gilbert HJ, Knox JP. *Understanding the biological rationale for the diversity of cellulose-directed carbohydrate-binding modules in prokaryotic enzymes.* J Biol Chem. 2006, 281:29321-29329.

Boisset C, Petrequin C, Chanzy H, Henrissat B, Schulein M. *Optimized mixtures of recombinant Humicola insolens cellulases for biodegradation of crystalline cellulose.* Biotechnol Bioeng. 2001, 72:339-45.

Bouzarelou D, Billini M, RoumeliotiK, Sophianopoulou V. *EglD, a putative endoglucanase with an expansin like domain is localized in the conidial cell wall of Aspergillus nidulans.* Fungal Genet Biol. 2008, 45:839-850.

Chanzy H, Henrissat B. *Unidirectional degradation of Valonia cellulose microcrystals subjected to cellulase action.* FEBS Lett. 1985, 184:285-288.

Cohen R; Suzuki MR, HammelKE. *Processive endoglucanase active in crystalline cellulose hydrolysis by the brown rot basidiomycete Gloeophyllum trabeum.* Appl Environ Microbiol. 2005, 71:2412-2417.

Cosgrove DJ. *Loosening of plant cell walls by expansins.* Nature. 2000, 407:321-326.

Cosgrove DJ: *Growth of the plant cell wall.* Nat Rev Mol Cell Biol 2005, 6:850-861.

Cosgrove DJ. *Use of GR2 proteins to modify cellulosic material and to enhance enzymatic and chemical modification of cellulose.* US Patent 2007, 2007/0166805 A1.

Coughlan MP. *The properties of fungal and bacterial cellulases with comment on their production and application.* Biotechnol gene engineer rev. 1985, 3:37-109.

Coward-Kelly G, Aiello-Mazzari C, Kim S, Granda C, Holtzapple M. *Suggested improvements to the standard filter paper assay used to measure cellulase activity.* Biotechnol Bioeng. 2003, 82:745-749.

Cox PM, Betts RA, Jones CD, Spall SA, Totterdell IJ. *Acceleration of global warming due to carbon-cycle feedbacks in a coupled climate model.* Nature. 2000, 408:184-187.

Dashtban M, Schraft H, Qin W. *Fungal bioconversion of lignocellulosic residues: opportunities and perspectives.* Int J Biol Sci. 2009, 5:578-595.

Decker SR, Adney WS, Jennings E, Vinzant TB, Himmel ME. *Automated filter paper assay for determination of cellulase activity.* Appl Biochem Biotechnol. 2003, 105-108:689-703.

Demain AL, Newcomb M, Wu JHD. *Cellulase, clostridia, and ethanol.* Microbiol Mol Biol Rev. 2005, 69:124-154.

Deshpande MV, Ericksson KE, *Patterson LG. Assay method for exoglucanase.* Annl Biochem. 1984, 138:841-842.

Din N, Damude HG, Gilkes NR, Miller RC, Warren RAJ, Kilburn DG. *C1-Cx revisited: intramolecular synergism in a cellulase.* Proc Natl Acad Sci USA. 1994, 91:11383-11387.

Ding SY, Himmel ME. *The maize primary cell wall microfibril: a new model derived from direct visualization.* J Agric Food Chem. 2006, 54:597-606.

Doi RH, Tamaru Y. *The Clostridium cellulovorans cellulosome: an enzyme complex with plant cell wall degrading activity.* Chem Rec. 2001, 1: 24-32.

Eriksson J, Malmsten M, Tiberg F, Callisen TH, Damhus T, Johansen KS. *Enzymatic degradation of model cellulose films.* J Colloid Interface Sci. 2005, 284:99-106.

Eyzaguirre J, Hidalgo M, Leschot A. *Beta-glucosidases from filamentous fungi: properties, structure, and applications.* In: Yarema KJ (ed.) Handbook carbohydrate engineering. London: Taylor & Francis Group. 2005, 645-685.

Fan LT, Lee Y-H, Beardmore DH. *Mechanism of the enzymatic hydrolysis of cellulose: effects of major structural features of cellulose on enzymatic hydrolysis.* Biotechnol Bioeng. 1980, 22:177-199.

Fields MW, Russell JB, Wilson DB. *The role of ruminal carboxymethylcellulases in the degradation of β-glucans from cereal grain.* FEMS Microbiol Ecol. 1998, 27:261-268.

Gama FM, Teixeira JA, Mota M. *Cellulose morphology and enzymatic reactivity: a modified solute exclusion technique.* Biotechonol Bioeng. 1994, 43:381-387.

García-Ochoa F, Santos VE, Casas JA, Gómez E. *Xanthan gum: production, recovery and properties.* Biotechnol Adv. 2000, 18:549-579.

Ghose TK. *Measurement of cellulase activities.* Pure Appl Chem. 1987, 59:257-68.

Gilkes RN, Jervis E, Henrissat B, Tekant B, Miller CR Jr, Warren JAR, Kilburn GD. *The adsorption of a bacterial cellulase and its two isolated domains to crystalline cellulose.* J Biol Chem. 1992, 267:6734-6749.

Han YJ, Chen HZ. *Synergism between corn stover protein and cellulase.* Enzyme Microb Technol. 2007, 41:638-645.

Hrmova M, De Gori R, BJ Smith, JK Fairweather, H Driguez, JN Varghese, GB Fincher, *Structural basis for broad substrate specificity in higher plant beta-D-glucan glucohydrolases.* Plant Cell. 2002, 14:1033-1052.

Hu G, Jr Heitmann JA, Rojas OJ. *Quantification of cellulase activity using the quartz crystal microbalance technique.* Ana Chem. 2009, 81: 1872-1880.

Karboune S, Geraert PA, Kermashat S. *Characterization of selected cellulolytic activities of multi-enzymatic complex system from Penicillium funiculosum.* J Agric Food Chem. 2008, 56:903-909.

Kerff F, Amoros A, Herman R, Sauvage E, Petrella S, Filee P, Charlier P, Joris B, Tabuchi A, Nikolaidis N, Cosgrove DJ. *Crystal structure and activity of Bacillus subtilis YoaJ (EXLX1), a bacterial expansin that promotes root colonization.* Proc Natl Acad Sci USA. 2008, 105:16876-16881.

Kim ES, Lee HJ, Bang WG, Choi IG, Kim KH. *Functional characterization of a bacterial expansin from Bacillus subtilis for enhanced enzymatic hydrolysis of cellulose.* Biotechnol Bioengineer., 2009, 102: 1342-1353.

Klyosov AA. *Cellulases of the third generation.* In: Aubert JP, Beguin P, Millet J (ed.) Biochemistry and genetics of cellulose degradation. London: Academic Press. 1988, 87-99.

Kongruang S, Han MJ, Breton CIG, Penner MH. *Quantitative analysis of cellulose-reducing ends.* Appl Biochem Biotechnol. 2004, 113-116:213-31.

Kopecny J, Hodrova B: *The effect of yellow affinity substance on cellulases of Ruminococcus flavefaciens.* Lett Appl Microbiol, 1997, 25:191-196.

Kosugi A, Arai T, Doi RH. *Degradation of cellulosome-produced cellooligosaccharides by an extracellular non-cellulosomal β-glucan glucohydrolase, BglA, from Clostridium cellulovorans.* Biochem Biophys Res Commun. 2006, 349:20-23.

Kosugi A, Murashima K, Doi RH. *Characterization of two noncellulosomal subunits, ArfA and BgaA, from Clostridium cellulovorans that cooperate with the cellulosome in plant cell wall degradation.* J Bacteriol. 2002, 184:6859-6865.

Lai TE, Pullammanappallil PC, Clarke WP. *Quantification of cellulase activity using cellulose-azure.* Talanta. 2006, 69:68-72.

Li X. *Streptomyces cellulolyticus sp. nov., a new cellulolytic member of the genus Streptomyces.* Int J syst Bacteriol. 1997, 47:443-445.

Li X, Gao PJ. *CMC-liquefying enzyme, a low molecular mass initial cellulose-decomposing cellulase responsible for fragmentation from Streptomyces sp. LX.* Appl Microbiol. 1997, 83:59-66.

Lykidis A, Mavromatis K, Ivanova N, Anderson I, Land M, Dibartolo G, Martinez M, Lapidus A, Lucas S, Copeland A, et al. *Genome sequence and analysis of the soil cellulolytic actinomycete Thermobifida fusca YX.* Bacteriology. 2007, 189:2477-2486.

Lynd L R, Weimer PJ, van Zyl WH, Pretorius IS. *Microbial cellulose utilization: fundamentals and biotechnology.* Microbiol Mol. Biol. Rev. 2002, 66:506-577.

Mansfield SD, Mooney C, Saddler JN. *Substrate and enzyme characteristics that limit cellulose hydrolysis.* Biotechnol Prog. 1999, 15:804-816.

Martinez D, Berka RM, Henrissat B, Saloheimo M, Arvas M, Baker SE, Chapman J, Chertkov O, Coutinho PM, Cullen D, et al. *Genome sequencing and analysis of the biomass-degrading fungus Trichoderma reesei (syn. Hypocrea jecorina).* Nat Biotechnol. 2008, 26:553-560.

Matthews JF, Skopec CE, Mason PE, Zuccato P, Torget RW, Sugiyama J, Himmel ME, Bradya JW. *Computer simulation studies of microcrystalline cellulose Iβ.* Carbohydr Res. 2006, 341:138-152.

Merino ST, Cherry J. *Progress and challenges in enzyme development for biomass utilization.* Adv Biochem Eng Biotechnol. 2007, 108:95-120.

Mingardon F, Chanal A, Tardif C, Bayer EA, Fierobe HP. *Exploration of new geometries in cellulosome-like chimeras.* Appl Environ Microbiol. 2007, 73:7138-7149.

Moser F, Irwin D, Chen SL, Wilson DB. *Regulation and characterization of Thermobifida fusca carbohydrate-binding module proteins E7 and E8.* Biotechnol Bioeng. 2008, 100:1066-1077.

Murashima K, Kosugi A, Doi RH. *Synergistic effects on crystalline degradation between cellulosomal cellulases from Clostridium cellulovorans.* J Bacteriol. 2002, 184:5088-5095.

Murashima K, Kosugi A, Doi RH. *Synergistic effects of cellulosomal xylanase and cellulases from Clostridium cellulovorans on plant cell wall degradation*. J Bacteriol. 2003, 185:1518-1524.

Nicholson MJ, Theodorou MK, Brookman JL. *Molecular analysis of the anaerobic rumen fungus Orpinomyces-insights into an AT-rich genome*. Microbiology. 2005, 151:121-133.

Nishiyama Y, Langan P, Chanzy H. *Crystal structure and hydrogen-bonding system in cellulose Iβ from synchrotron X-ray and neutron fiber diffraction*. J Am Chem Soc. 2002, 124:9074-9082.

Nordmark TS, Bakalinsky A, Penner MH. *Measuring cellulase activity*. Appl Biochem Biotechnol. 2007, 136-140:131-139.

Notley SM, Pettersson B, Wagberg L. *Direct measurement of attractive van der Waals' forces between regenerated cellulose surfaces in an aqueous environment*. J Am Chem Soc. 2004, 126:13930-13931.

Okada H, Tada K, Sekiya T, Yokoyama K, Takahashi A, Tohda H, Kumagai H, Morikawa Y. *Molecular characterization and heterologous expression of the gene encoding a low-molecular-mass endoglucanase from Trichoderma reesei QM9414*. Appl Environ Microbiol. 1998, 64:555-563.

Qian F, An L, Wang M, Li C, Li X. *Isolation and characterization of a xanthan-degrading Microbacterium sp. strain XT11 from garden soil*. J Appl Microbiol. 2007, 102:1362-1371.

Rabinovich ML, Melnik MS, Bolobova AV. *Dedicated to the memory of Berez IV and Feniksova RV microbial cellulases*. J Appl Biochem Microbiol. 2002, 38:305-322.

Ramirez-Ramirez N, Romero-Garcia ER, Calderon VC, Avitia CI, Tellez-Valencia A, Pedraza-Reyes M. *Expression, characterization and synergistic interactions of Myxobacter sp AL-1 Cel9 and Cel48 glycosyl hydrolases*. Int J Mol Sci. 2008, 9:247-257.

Saloheimo M, Paloheimo M, Hakola S, Pere J, Swanson B, Nyyssonen E, Bhatia A, Ward M, Pentilla M. Swollenin, *A Trichoderma reesei protein with sequence similarity to the plant expansins, exhibits disruption activity on cellulosic materials*. Eur J Biochem. 2002, 269:4202-4211.

Samejima M, Sugiyama J, Igarashi K, *Eriksson KEL Enzymatic hydrolysis of bacterial cellulose*. Carbohydr Res. 1998, 305:281-288.

Sanchez C. *Lignocellulosic residues: biodegradation and bioconversion by fungi*. Biotechnol Adv. 2009, 27:185-194.

Sánchez MM, Irwin DC, Pastor FIJ, Wilson DB, Diaz P. *Synergistic activity of Paenibacillus sp. BP-23 cellobiohydrolase Cel48C in association with the contiguous endoglucanase Cel9B and with endo or exo-acting glucanases from Thermobifida fusca*. Biotechnol Bioengineer. 2004, 87:161-169.

Sandford PA, Pittsley JE, Knutson CA, Watson PR, Cadmus MC, Janes A. *Variation in Xanthomonas campestris NRRL B-1459: characterization of xanthan products of differing pyruvic acid content*. In: Sandford PA, Laskin A (eds.), Extracellular microbial polysaccharides. Washington, DC: American Chemical Society. 1977, 192-210.

Schwarz WH. *The cellulosome and cellulose degradation by anaerobic bacteria*. Appl Microbiol Biotechnol. 2001, 56:634-649.

Setlow B, Cabrera-Hernandez A, Cabrera-Martinez RM, Setlow P. *Identification of aryl-phospho-beta-D-glucosidases in Bacillus subtilis*. Arch Microbiol. 2004, 181:60-7.

Shoseyov O, Shani Z, Levy I. *Carbohydrate binding modules: biochemical properties and novel applications*. Microbiol Mol Biol Rev. 2006, 70:283-295.

Taylor LE, Henrissat B, Coutinho PM, Ekborg NA, Hutcheson SW, Weiner RM. *Complete cellulase system in the marine bacterium Saccharophagus degradans strain 2-40T*. J Bacteriol. 2006, 188:3849-3861.

Tokuda G, Watanabe H, Matsumoto T, Noda H. *Cellulose digestion in the wood-eating higher termite, Nasutitermes takasagoensis (Shiraki): distribution of cellulases and properties of endo-β-1,4-glucanase*. Zool. Sci. 1997, 14:83-93.

Toyama H, M Nakano, Satake Y, Toyama N. *Rapid isolation of the Trichoderma strain with higher degrading ability of a filter paper and superior proliferation characteritics using avicel plates and the double-layer selection medium*. Appl biochem biotechnol. 2008, 145:23-28.

Wang LS, Zhang YZ, Gao PJ. *A novel function for the cellulose binding module of cellobiohydrolase I*. Sci China Ser C. 2008, 51:620-629.

Wang TH, Liu T, Wu ZH, Liu SL, Lu Y, Qu YB. *Novel cellulase profile of Trichoderma reesei strains constructed by cbh1 gene replacement with eg3 gene expression cassette*. Acta Biochimica et Biophysica Sinica. 2004, 36:667-672.

Watanabe H, Nakashima K, Saito H, Slaytor M. *New endo-β-1,4-glucanases from the parabasalian symbionts, Pseudotrichonympha grassii and Holomastigotoides mirabile of Coptotermes termites*. Cell Mol Life Sci. 2002, 59:1983-1992.

Weimer PJ, Hackney JM, Jung HJG, Hatfield RD. *Fermentation of a bacterial cellulose/xylan composite by mixed ruminal microflora: implications for the role of polysaccharide matrix interactions in plant cell wall biodegradability*. J Agric Food Chem. 2000, 48:1727-1733.

Wilson DB. *Studies of Thermobifida fusca plant cell wall degrading enzymes*. Chem Rec. 2004, 4:72-82.

Wilson DB. *Three microbial strategies for plant cell wall degradation*. Ann. NY Acad Sci. 2008, 1125:289-297.

Wood TM. *Fungal cellulases*. In: Haigler CH, Weimer PJ (ed.) Biosynthesis and biodegradation of cellulose. New York, NY: Marcel Dekker Inc. 1991, 491-533.

Wood TM, Bhat KM. *Methods of measuring cellulase activities*. Method Enzymol. 1988, 160: 87-112.

Wyman CE. *What is (and is not) vital to advancing cellulosic ethanol*. Trend Biotechnol. 2007, 25:153-157.

Xie G, Bruce DC, Challacombe JF, Chertkov O, Detter JC, Gilna P, Han CS, Lucas S, Misra M, Myers GL, et al. *Genome sequence of the cellulolytic gliding bacterium Cytophaga hutchinsonii*. Appl Environ Microbiol. 2007, 73:3536-3546.

Yao Q, Sun TT, Liu WF, Chen GJ. *Gene cloning and heterologous expression of a novel endoglucanase, swollenin from Trichoderma pseudokoningii S38*. Biosci Biotechnol Biochem. 2008, 72:2799-2805.

Yaoi K, Kondo H, Hiyoshi A, Noro N, Sugimoto H, Tsuda S, Mitsuishi Y, Miyazaki K. *The structural basis for the exo-mode of action in GH74 oligoxyloglucan reducing end-specific cellobiohydrolase*. J Mol Biol. 2007, 370:53-62.

Zhang YHP, Lynd LR. *Toward an aggregated understanding of enzymatic hydrolysis of cellulose: noncomplexed cellulose systems*. Biotechnol Bioeng. 2004, 88:797-824.

Zhang YHP, Lynd LR. *Regulation of cellulase synthesis in batch and continuous cultures of Clostridium thermocellum*. J Bacteriol. 2005, 187:99-106.

Zhang YH, Ding SY, Mielenz JR, Cui JB, Elander RT, Laser M, Himmel ME, McMillan JR, Lynd LR. *Fractionating recalcitrant lignocellulose at modest reaction conditions.* Biotechnol Bioeng. 2007, 97:214-223.

Zhou J, Wang YH, Chu J, Zhuang YP, Zhang SL, Yin P. *Identification and purification of the main components of cellulases from a mutant strain of Trichoderma viride T100-14.* Bioresource Technol. 2008, 99:6826-6833.

Zhou X, Chen H, Li Z. *CMCase activity assay as a method for cellulase adsorption analysis.* Enzyme and Microbial Technol. 2004, 35:455-459.

Zverlov VV, Schantz N, Schwarz WH. *A major new component in the cellulosome of Clostridium thermocellum is a processive endo-beta-1,4-glucanase producing cellotetraose*. FEMS Microbiol Lett. 2005, 249:353-358.

In: Cellulase: Types and Action, Mechanism, and Uses ISBN: 978-1-61761-983-0
Editor: Adam E. Golan

Chapter 9

CELLULASE: TYPES, ACTIONS, MECHANISMS AND USES

Tzi bun Ng * ***and Randy Chi Fai Cheung***
School of Biomedical Sciences, Faculty of Medicine,
The Chinese University of Hong Kong. P.R. of China

Cellulases are celluloytic enzymes (EC 3.2.1.4) produced mainly by microbes including fungi, bacteria, and also by protozoans. However, plants and animals also produce cellulases. Several different kinds of cellulases differing in structure and mechanism of action are known. Cellulases catalyze the hydrolysis of 1, 4-beta-D-glycosidic linkages in cellulose, lichenin and cereal beta-D-glucans. Other names of cellulase are endoglucanase, endo-1,4-beta-glucanase, carboxymethyl cellulase (CMCase), endo-1,4-beta-D-glucanase, beta-1,4-glucanase, beta-1,4-endoglucan hydrolase and celludextrinase. Excocellulases and beta-glucosidases are other types of cellulases. Avicelase refers to the total cellulase activity of a given sample of the enzyme(s). The cellulase activity may be the consequence of the action of more than one type of enzymes.

TYPES

There are five general types of cellulases based on the type of reaction catalyzed:

1) Endo-cellulases break internal bonds to disrupt the crystalline structure of cellulose and expose individual polysaccharide chains
2) Exo-cellulases remove 2-4 units from the ends of the exposed chains produced by endocellulase, producing tetrasaccharides or disaccharides such as cellobiose. There are two main types of exo-cellulases (or cellobiohydrolases) working

* Corresponding author (Tel: 852-26098031, Fax: 852-26035123, Email: b021770@mailserv.cuhk.edu.hk

processively from the reducing end and from the non-reducing end of cellulose, respectively.

3) Cellobiases or beta-glucosidases hydrolyse the exo-cellulase product into individual monosaccharides.
4) Oxidative cellulases depolymerize cellulose by radical reactions, e.g. cellobiose dehydrogenase (acceptor).
5) Cellulose phosphorylases depolymerize cellulose using phosphates rather than water.

Cellulases can also be classified into three types of enzymes as follows: (1) endoglucanases (EC 3.2.1.4), which catalyze the cleavage of internal β-1, 4-glucosidic bonds; (2) exoglucanases (EC 3.2.1.91), which processively act on the reducing and non-reducing ends of cellulose chains to liberate short-chain cello-oligosaccharides; and (3) β-glucosidases (EC 3.2.1.21), that catalyzes the hydrolysis of soluble cellooligosaccharides (e.g. cellobiose) to glucose. Hemicellulases comprise enzymes that break down both β-1, 4- xylan (xylanases EC 3.2.1.8 and β-xylosidases EC 3.2.1.37) and various side chains (α-l-arabinofuranosidases EC 3.2.1.55, α-glucuronidases EC 3.2.1.139, acetyl xylan esterases EC 3.1.1.72, ferulic acid esterases EC 3.1.1.73, and α-galactosidases EC 3.2.1.22). New cellulases and hemicellulases from bacteria and fungi have been purified (Shallom and Shoham, 2003; Hilden and Johansson, 2004), and the regulation of microbial production of cellulase has been studied (Aro et al., 2005; Foreman et al., 2003). Furthermore, significant progress has been made to reduce the cost of cellulases. For instance, cellulases from the aerobic fungus *Trichoderma reesei* have been in the limelight of research for half a century and are the most frequently used cellulases in laboratory and pilot-scale bioethanol production. A cost reduction of greater than 10-fold has been achieved for *T. reesei* cellulases (Greer, 2005), resulting in an enzyme cost of around 10 to 20 cents per gallon of ethanol produced. Cost reduction was made possible by enzyme engineering and fermentation process development.

Actions

Cellulase is produced mainly by symbiotic bacteria in the gastric chambers of ruminants to degrade cellulose to beta-glucose. Most animals (including humans) other than ruminants, do not produce cellulase and are unable to use most of the energy contained in plant materials. Enzymes that hydrolyze hemicellulose are often called hemicellulases and classified under cellulases. Lignin degrading enzymes are sometimes classified as cellulases, though this classification is usually considered inappropriate.

Within the above types there are also progressive (also known as processive) and non-progressive types. Progressive cellulases will continue to interact with a single polysaccharide strand. Non-progressive cellulases will interact once, then disengage, and engage another polysaccharide strand.

The majority of fungal cellulases exhibit a two-domain structure composed of a catalytic domain and a cellulose binding domain, linked by a flexible linker. This structure facilitates enzymatic action on an insoluble substrate and it enables the enzyme to diffuse two-dimensionally on a surface like a caterpillar. However, cellulases (mostly endoglucanases) devoid of a cellulose binding domain, which probably have a swelling function, are present.

MECHANISM

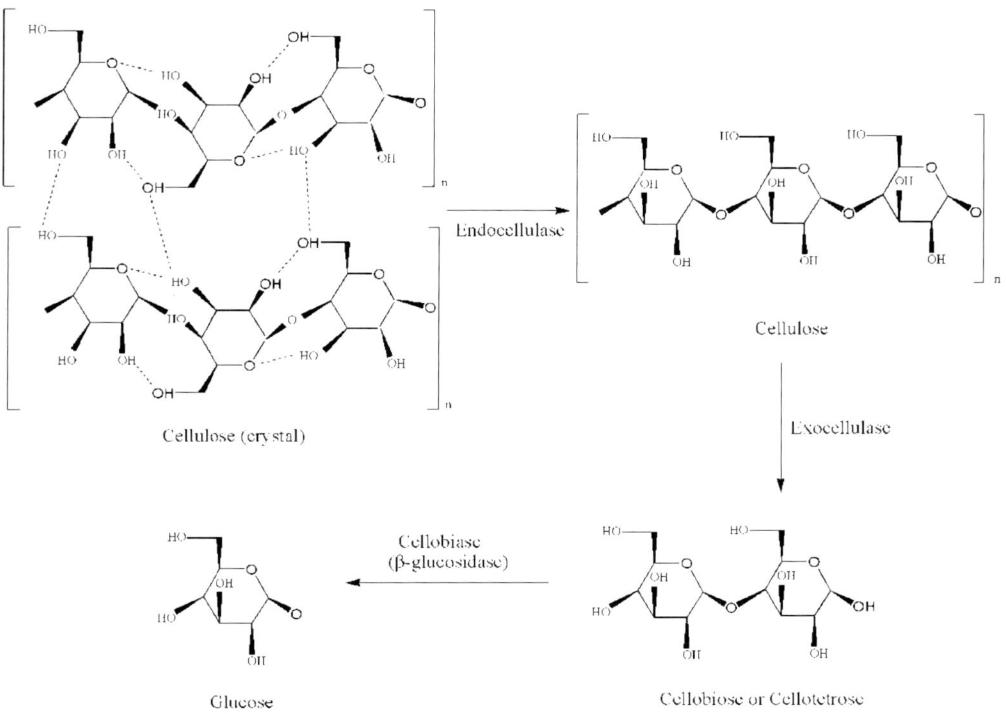

The following are three types of reaction catalyzed by cellulases:

1) Breakage of the non-covalent interactions present in the crystalline structure of cellulose (endo-cellulase)
2) Hydrolysis of the individual cellulose fibers to break it down into smaller sugars (exo-cellulase)
3) Hydrolysis of disaccharides and tetrasaccharides into glucose (beta-glucosidase).

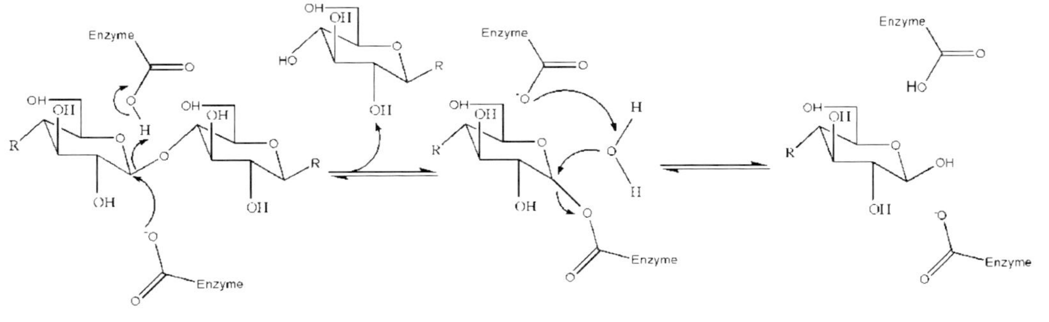

Mechanistic Details of Beta-Glucosidase Activity of Cellulase

In some cellulolytic microorganisms, several types of cellulases are assembled into a multifunctional supramolecular complex, referred to as the cellulosome, with subunits consisting of interacting functional domains. One of these subunits is a class of noncatalytic scaffolding polypeptide, which integrates the cellulase and xylanase subunits into the cohesive complex. Application of cellulosome hybrids and chimeric constructs of cellulosomal domains may promote use of cellulosic biomass and economical conversion of cellulosic biomass to biofuels and may facilitate diverse applications in research, medicine and industry. An array of chemical, crystallographic, microscopic, biochemical and molecular genetic techniques facilitates new insights into both the structure of cellulose and the mechanisms of its hydrolysis (Bayer et al., 1994, 1998).

Cellulolytic proteins form a complex of enzymes that collaborate to break down cellulose into the soluble products cellobiose and glucose. Studies on the molecular mechanisms of cellulose action have been expedited by advances in molecular biology. Homology and similarity in mechanisms of action exist between cellulases from various microorganisms (Mosier et al., 1999). A more comprehensive picture of the cellulolytic action of cellulases has appeared and combines the physical and chemical characteristics of solid cellulose substrates with the specialized structure and function of the cellulases that degrade it.

Amino acid sequence comparison reveals that the catalytic cores of cellulases belong to several families. The enzymes in each family have similarities in folding pattern, catalytic residues, and reaction mechanism that involves either single substitution with inversion of configuration or double substitution leading to retention of the beta-configuration at the anomeric carbon. Other than catalytic domains, many cellulolytic enzymes exhibit domains that are not involved in catalysis, but play a role in substrate binding, multi-enzyme complex formation, or attachment to cell surface. These domains may facilitate the degradation of crystalline cellulose by inhibiting enzyme detachment from the substrate surface, by focusing hydrolysis on specific regions in which the substrate is destabilized by multiple cutting events, and by facilitating recovery of the soluble degradation products. In many cellulolytic organisms, cellulase synthesis is suppressed by soluble metabolizable carbon sources and stimulated by cellulose (Béguin and Aubert, 1994).

USES

Cellulase is utilized in commercial food processing in coffee. It hydrolyses cellulose during the drying of beans. Cellulase is used for fermenting biomass into biofuels at a relatively experimental stage. Cellulase is used for treating Phytobezoars, a form of cellulose bezoar found in the human stomach. The main potential applications are in food, animal feed, textile, fuel and chemical industries (Beguin and Anbert, 1993; Coughlan, 1985; Mandels, 1985). Other areas of application encompass paper and pulp industry, waste management, medical/ pharmaceutical industry, protoplast production, genetic engineering and pollution treatment (Beguin and Anbert, 1993; Coughlan, 1985; Mandels, 1985). In textile industry, cellulases are exploited for (a) eliminating excessive dye from the denim fabric in pre-faded blue jeans (biostoning), (b) removing microfibrils projective from cotton fabrics following several washing cycles, and (c) reinstating the softness and color brightness of cotton fabrics (Beguin and Anbert, 1993; Mandels, 1985). In addition, either cellulases or mixture of glucanases have been utilized for the production of plant and fungal protoplasts, in producing hybrid strains as well as in other genetic engineering experiments (Banehop, 1981).

BIOFUEL

There has been a search for alternatives to petroleum-derived fuels in an attempt to decrease the dependence on non-renewable resources. The most common renewable fuel is ethanol obtained from corn grain (starch) and sugar cane (sucrose). It is anticipated that in the near future, the supply of raw materials will be limited, hence lignocellulosic biomass, comprising agricultural residues, wood, municipal solid waste and energy crops may be regarded as supplies of ethanol in the future (Gray et al., 2006). Development of technologies that will enable cost-effective conversion of biomass into fuels and chemicals is in progress. These technologies include low-cost thermochemical pretreatment, highly effective cellulases and hemicellulases and efficient and robust fermentative microorganisms.

Biomass is composed of cellulose (40–50%), hemicellulose (25–35%) and lignin (15–20%). Cellulose is made up of glucose connected via β-1,4 glycosidic bonds. Due to the β-1,4 linkage, cellulose is highly crystalline and compact confering on it high resistance to biological attack. In general, hemicellulose is comprised of a main chain xylan backbone (β-1,4 linkages) and branches of arabinose, galactose, glucuronic acid, mannose, etc. The extent of branching and type of minor sugars in hemicellulose change with the type of plants. Furthermore, lignin can be covalently linked to hemicellulose by ferulic acid ester bonds. It is much more difficult to enzymatically degrade lignocellulose than starch to fermentable sugars because of the compactness and complexity of the former. Thus, it is more expensive to produce ethanol from biomass than production from starch (Wyman, 2003). In order to compete with grain-derived ethanol, the enzymes used to hydrolyse biomass have to be more efficient and less costly. The presence of non-glucose sugars in the feedstock complicates the fermentation process since conversion of pentose sugars into ethanol is less efficient than conversion of the hexose sugars.

Several approaches have been adopted to enhance cellulase performance and reduce the quantity of enzyme required to saccharify biomass substrates. The primary target for cellulase engineering has been the cellobiohydrolases, as they make up 60–80% of natural cellulase systems (Lynd et al., 2002). Teter et al. (2004) employed site-directed mutagenesis, site-saturation mutagenesis, error-prone PCR and DNA shuffling to generate variants of *T. reesei* Cel7A. The mutants, which were expressed in *S. cerevisiae,* were screened for enhanced thermal stability and thermal activity. One of the variants identified was expressed in *T. reesei* instead of wild-type Cel7A. The recombinant strain had higher efficacy than the parent *T. reesei* with regard to hydrolysis of pretreated corn stover. Site-directed mutagenesis was employed by Day et al. (2003) to produce *Hypocrea jecorena* cellobiohydrolase Cel7A (CBH1) variants. Several mutants expressed in *T. reesei* demonstrated better thermostability and reversibility. An alternative strategy is to introduce heterologous enzymes into an existing system such as *T. reesei* so that improvement of the overall performance of the system can be achieved. Bower (2005) introduced several bacterial endoglucanases into *T. reesei.* One of them, GH5A from *Acidothermus cellulolyticus*, was fused to *T. reesei* cellobiohydrolase CBH1. The fusion protein expressed in *T. reesei* exhibited higher efficacy in the saccharification of pretreated corn stover. It required 6 hours to achieve 20% cellulose conversion, compared with 10 hours for the same conversion with the parent cellulase. Various CBH I and II homologs from a variety of fungi that may be useful in biomass hydrolysis have been discovered (Wu et al., 2003; Lange et al., 2002).

Cellulases possess carbohydrate-binding modules (CBMs) which facilitate the interaction with the substrate surface. CBMs are placed in different families according to amino acid sequence and crystal structure. Very different ligand specificity can arise from slight structural differences in CBMs (Boraston et al., 2004). By binding to the surface of crystalline cellulose, CBMs target their cognate catalytic domains to specific substrates and augment catalytic efficiency by raising the effective concentration at the surface. Other than targeting, CBMs may disrupt the structure of polysaccharides and thus increase the hydrolytic rate. Vaaje-Kolstad et al. (2005a. b) demonstrated that efficient chitin degradation is reliant on the presence of a small non-catalytic protein, CBP21, that binds to the crystalline substrate. CBP21 binding brings about structural changes in the substrate and hydrolysis is enhanced. The performance of cellulase may be increased with the help of similar CBMs that also display disruptive function. The lignin component of the biomass material poses challenges for enzymatic degradation owing to its nonproductive binding and inactivation of cellulases. Berlin et al. (2005) proposed a new strategy to enhance activity of cellulases for lignocellulosic hydrolysis by employing weak lignin-binding enzymes. Naturally occurring cellulases with similar catalytic activity on a model cellulosic substrate differ considerably in affinity for lignin, hence influencing their performance on native substrates. Palonen (2004) noted that the location and structure of lignin affect the enzymatic hydrolysis more than the absolute amount of lignin. Modification of lignin surfaces by oxidative treatments with laccase alone and delignification treatment with a laccase-mediator system results in enhanced lignocellulose hydrolysis.

The efficient degradation of hemicellulose necessitates the interplay of various enzymes. Hemicellulases promote cellulose hydrolysis by exposing and making cellulose fibers more accessible (Shallom and Shoham, 2003). Commercial development of hemicellulases for lignocellulose hydrolysis is not at par with cellulases since commercial preparations have been developed on biomass pretreated with dilute-acid from which hemicellulose has been removed before saccharification. On the other hand, for non-acid pretreatment methods (Mosier et al., 2005) in which the hemicellulose fraction remains intact, there is a need for hemicellulases. Cellulases like those from *T. reesei* have low hemicellulase activity and cannot achieve complete conversion to monomer sugar. Progress has been made in understanding the structure/function of xylanases (Kaneko et al., 2004; Pell et al., 2004a), enzyme specificity (Pell et al., 2004b; Vardakou et al. 2005), and hemicellulase CBMs (Boraston et al., 2004; Levasseur et al., 2005). Development of inexpensive, commercial hemicellulases that synergize with cellulases in production of bioethanol is an important goal.

Significant progress has been made in all aspects of lignocellulosic conversion to ethanol. It is important to cut capital and operating costs of each of the unit operations. Enzyme costs have been decreased by a combination of protein engineering and process development. However, further cost cuts are needed and will more than likely come from novel, tailored cocktails of enzymes with higher specific activities than current commercial enzymes. A commercial organism must, however, be able to withstand the rigors of a full-scale process, including potential toxic compounds present in the sugar hydrolysate.

As the largest developing country, China encounters the challenge in meeting its need for immense amounts of energy resources, in particular for liquid fuel. As China has an enormous population, the Chinese government has commenced a bioethanol project, and has generated approximately 1 million tons of ethanol fuel from corn and wheat in 2005. Research projects in China on lignocellulose biodegradation and biotransformation have been implemented. A corncob biorefinery process has been developed in Shandong University. By combining cellulase and ethanol production with production of xylose-related products, the total production cost can be cut (Yinbo et al., 2006).

WASTE TREATMENT

Lignocellulose is the major plant cell wall component of the biosphere and gives rise to the largest volume of waste. Hence, the landfills are rapidly filling up. Microorganisms that produce cellulosomes are capable of transforming lignocelluose to microbial cell mass and products (e.g. ethanol). The combination of designer cellulosomes with new production concepts could in the future provide the breakthroughs essential for economical conversion of cellulosic biomass to biofuels (Bayer et al., 2007).

Municipal solid waste (MSW) comprises essentially cellulose in various forms including cardboard, newspaper and wood which are mostly biodegradable (Rathje, 1991), the bulk of which is dumped in landfills. The uncontrollable landfill environment contains variable microbial populations and/or enzyme systems, as well as suboptimal environments, leading to the slow rate of anaerobic degradation in landfills. Moreover, landfills contaminate groundwater and rapidly fill to capacity. The conversion of cellulosic waste from MSW and landfills to biofuels has long been considered to be a worthwhile undertaking. The benefits would be two-fold: firstly, the amount of cellulose waste (the largest single waste byproduct of our society) would be reduced and its environmental effects will be lowered, and secondly the pollutant would be changed to an energy source to lessen the reliance on fossil fuels.

In natural anaerobic digestion processes in MSW landfills, some micro-organisms collectively produce fermentable sugars from polysaccharides and others convert sugars to methane and carbon dioxide. Such mixed fermentations are difficult to establish and maintain at large scale. MSW, herbaceous crops and woody biomass share the same rate-limiting step for bioconversion processes: the hydrolysis of complex polysaccharides to fermentable sugars (Horton et al., 1980). The main biodegradable polymer, cellulose, is usually protected by lignin, a relatively inactive, polyphenylpropane, three-dimensional polymer (Fengel and Wegener, 1984), and by hemicelluloses (Grohmann et al., 1985). During the aerobic degradation of lignocellulose, lignocellulosic sugars are liberated by thermal and chemical pretreatment, and then by aerobic enzymatic hydrolysis of chopped or milled biomass. The pretreated soluble fraction of biomass is referred to as the hydrolysate, and the hydrolysate which contains insoluble material is known as the slurry. In diluted acid pretreatment, the bulk of the hemicellulosic sugars (xylose, arabinose, galactose and mannose) is solubilized. However, the glucose component stays as cellulose in the solid form, and it is depolymerized by cellulases. In the case where enzymes are added to the slurry and the saccharification process is permitted to proceed independently of fermentation, the process is known as separate hydrolysis and fermentation (SHF). When cellulases are combined with anaerobic fermentation (usually yeast) to relieve the enzymes from product inhibition, the process is referred to as simultaneous saccharification and fermentation (SSF). A process based on the fermentation of pentose sugars (derived from the hydrolysate), combined with the saccharification of cellulose and fermentation of glucose (derived from cellulose), is called simultaneous saccharification and cofermentation (SSCF). Alternatively, a hybrid process with partial enzymatic hydrolysis to obtain high cellulose hydrolysis rates by operating at high temperature and co-fermentation can be used to achieve high overall conversion rates of biomass sugars to ethanol. This process is referred to as hybrid saccharification and fermentation (HSF). It exploits thermostable enzymes capable of functioning under conditions in which known ethanologens fail to function (i.e. >75 ^{o}C). After several days of saccharification and fermentation, the bulk of the major and minor sugars will have been converted to ethanol.

TREATMENT FOR PHYTOBEZOARS

A bezoar is a hard lump of undigested foreign matter in the gastrointestinal tract. The symptoms of bezoars comprise distension, pain and vomiting. In this case, ulceration, gastric bleeding and even perforation may be present. If untreated, a significant mortality would ensue. The most common bezoar is the phytobezoar, made up of vegetable fibers. The fiber content (cellulose, hemicellulose, lignin, and tannins) in phytobezoars has a large amount of polymerized tannins consisting of leucoanthocyanins and catechins predominantly. Unripe fruits have a high concentration of tannin monomers. The formation of phytobezoars is attributed to polymerization of these tannin monomers. Upon ingestion of unripe fruits, gastric hydrochloric acid initiates the polymerization, leading to the formation of a tannin-cellulose-hemicellulose-protein complex. Cellulase is administered by a nasogastric tube, No adverse effects have been reported (Fernández Morató et al., 2009). Enzymes, especially cellulose, help to break up the mass which it can then be aspirated or allowed to pass on. Treatment of patients with phytobezoars with cellulose has been proven to be simple, safe and effective. Stanten and Peters (1975) presented a review of phytobezoars that focused on medical treatment using enzymatic dissolution for bezoars found in the gastric pouch. Clinical and *in vitro* investigations disclosed the effectiveness of both papain and cellulase. It was suggested that a combination of them be administered since each acts on a different component of the bezoar. There were no complications reported by the treated patients. The review by Walker-Renard (1993) provides clinicians with information concerning available medicinal agents for the management of phytobezoars. Data sources were accquired by a Medline search from 1966 to 1993. All citations containing references to patients with a phytobezoar treated with medicinal agents were chosen and reviewed for the treatment regimen, number of patients treated, duration of therapy, success rate, and adverse effects. A total of 36 patients with phytobezoars were reviewed. Papain was efficacious in treating 87% (13 of 15) and cellulase in 100% (19 of 19) of the patients. Adverse effects reported in the papain-treated group included gastric ulcer, esophageal perforation, and hypernatremia. No adverse effects were reported by the cellulose-treated group. The review reveals the efficacy of papain and cellulase in the management of phytobezoars in the small number of patients studied. However, controlled clinical trials are needed to compare the safety and efficacy of the two agents.

DETERGENTS

Alkaliphilic *Bacillus* species capable of producing alkaline exoenzymes have been isolated and a possible application of alkaline cellulase (carboxymethylcellulase) as an additive to improve the efficiency of detergents has been found (Ito, 1997). The enzymatic properties of some candidate cellulases indicate that they would be suitable for use as laundry detergents. The characteristics and possible catalytic mechanism of the hydrolytic reaction and the gene for the industrial alkaline cellulase produced by one of the isolates, *Bacillus* sp KSM-635, were reported. When a colored cotton garment is washed repeatedly, it gradually loses the brightness of its color. The effects are caused by microfibril formation from cotton fibers. The larger surface area results in more light reflection, causing the brightness of the fabric color to diminish. The cellulase molecule binds to and hydrolyzes an exposed fibril on the yarn surface by acting on the beta-1, 4-glucosidic bond. The enzyme leaves the interior part of the cotton fibre in the yarn intact. Cellulases are useful cleaning additives. They remove fine surface fuzz and fibrils from cotton textiles and inhibit the accumulation of new pills on the textile surface. These effects confer a more shiny look to colored textiles, in spite of the wear and tear from frequent washing.

TEXTILE INDUSTRY

Cellulase expedites the degradation of cellulose, present in jeans as the chief constituent of cotton and other natural plant fibers. The surface cellulose binds to the active sites of cellulase, disrupting bonds and liberating indigo dye particles from the jeans surface. After the reaction, only the dye has encapsulated. The jeans remain intact and become faded. Among the various techniques used in the jean industry, biowashing with cellulase has the highest popularity due to its environment-friendliness (different form the use of pumice stones or acid, enzymes can be recycled and do not constitute health hazard) and effectiveness (since cellulose is influenced by temperature and pH which in turn can be manipulated to regulate the different levels of stonewashing.)

PULP AND PAPER INDUSTRY

Cellulases have been used in the pulp and paper industry for biomechanical pulping to modify the coarse mechanical pulp and hand sheet strength properties, deinking of recycling fibers and to improve drainage and runnability of paper mills. Cellulases are used to remove inks, coating and toners from paper, bio-characterize of pulp fiber, prepare easily biodegradable cardboard, and manufacture soft paper including paper towels and sanitary paper.

Food and Animal Feed

Cellulases are employed in the food industry for (a) extracting fruit juices and oil from seeds, (b) clearing fruit juices, (c) improving the soaking efficiency and homogeneous water absorption of cereals, (d) removing external soybean coat during production of fermented soybean foods such as soysauce and miso, (e) isolating proteins from soybean and coconut, (f) isolating starch from corn and sweet potato, (g) gelatinizing seaweeds to enhance digestibility, (h) extracting agar from seaweeds, and (i) digesting ball-milled lignocellulose which can be employed as food additive (Beguin and Anbert, 1993; Coughlan, 1985; Mandels, 1985). Cellulases can also be utilized for (a) elevating the nutritive quality of fermented foods, (b) enhancing the rehydrability of dried vegetables and soup mixtures, (c) producing cello-oligosaccharides, glucose and other soluble sugars from cellulosic wastes, and (d) removing cell wall which will expedite the liberation of flavors, enzymes, polysaccharides and proteins (Mandels, 1985). In brewing and wine industries, cellulases are utilized for (a) hydrolyzing 13-1, 3 and 13-1, 4 glucan present in barley of a low grade and facilitate the filtration of beer, and (b) strengthening wines aroma. The recombinant yeasts producing 13-1, 3 and 13-1, 4-glucanases have been exploited in brewing industry (Beguin and Anbert, 1993). In animal feed industry, the cellulases are employed (a) as a feed supplement for ruminants and monogastric animals, (b) in pretreatment of lignocellulosic material, dehulling of cereal grains, treatment of silage to increase the digestibility of ruminants and monogastric animals (Mandels, 1985). Another application is the cloning of cellulase genes to yield transgenic animals capable of secreting the desired cellulases into the gastrointestinal tract and facilitate roughage digestion (Beguin and Anbert, 1993).

Conclusion

From the foregoing account, it can be seen that cellulases are enzymes that have many applications, continuing research may reveal sources of cellulases with more desirable characteristics for application.

Acknowledgments

We thank Dr. Chuan-hao Li, for diagram drawing.

References

Aro N, Pakula T, Penttilae M. (2005) *Transcriptional regulation of plant cell wall degradation by filamentous fungi.* FEMS Microbiol Rev 29: 719-739.

Banehop T. (1981) *The anaerobic fungi in rumen fiber digestion.* Agric. Environ. 6: 339-348.

Bayer EA, Chanzy H, Lamed R, Shoham Y. (1998) *Cellulose, cellulases and cellulosomes.* Curr Opin Struct Biol. 8(5):548-57.

Bayer EA, Lamed R, Himmel ME. (2007) *The potential of cellulases and cellulosomes for cellulosic waste management.* Curr Opin Biotechnol. 18(3):237-45.

Bayer EA, Morag E, Lamed R. (1994) *The cellulosome-a treasure-trove for biotechnology.* Trends Biotechnol. 12(9):379-86.

Beguin P, Anbert JP. (1993) *The biological degradation of cellulose.* FEMS Microbiol. Rev. 13: 25-58.

Béguin P, Aubert JP. (1994) *The biological degradation of cellulose.* FEMS Microbiol Rev. 13(1):25-58.

Berlin A, Gilkes N, Kurabi A, Bura R, Tu M, Kilburn D, Saddler J. (2005) *Weak lignin-binding enzymes. A novel approach to improve activity of cellulases for hydrolysis of lignocellulosics.* Appl Biochem Biotechnol 121-124: 163-170.

Boraston AB, Bolam DN, Gilbert HJ, Davies GJ. (2004) *Carbohydrate binding modules: fine-tuning polysaccharide recognition.* Biochem J 382:769-781.

Bower BS (Genencor International IU). (2005) *Fusion proteins of an exocellobiohydrolase and an endoglucanase for use in the saccharification of cellulose and hemicellulose.* Patent 2005093073.

Coughlan MP. (1985) *Cellulases: Production properties and applications.* Biochem. Soc. Trans. 13: 405-406

Day AG (Genencor International IU). (2003) *Novel variant Hypocrea jecorina CBH I cellulases, their production with recombinant cells, and their uses.* patent 2004016760.

Fengel D, Wegener G. (1984) *Wood: Chemistry, Ultrastructure, Reactions.* Berlin: Walter de Gruyter.

Fernández Morató J, Ilzarbe Sánchez L, Bessa Caserras J, Mateu de Antonio J. (2009) *Cellulase treatment in 3 cases of large phytobezoars.* Farm Hosp. 33: 100-3.

Foreman PK, Brown D, Dankmeyer L, Dean R, Diener S, Dunn-Coleman NS, Goedegebuur F, Houfek TD, England GJ, Kelley AS. (2003) *Transcriptional regulation of biomass degrading enzymes in the filamentous fungus Trichoderma reesei.* J Biol Chem 278: 31988-31997.

Gray KA, Zhao L, Emptage M. (2006) *Bioethanol. Curr* Opin Chem Biol. 10: 141-6.

Greer D. (2005) *Spinning straw into fuel.* Biocycle 46: 61-65.

Grohmann K, Torget R, Himmel ME. (1985) *Optimization of dilute acid pretreatment of biomass.* Biotechnol Bioeng Symp 15: 59-80.

Hilden L, Johansson G. (2004) *Recent developments on cellulases and carbohydrate-binding modules with cellulose affinity.* Biotechnol Lett 26: 1683-1693.

Horton GL, Rivers DB, Emert GH. (1980) *Preparation of cellulosics for enzymatic conversion.* Ind Eng Chem Prod Res Dev 19: 422-429.

Ito S. (1997) *Alkaline cellulases from alkaliphilic Bacillus: enzymatic properties, genetics, and application to detergents.* Extremophiles. 1(2):61-6.

Kaneko S, Ichinose H, Fujimoto Z, Kuno A, Yura K, Go M, Mizuno H, Kusakabe I, Kobayashi H. (2004) *Structure and function of a family 10 β-xylanase chimera of Streptomyces olivaceoviridis E-86 FXYN and Cellulomonas fimi cex.* J Biol Chem 279: 26619-26626.

Lange L, Wu W, Aubert D, Landvik S, Schnorr KM, Clausen IG (Novozymes A/S D). (2002) *Cloning and sequences of microbial cellobiohydrolase I for use in DNA shuffling, ethanol prodn. transgenic plants, and detergents.* patent 2003000941.

Levasseur A, Navarro D, Punt PJ, Belaich JP, Asther M, Record E. (2005) *Construction of engineered bifunctional enzymes and their overproduction in Aspergillus niger for improved enzymatic tools to degrade agricultural by-products.* Appl Environ Microbiol 71: 8132-8140.

Lynd LR, Weimer PJ, van Zyl WH, Pretorius IS. (2002) *Microbial cellulose utilization: fundamentals and biotechnology.* Microbiol Mol Biol Rev 66: 506-577.

Mandels M. (1985) *Applications of cellulases.* Biochem. Soc. Trans. 13: 414-415.

Mosier N, Wyman C, Dale B, Elander R, Lee YY, Holtzapple M, Ladisch M. (2005) *Features of promising technologies for pretreatment of lignocellulosic biomass.* Bioresour Technol 96: 673-686.

Mosier NS, Hall P, Ladisch CM, Ladisch MR. (1999) *Reaction kinetics, molecular action, and mechanisms of cellulolytic proteins.* Adv Biochem Eng Biotechnol. 65:23-40.

Palonen H. (2004) *Role of lignin in the enzymatic hydrolysis of lignocellulose.* VTT Publications 520: 1-80.

Pell G, Szabo L, Charnock SJ, Xie H, Gloster TM, Davies GJ, Gilbert HJ. (2004a) *Structural and biochemical analysis of Cellvibrio japonicus xylanase 10C: how variation in substrate-binding cleft influences the catalytic profile of family GH-10 xylanases.* J Biol Chem 279: 11777-11788.

Pell G, Taylor EJ, Gloster TM, Turkenburg JP, Fontes CMGA, Ferreira LMA, Nagy T, Clark SJ, Davies GJ, Gilbert HJ. (2004b) *The mechanisms by which family 10 glycoside hydrolases bind decorated substrates.* J Biol Chem 279: 9597-9605.

Rathje WL. (1991) *Once and future landfills.* Natl Geogr Mag 179: 116-134.

Sarkanen KV, Ludwig CH. (1971) *Lignins: Occurrence, Formation, Structure and Reactions.* New York: Wiley-Interscience.

Shallom D, Shoham Y. (2003) *Microbial hemicellulases.* Curr Opin Microbiol 6: 219-228.

Stanten A, Peters HE Jr. (1975) *Enzymatic dissolution of phytobezoars.* Am J Surg. 130(2): 259-61.

Teter S, Cherry J, Ward C, Jones A, Harris P, Yi J (Novozymes Biotech IU). (2004) *Variants of cellobiohydrolase I from Trichoderma reesei with improved properties.* 2005048619.

Vaaje-Kolstad G, Horn SJ, van Aalten DMF, Synstad B, Eijsink VGH. (2005b) *The non-catalytic chitin-binding protein CBP21 from Serratia marcescens is essential for chitin degradation.* J Biol Chem 280: 28492-28497.

Vaaje-Kolstad G, Houston DR, Riemen AHK, Eijsink VGH, van Aalten DMF. (2005a) *Crystal structure and binding properties of the Serratia marcescens chitin-binding protein CBP21.* J Biol Chem 280: 11313-11319.

Vardakou M, Flint J, Christakopoulos P, Lewis RJ, Gilbert HJ, Murray JW. (2005) *A family 10 Thermoascus aurantiacus xylanase utilizes arabinose decorations of xylan as significant substrate specificity determinants.* J Mol Biol 352: 1060-1067.

Walker-Renard P. (1993) *Update on the medicinal management of phytobezoars.* Am J Gastroenterol. 88(10): 1663-6.

Wu W, Lange L, Skovlund DA, Liu Ye (Novozymes A/S D). (2003) *Cloning and sequences of cellobiohydrolase II from thermophilic fungi and their use for ethanol production and in detergents*. patent 2004056981.

Wyman CE. (2003) *Potential synergies and challenges in refining cellulosic biomass to fuels, chemicals, and power*. Biotechnol Prog 19: 254-262.

Yinbo Q, Zhu M, Liu K, Bao X, Lin J. (2006) *Studies on cellulosic ethanol production for sustainable supply of liquid fuel in China*. Biotechnol J. 1(11):1235-40.

In: Cellulase: Types and Action, Mechanism, and Uses
Editor: Adam E. Golan
ISBN: 978-1-61761-983-0

Chapter 10

SYNERGISTIC EFFECTS OF SNAIL AND TRICHODERMA REESEI CELLULASES ON ENZYMATIC HYDROLYSIS AND ETHANOL FERMENTATION OF LIGNOCELLULOSE

Ding Wenyong and Chen Hongzhang*
National Key Laboratory of Biochemical Engineering, Institute of Process Engineering, Chinese Academy of Sciences, Beijing 100190, PR China

ABSTRACT

To evaluate the synergism of cellulases from animal and microorganism, mixture of cellulases from snail (CES) and *Trichoderma reesei* (CET) was used to enzymatic hydrolysis and ethanol fermentation of lignocellulose. When the mixed cellulase was used to enzymatically hydrolyze *Pennisetum hydridum*, the optimal ratio of CES and CET was 3:1, and the glucose yield using the mixed enzyme was 100.3% and 50.2% higher than that produced individually by CES and CET, respectively. For ethanol fermentation of lignocellulose, the optimal ratio of CES and CET was 1:3, the ethanol yield using the mixed enzyme was 42.5% and 20.1% higher than that produced individually by CES and CET, respectively. Our results showed that mixed cellulase from animal and microorganism is a potential approach for improving enzymatic hydrolysis and ethanol fermentation of lignocellulose.

Keywords: cellulase; synergism; animal; microorganism; Simultaneous saccharification and fermentation (SSF)

*Corresponding author. Tel: +86-10-82627067; fax: +86-10-82627071. E-mail address: hzchen@home.ipe.ac.cn (Chen Hongzhang).

1. Introduction

Cellulase is a multicomponent enzyme with three components: endo-1,4-glucanase (EC 3.2.1.4, Cx, CMCase), exo-1,4-glucanase (EC 3.2.1.91, C1, cellobiohydrolase) and 1,4-β-glucosidase (EC 3.2.1.21, cellobiase) (Lynd et al., 2002; Wood, 1992). The three enzymes exert a synergic effect to fully hydrolyze cellulose into glucose (Henrissat et al., 1985; Mosier et al., 1999; Reese et al., 1950). However, the proportions of these three components differ among cellulases from different sources (Bravo et al., 2001; Wood et al., 1994). As the proportion of the three components from a single source are usually not at an optimal ratio, the enzyme is not optimum for cellulose degradation. The efficiency of cellulase is accordingly depressed, and wastage of the enzyme also unavoidable. Therefore, mixtures of enzymes containing different kinds of cellulase have been used to degrade cellulose. Such mixtures result in a higher sugar yield than that obtained using cellulase from a single source (Berlin et al., 2007; Gusakov et al., 2007).

At present, microbial fermentation of *Trichoderma* is the most common method for cellulase production (Chandra et al., 2009; Kovács et al., 2008; Latifian et al., 2007). In previous studies, cellulases were obtained mostly from mixed cultures or a mixture of cultures of *Trichoderma* and *Aspergillus niger* (Ahamed and Vermette, 2008; Imai et al., 2004). Cellulases are also present in the gut of some animals (Watanabe and Tokuda, 2001), such as termite and snail (Destevens, 1955; Warnecke et al., 2007). To achieve industrial cellulases production by imitating the micro-bioreactor of termite, Warnecke et al., (2007) analyzed the genome sequences of microorganisms derived from the termite hindgut. The scale of cellulase source would extended if the cellulase from animal mixed with existing cellulase from microorganisms can be used effectively. However, enzymatic hydrolysis and ethanol fermentation of lignocellulose using mixed cellulases from animals and microorganisms has never been reported. In this chapter, we examined the synergistic effects of cellulases from snail and *Trichoderma reesei* on enzymatic hydrolysis and ethanol fermentation of lignocellulose, and demonstrated the feasibility of using mixed cellulases from animal and microorganism.

2. Materials and Methods

2.1 Materials

Steam-exploded *Pennisetum hydridum* (Guangdong, China) was used as substrate in this chapter. The steam explosion pretreatment was performed in a 7.50 L vessel (Weihai Automatic Control Reactor Ltd., China) with pressure 1.70 MPa for 7 min (Chen and Liu, 2007). After pretreatment, the material was dried at ambient temperature and kept at 4℃. Dry solids content (cellulose, hemicellulose and lignin) was estimated according to the procedures of Goering and Van Soest (Goering and Van Soest, 1970). The composition of the *Pennisetum hydridum* after pretreatment was 37.6 % (w/w) cellulose, 18.6 % (w/w) hemicellulose and 13.3 % (w/w) lignin.

The cellulases extracted from snail and *Trichoderma reesei* were obtained from Meijing Co. Ltd. (Fujian, China) and Xiasheng Co. Ltd. (Ningxia, China), respectively. Snail was the representative herbivorous animal, so the cellulase of which was selected .The cellulase from *Trichoderma reesei* was selected because it was representative commercial cellulase preparations. Commercial instant active dry yeast (*Saccharomyces cerevisiae*) was obtained from Angel Yeast Co. Ltd., Hubei, China. All other chemicals used in this chapter were of analytical grade and purchased from Beijing Chemical Reagent Corp., China.

2.2 Enzymatic Hydrolysis

The enzymatic hydrolysis experiments were carried out at 50℃ with 3 g (dry weight, DW) of lignocellulose suspended in 0.2 mol/L acetic acid buffer (pH 4.8) with final volume of 90 mL in a 500 mL flask. The slurry was added to mixture of CES and CET. The flask was sealed with a lid, and the hydrolysis was carried out in shaker at 150 rpm for 72h. Samples were withdrawn periodically for sugar analysis by HPLC. All experiments were performed in triplicate.

2.3 Simultaneous Saccharification and Fermentation (SSF) of Lignocellulose to Ethanol

2.3.1 Preparation of Yeast Inoculum

For dissolution and activation, 1 g dry yeast was added to 20 mL sterile water containing 2% glucose, and incubated at 37℃for 1 h.

2.3.2 SSF

The fermentation experiments were carried out at 37℃ in a 500 mL flask with volume of 90 mL, and 3 g (DW) lignocellulose was suspended in medium. The mixture of CES and CET was added to the slurry . The medium composition consisted of 2 g/L $(NH_4)_2SO_4$, 5 g/L KH_2PO_4, 0.4 g/L $MgSO_4 \cdot 7H_2O$, 0.2 g/L $CaCl_2$ and 2 g/L yeast extract. The substrate was autoclaved for 15 min at 121℃ before adding enzymes and inoculum. The amount of inoculum was 10% (v/v) of the SSF medium. The flask was sealed with a lid, and the fermentation was carried out in shaker at 150 rpm for 72h. Samples were withdrawn periodically for sugar and ethanol analysis by HPLC. All experiments were performed in triplicate.

2.4 Analytical Methods

2.4.1 Enzymatic Activity Assays

Two kinds of commercail cellulase were applied in the experiment. Solid powder of cellulase from snail was dissolved to determined enzymatic activity. Liquid of cellulase from *Trichoderma reesei* was diluted directly to determined enzymatic activity. CMCase, cellobiohydrolase and β-glucosidase activity were determined using the Mandels method,

Corbett method and the Sternberg method, respectively (Corbett et al., 1963; Mandels et al., 2009; Sternberg et al., 1977). Filter paper activity (FPA) , the synthetic enzymatic activity of CMCase, cellobiohydrolase and β-glucosidase, was determined as recommended by Ghose (Ghose, 1987) and the enzymatic activity unit was denoted as FPU. Reducing sugars were analyzed by the Miller method (Miller, 1959). All activities were expressed in International Units, i.e. one unit of activity corresponded to the quantity of enzyme hydrolyzing one μmol of substrate or releasing one μmol of reducing sugars (in glucose equivalents) per min.

2.4.2 Analysis of Sugar and Ethanol by HPLC

Samples of the hydrolysis and fermentation liquid were centrifuged at 14,000 rpm for 5 min. The supernatant was filtered with 0.45 μm sterile filters. The concentrations of glucose, cellobiose and ethanol were determined using a HPLC (Agilent technology 1200 series, Palo Alto, CA). The separation was performed on an Aminex Hpx-87H ion exclusion column (300 mm × 7.8 mm) at 35℃ with a refractive index detector. The eluent used was 5 mM H_2SO_4 with a flow rate of 0.6 mL/min.

3. Results and Discussion

3.1. Enzymatic Activity of CES and CET

Enzymatic activities of CES and CET are shown in Table 1. To compare units of enzymatic activities of CES and CET, which are expressed as IU/g and IU/mL, respectively, the FPU of CES and CET was transformed to the same value (assigned a value of 1), and other enzymatic activities are expressed relative to the FPU (Table 1). As shown in Table 1, the CES and CET cellulases both have three components. However, the enzymatic activity of each component differed between CES and CET. Endoglucanase (CMCase) activity in CET was 5.6 times that in CES, while the β-glucosidase and exoglucanase (C1) activities in CES were 14.3 times and 1.5 times, respectively, those in CET.

Table 1. Enzymatic Activities of CES and CET.

	FPA	CMCase	β-Gase	C_1
Activity of CES (IU/g)	29.0±1.4	234.0±10.0	107.9±4.9	5.9±0.3
Activity of CET (IU/ml)	110.2±4.8	4980.0±230.0	29.1±1.3	14.3±0.7
Ratio of CES activity to FPU	1	8.07	3.72	0.20
Ratio of CET activity to FPU	1	45.27	0.26	0.13

3.2 Enzymatic Hydrolysis

3.2.1 Enzymatic Hydrolysis using Equal Activities of CES and CET

Different enzyme activities (10, 20, 30 FPU/g substrate) of either CES or CET were used for enzymatic hydrolysis of substrate. The concentration and production velocity of glucose from CET was higher than that of CES at the same enzyme applied (Fig. 1a). After 16 h's hydrolysis with CET, the glucose concentration all reached or exceeded the production level by CES after 72 h with enzyme loading of 10, 20 or 30 FPU/g. The final glucose concentrations after 72 h hydrolysis with CET were 51.2, 33.3 and 13.4% higher than those produced using CES at enzyme loading of 10, 20 and 30 FPU/g, respectively (Fig. 1a). The advantage of CET to CES was more significant at low cellulase loadings. Because of the higher β-glucosidase activity in CES, the cellobiose produced during the enzymatic hydrolysis of substrate was quickly degraded by β-glucosidase, which resulted in no detectable cellobiose during the reaction period (data not shown). However, cellobiose concentration increased with increasing CET loadings during the hydrolysis period (Fig. 1b). This was due to incomplete degradation of cellobiose that resulted from the low β-glucosidase activity.

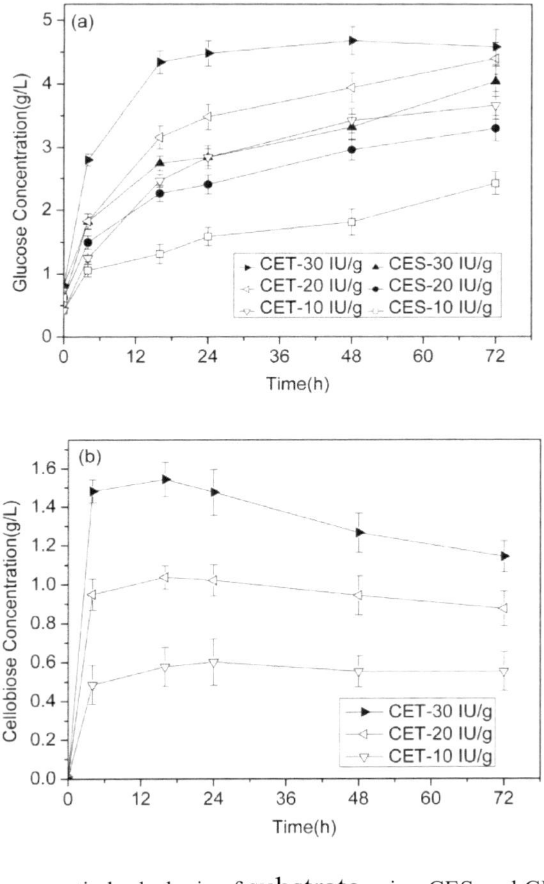

Figure 1. Comparison of enzymatic hydrolysis of substrate using CES and CET. (a) Glucose concentration. (b) Cellobiose concentration.

Because of its high β-glucosidase activity, CES complemented the activity of CET. The various ratios of CES and CET had different enzymatic activities. All mixtures showed higher FPU values than either enzyme individually (Table 2). For example, the total FPU of mixed cellulases increased by 20, 30 and 47.5%, at ratio of 10/30, 20/20 and 30/10 FPU of CES to

CET, respectively (Table 2). These increased FPU values suggest that the ratio of the three components of the mixture can be optimized, to improve the synergistic effects of the three components. As shown in Tables 2, the ratios of β-glucosidase activity to FPA activity were calculated as 0.83, 1.53, and 2.02 in cellulase mixtures containing 10/30, 20/20 and 30/10 FPU of CES to CET, respectively.

Table 2. Changes in FPU after mixing CES and CET.

CES (FPU)	CET (FPU)	Mixed cellulases (IU)	
		FPA	β-Gase :FPA
10	30	48.0±2.0	0.83
20	20	52.0±2.5	1.53
30	10	59.0±2.3	2.02

3.2.2 Enzymatic Hydrolysis using 1:1 CES and CET Mixtures

The mixtures of CES to CET with ratio of 1:1 at different loadings (5/5, 10/10, 15/15 FPU/g substrate) were used for the enzymatic hydrolysis of substrate. Mixed enzymes resulted in increased glucose concentration compared with hydrolysis using either of the enzymes individually (Fig. 2). The glucose concentration from the 10/10 FPU/g mixture was equivalent to the 72-hr concentration of 20 FPU/g CES at 11-hr and to the 72-hr concentration of 20 FPU/g CET at 21-hr of hydrolysis. The final glucose concentration after 72-hr hydrolysis from the 10/10 FPU/g mixture was 96.7% higher than that produced by 20 FPU/g CES alone, and 47.5% higher than that produced by 20 FPU/g CET alone (Fig. 2). This showed that the 1:1 mixture required a lower enzyme load and less treatment time to yield the same amount of glucose as either individual enzyme. As shown in Table 2, the activity of the 20/20 mixed cellulase is 52 FPU/g, it can concluded that the activity of the 10/10 mixed cellulase is 26 FPU/g. This is higher than the 20 FPU/g (substrate) of individual CES and CET. Therefore, the 10/10 FPU/g of mixed cellulase has better enzymatic hydrolysis effect than either CES or CET of 20 FPU/g individually. The glucose concentration from the 5/5 FPU/g mixture was equivalent to the 72-hr concentration of 20 FPU/g CES at 18-hr and to the 72-hr concentration of 20 FPU/g CET at 35-hr of hydrolysis.The final glucose concentration after 72-hr hydrolysis from the 5/5 FPU/g mixture was 67.3% higher than that achieved using 20 FPU/g CES alone, and 25.4% higher than that achieved using 20 FPU/g CET alone (Fig. 2). As shown in Table 2, it also can be concluded that the activity of the 5/5 mixed cellulase was 13 FPU/g. The FPA of 5/5 mixed cellulase was lower than 20 FPU/g of individual CES and CET, but the enzymatic hydrolysis of mixed cellulase was still more efficient than that using either CES or CET individually. This result showed that enzymatic hydrolysis can be improved using mixed cellulases, with the ratio of the three components be optimized. Cellobiose was only detected when CET was used individually (Fig. 1b), and was undetectable using CES or mixed cellulases (data not shown). This showed that the high β-glucosidase activity of CES could compensate for the low β-glucosidase activity of CET in mixed cellulase.

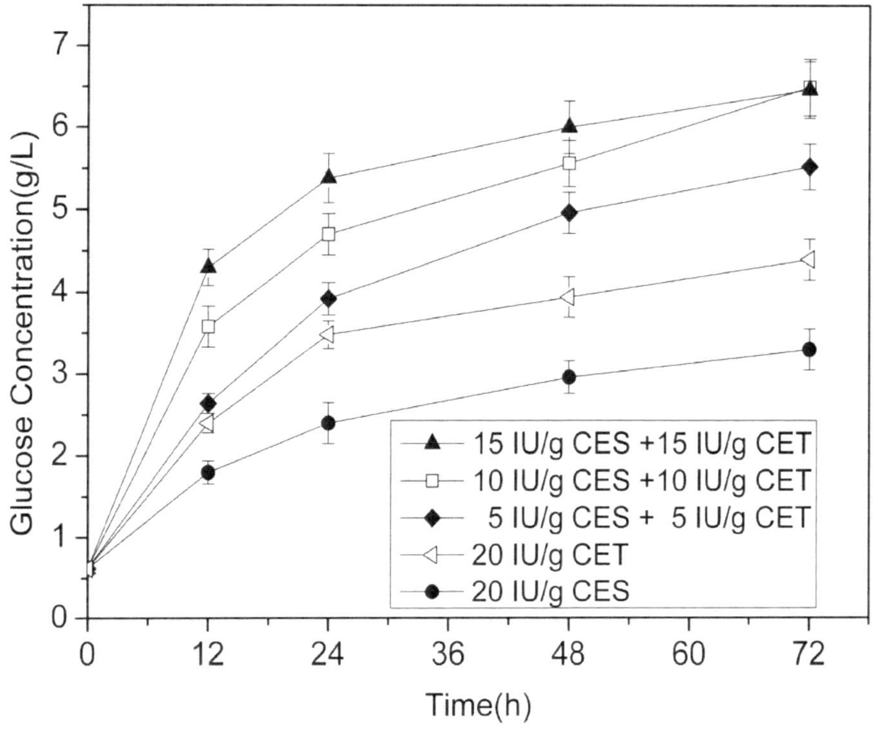

Figure 2. Comparison of enzymatic hydrolysis of substrate using 1:1 mixture of CES and CET.

3.2.3 Enzymatic Hydrolysis using Various Ratios of CES and CET

Cellulase mixtures of various ratios of CES to CET (5/15, 10/10, 15/5 FPU/g substrate) were used for enzymatic hydrolysis of substrate. The final glucose and cellobiose concentrations were determined after enzymatic hydrolysis for 72 h. All mixtures of the two enzymes resulted in higher concentrations of glucose compared with single enzyme treatments (Fig. 3). When the ratios of CES to CET of the enzyme mixture were 5/15, 10/10, 15/5 FPU/g, the final glucose concentrations were 84.8, 96.7, and 100.3% higher than that of using CES alone, and 38.6, 47.5, and 50.2% than that of using CET alone, respectively (Fig. 3). As shown in Table 2, the activity of the 10/30, 20/20, 30/10 FPU/g mixed cellulase is 48, 52, 59 FPU/g, it can be concluded that the activity of the 5/15, 10/10, 15/5 FPU/g mixed cellulase was 24, 26, 29.5 FPU/g. Glucose concentrations increased with increasing actual FPA. The ratio of β-glucosidase activity to FPA was 2.02 in the mixture (15/5 FPU/g) yielding the highest final glucose concentration. It has been reported that the ideal ratio of β-glucosidase activity to FPA ranges from 0.12 to 1.5, depending on the source of enzyme and the type of substrate (Duff and Murray, 1996). Differences in the ideal ratio may arise from the differences between animal and microorganism cellulases. The final cellobiose concentration was 1.4 g/L when CET alone was used to hydrolyze the substrate. The final cellobiose was not detected when CES was added to the enzyme mixture in any of the ratios tested. This result indicated that the effects of the low β-glucosidase activity of CET were alleviated by CES addition in cellulase mixtures.

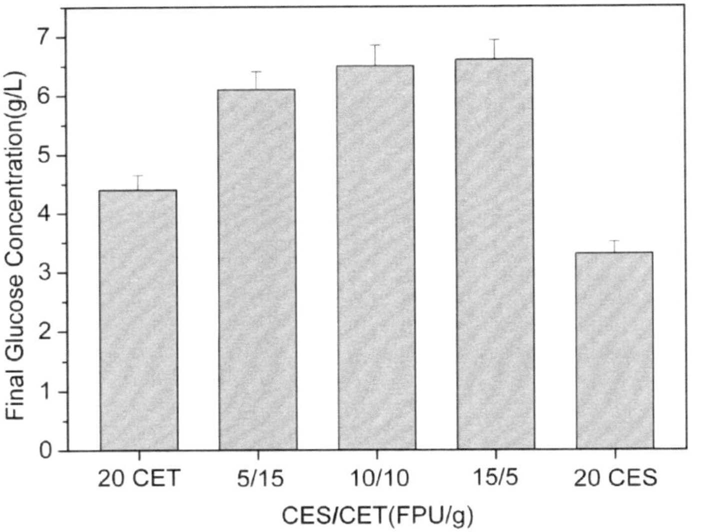

Figure 3. Comparison of enzymatic hydrolysis of substrate using mixtures with different ratios of CES and CET.

3.3 SSF

3.3.1 SSF using Equal Activities of CES and CET

Equal enzyme activities of either CES or CET of 20 FPU/g were used for ethanol fermentation of substrate. At the same FPU, CET produced ethanol more rapidly and the final ethanol concentration was higher than that of CES (Fig. 4). This was consistent with the results of glucose production by CET and CES. The final ethanol concentration after 48-hr fermentation of 20 FPU/g CET was 18.6% higher than that of 20 FPU/g CES (Fig. 4).

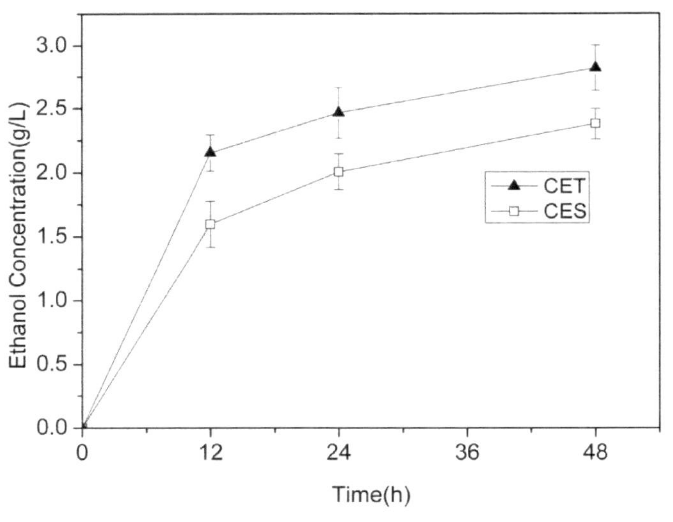

Figure 4. Comparison of ethanol fermentation of substrate using CES and CET.

3.3.2 SSF using 1:1 CES and CET Mixtures

The 1:1 mixtures of CES to CET at different loadings (5/5, 10/10, 15/15 FPU/g substrate) were used for ethanol fermentation of substrate. The final ethanol concentration was determined after ethanol fermentation for 48 h. Mixed enzymes resulted in increased ethanol concentration compared with fermentation using either of the enzymes individually (Fig. 5). The final ethanol concentrations of 5/5, 10/10, 15/15 FPU/g mixture was 9.4%, 39.2%, 48.4% higher than that using 10, 20, 30 FPU/g CES alone, and 7.2%, 17.4%, 7.73% higher than that using 10, 20,30 FPU/g CET alone (Fig. 5). Furthermore, the final ethanol concentrations using CET alone were 2.0, 18.6, and 37.7% higher than those obtained using CES alone (Fig. 5), at loadings of 10, 20, 30 FPU/g, respectively. The advantage of CET to CES in ethanol fermentation was more remarkable at higher activities, but this was not the case in enzymatic hydrolysis. This might be conscribed to the glucose was fermented to ethanol quickly, resulting in low concentrations of glucose. In turn, this weakened the feedback inhibition of glucose on β-glucosidase during CET-catalyzed ethanol fermentation. This compensates for the low activity of β-glucosidase in CET, and thus, increases enzyme utilization efficiency of CET and alleviates the disadvantage of CET compared with mixed cellulases. This could also be explained by the absence of cellobiose when CET was used for ethanol fermentation (data not shown). The fact that cellobiose was degraded to glucose quickly indicated that the function of β-glucosidase of CET was improved in ethanol fermentation.

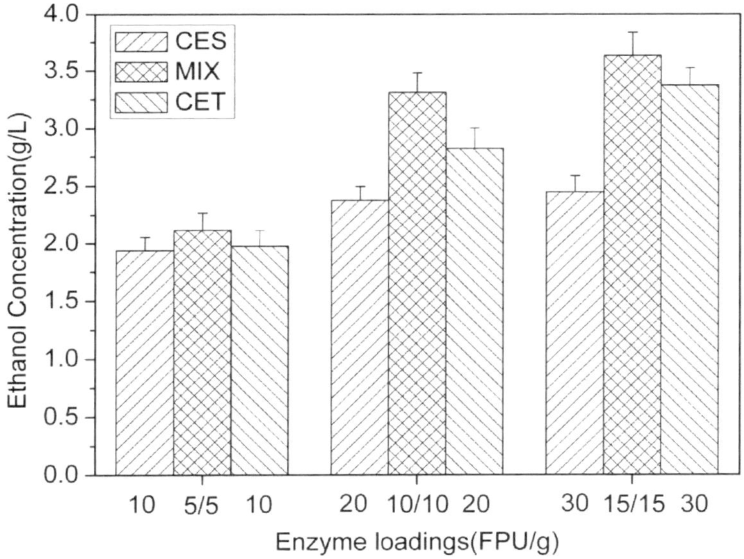

Figure 5. Comparison of ethanol fermentation of substrate using 1:1 mixture of CES and CET. MIX, mixed cellulase.

3.3.3 SSF using Various Ratios of CES and CET Mixtures

Mixtures of various ratios of CES to CET (5/15, 10/10, 15/5 FPU/g substrate) were used for ethanol fermentation of substrate. The final ethanol concentration was determined at the end of ethanol fermentation for 48 h. The ethanol concentrations resulting from mixed cellulases were higher than those resulting from either CES or CET individually (Fig. 6). When CES to CET were 5/15, 10/10, 15/5 FPU/g of the enzyme mixture, the final ethanol concentrations were 42.5, 39.2, and 38.2% higher, respectively, than that of 20FPU/g CES alone, and 20.1, 17.4, and 16.5% higher, respectively, than that of 20FPU/g CET alone (Fig. 6). The highest final ethanol concentration was obtained using an enzyme mixture of 5/15 FPU/g (Fig. 6), but the highest final glucose concentration in enzymatic hydrolysis was obtained using an enzyme mixture of 15/5 FPU/g (Fig. 3). The high endoglucanase activity of CET resulted in more complete degradation of the substrate. During ethanol fermentation, this alleviated the disadvantage of low β-glucosidase activity of CET compared with that of mixed cellulases. A higher proportion of CET in the enzyme mixture improved the efficiency of ethanol fermentation. According to Table 2, the optimal ratio of β-glucosidase activity to FPA was 0.83 at the highest final ethanol concentration of 5/15 FPU/g. Compared with the optimal ratio 2.02 in enzymatic hydrolysis, a lower ratio of β-glucosidase:FPA is required for optimal ethanol fermentation.

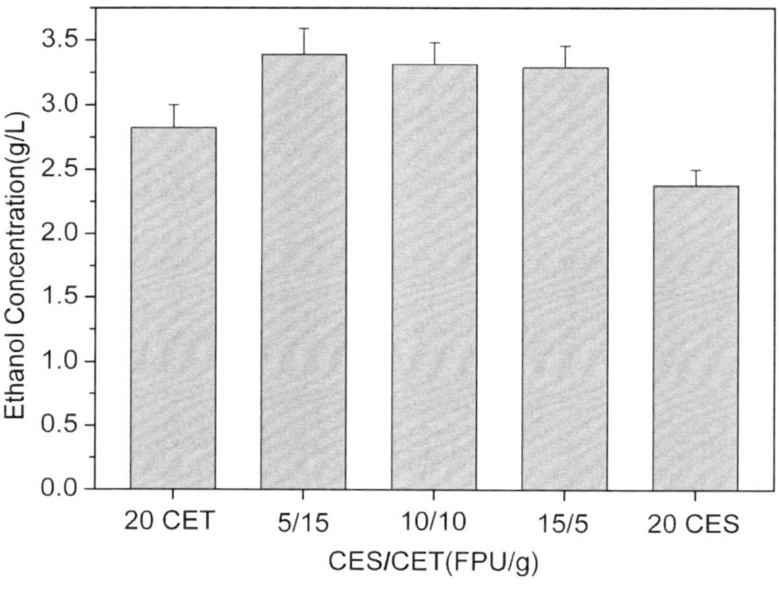

Figure 6. Comparison of ethanol fermentation of substrate using mixtures with different ratios of CES and CET.

CONCLUSION

Our results demonstrate that a mixture of cellulases from snail and *Trichoderma reesei* increased the efficiency of enzymatic hydrolysis and ethanol fermentation of lignocellulose. This result indicates that there are significant synergistic effects when cellulases from animals and microorganisms are combined. The proportion of the three components was optimized in a cellulase mixture. The similar characteristics of the cellulases from animal and microorganism indicated animal tissues could be a new source of cellulases for enzymatic hydrolysis and lignocellulosic ethanol fermentation. Moreover, the mixed use of these two kinds of cellulases will improve the efficiency of enzymatic hydrolysis and fermentation performance as well.

ACKNOWLEDGMENTS

This work was financially supported by National Basic Research Program of China (973 Project, No. 2004CB719700) and Knowledge Innovation Program of CAS (KSCX1-YW-11A; KGCX2-YW-328).

REFERENCES

Ahamed, A.; Vermette, P. *Enhanced enzyme production from mixed cultures of Trichoderma reesei RUT-C30 and Aspergillus niger LMA grown as fed batch in a stirred tank bioreactor.* Biochem. Eng. J. 2008, 42, 41-46.

Berlin, A.; Maximenko, V.; Gilkes, N.; Saddler, J. *Optimization of enzyme complexes for lignocellulose hydrolysis.* Biotechnol. Bioeng. 2007, 97, 287-296.

Bravo, V.; Páez, M.P.; El-Hadj, M.A.; Reyes, A.; García, A.I. *Hydrolysis of carboxymethylcellulose with mixtures of cellulase and β-1,4-glucosidase.* J. Chem. Technol. Biotechnol. 2001, 77, 15-20.

Chandra, M.; Kalra, A.; Sangwan, N.S.; Gaurav, S.S.; Darokar, M.P.; Sangwan, R.S. *Development of a mutant of Trichoderma citrinoviride for enhanced production of cellulases.* Bioresour. Technol. 2009, 100, 1659-1662.

Corbett, W.M.; Green, J.W.; Bemiller, J.N. *Purification of Cotton Cellulose.* Methods. Carbohydr. Chem. 1963, 3, 3-4.

Destevens, G. *Cellulase preparation from Helix pomatia (snails).* Methods. Enzymol. 1955, 1, 173-178.

Duff, S.J.B.; Murray, W.D. *Bioconversion of forest products industry waste cellulosics to fuel ethanol: a review.* Bioresour. Technol. 1996, 55, 1-33.

Ghose, T.K. *Measurement of cellulase activities.* Pure. Appl. Chem. 1987, 59, 257-268.

Goering, H.K.; Van Soest, P.J. *Forage fiber analyses (apparatus, reagents, procedures, and some applications).* In: Agriculture Handbook, ARS, USDA, Washington, DC, 1970.

Gusakov, A.V.; Salanovich, T.N.; Antonov, A.I.; Ustinov, B.B.; Okunev, O.N.; Burlingame, R.; Emalfarb, M.; Baez, M.; Sinitsyn, A.P. *Design of highly efficient cellulase mixtures for enzymatic hydrolysis of cellulose.* Biotechnol. Bioeng. 2007, 97, 1028-1038.

Henrissat, B.; Driguez, H.; Viet, C.; Schülein, M. *Synergism of cellulases from Trichoderma reesei in the degradation of cellulose.* Nat. Biotechnol. 1985, 3, 722-726.

Chen, H.Z.; Liu, L.Y. *Unpolluted fractionation of wheat straw by steam explosion and ethanol extraction.* Bioresour. Technol. 2007, 98, 666-676.

Imai, M.; Ikari, K.; Suzuki, I. *High-performance hydrolysis of cellulose using mixed cellulase species and ultrasonication pretreatment.* Biochem. Eng. J. 2004, 17, 79-83.

Kovács, K.; Megyeri, L.; Szakacs, G.; Kubicek, C.P.; Galbe, M.; Zacchi, G. *Trichoderma atroviride mutants with enhanced production of cellulase and β-glucosidase on pretreated willow.* Enzyme. Microb. Technol. 2008, 43, 48-55.

Latifian, M.; Hamidi-Esfahani, Z.; Barzegar, M. *Evaluation of culture conditions for cellulase production by two Trichoderma reesei mutants under solid-state fermentation conditions. Bioresour.* Technol. 2007, 98, 3634-3637.

Lynd, L.R.; Weimer, P.J.; van Zyl, W.H.; Pretorius, I.S. *Microbial cellulose utilization: fundamentals and biotechnology*. Microbiol. Mol. Biol. R. 2002, 66, 506-577.

Mandels, M.; Andreotti, R.; Roche, C. *Measurement of saccharifying cellulase*. Biotechnol. Bioeng. Symp. 1976, 6, 21-33.

Miller, G.L. *Use of dinitrosalicylic acid reagent for determination of reducing sugar*. Anal. Chem. 1959, 31, 426-428.

Mosier, N.S.; Hall, P.; Ladisch, C.M.; Ladisch, M.R. *Reaction kinetics, molecular action, and mechanisms of cellulolytic proteins*. Adv. Biochem. Eng. Biotechnol. 1999, 65, 23-40.

Reese, E.T.; Siu, R.G.H.; Levinson, H.S. *The biological degradation of soluble cellulose derivatives and its relationship to the mechanism of cellulose hydrolysis*. J. Bacteriol. 1950, 59, 485-497.

Sternberg, D.; Vijayakumar, P.; Reese, E.T.*β-Glucosidase: microbial production and effect on enzymatic hydrolysis of cellulose*. Can. J. Microbiol. 1977, 23, 139-147.

Warnecke, F.; Luginbuhl, P.; Ivanova, N.; Ghassemian, M.; Richardson, T.H.; Stege, J.T.; Cayouette, M.; McHardy, A.C.; Djordjevic, G.; Aboushadi, N. *Metagenomic and functional analysis of hindgut microbiota of a wood-feeding higher termite*. Nature. 2007, 450, 560-565.

Watanabe, H.; Tokuda, G. *Animal cellulases*. Cell. Mol. Life. Sci. 2001, 58, 1167-1178.

Wood, T.M. *Fungal cellulases*. Biochem. Soc. Trans. 1992, 20, 46-53.

Wood, T.M.; Wilson, C.A.; McCrae, S.I. *Synergism between components of the cellulase system of the anaerobic rumen fungus Neocallimastix frontalis and those of the aerobic fungi Penicillium pinophilum and Trichoderma koningii in degrading crystalline cellulose*. Appl. Microbiol. Biotechnol. 1994, 41, 257-261.

In: Cellulase: Types and Action, Mechanism, and Uses
Editor: Adam E. Golan

ISBN: 978-1-61761-983-0

Chapter 11

ENGINEERING THERMOBIFIDA FUSCA CELLULASES: CATALYTIC MECHANISMS AND IMPROVED ACTIVITY

Thu V. Vuong and David B. Wilson
Department of Molecular Biology and Genetics, Cornell University, Ithaca, New York, USA

ABSTRACT

The importance of cellulases in the production of fuels from biomass makes understanding their catalytic mechanisms on crystalline cellulose important in order to design more active enzymes. Seven modular cellulases from Thermobifida fusca have been purified and characterized; of which, three inverting cellulases: endocellulase Cel6A, exocellulase Cel6B and processive endocellulase Cel9A have been studied extensively. Each one has an atypical catalytic mechanism: two Asp residues hold the nucleophilic water in Cel9A while no single catalytic base was found in the family-6 enzymes, suggesting that several residues might be involved in catalysis and form a network that functions as the catalytic base in these enzymes. Site-directed mutagenesis and removal of domains demonstrate the important role of cellulose-binding modules in crystalline substrate hydrolysis and processivity. To investigate if independent enzymes could function effectively in a cellulosome, the catalytic domains of the two family-6 T. fusca cellulases were attached to dockerin domains and then the chimeric enzymes were used to form designer cellulosomes. Additionally, Cel6B enzymes have been fluorescence-labeled, providing another way to measure binding and processivity. These studies have created several enzymes with higher activity on crystalline cellulose; however, better strategies are necessary to produce more active engineered cellulases that will be able to lower the cost of cellulases for biomass hydrolysis.

INTRODUCTION

An important step towards enhancing the economic competitiveness of biofuels is to lower the cost of enzymes used to hydrolyze cell wall polymers to sugars. Engineering cellulases with improved activity will help reduce the amount of enzymes required and the time needed for cellulose hydrolysis in biofuel production. Site-directed mutagenesis and directed evolution have been used to try to engineer cellulases with higher activity on crystalline substrates [1]. Site-directed mutagenesis is a useful approach as it can reveal the specific role of an amino acid of interest, facilitating an understanding of the detailed catalytic mechanisms of cellulases [2, 3]. This knowledge will allow rational design of improved cellulases.

Microorganisms have developed several different types of enzyme systems to degrade cellulose efficiently, ranging from synergistic mixtures of free enzymes, found mainly in aerobic species [4] to large and complex structures, called cellulosomes, commonly found in anaerobic bacteria [5]. Several other less studied mechanisms also are used by some cellulolytic microorganisms [6, 7]. A typical free cellulase consists of a catalytic domain joined to a carbohydrate-binding module (CBM) via a linker. Catalytic domains and binding modules are grouped into different families based on sequence similarities (*www.cazy.org*) [8]. Cellulolytic microorganisms can produce multiple cellulases with catalytic domains belonging to several glycoside hydrolases (GH) families. The thermophilic, aerobic bacterium *Thermobifida fusca* secretes seven distinct cellulases [4, 9]: two in GH family 5 (Cel5A and Cel5B), two in GH family 6 (Cel6A and Cel6B), two in GH family 9 (Cel9A and Cel9B) and one in GH family 48 (Cel48A).

Three of these *T. fusca* cellulases, Cel6A, Cel6B and Cel9A are good candidates for rational design, as they have been characterized intensively, cloned, and over-expressed in both *Escherichia coli* and *Streptomyces lividans,* and the x-ray structures of the Cel6A catalytic domain and Cel9A-68 were solved [10, 11]. These *T. fusca* cellulases represent all three known modes of action: Cel6A is an endocellulase (EC 3.2.1.4) while Cel6B is an exocellulase (EC 3.2.1.91) and Cel9A is a processive endocellulase. The difference in their function is reflected in their structures (Figure 1).

The x-ray structure of the Cel6A catalytic domain shows a modified α/β barrel with an open active site cleft [11]. Structural models of exocellulase Cel6B [2, 12], built from the *Humicola insolens* Cel6A catalytic domain (1OCB) and the *Trichoderma reesei* (also known as *Hypocrea jecorina*) Cel6A catalytic domain (1QK2) showed that the active site is enclosed by two long loops, forming a tunnel, which allows processive movement on a cellulose chain. Processive endocellulase Cel9A is the most active *T. fusca* cellulase on crystalline substrates. Its x-ray structure (4TF4) shows a catalytic domain with an open active site cleft, rigidly attached to a family-3c CBM by a short linker, allowing both modules to bind to a single cellulose chain [10].

Instead of secreting individual cellulases, anaerobic bacteria such as *Clostridium thermocellum* [5] and *C. cellulovorans* [13] have developed a different strategy to break down plant cell walls by producing cellulosomes, where a number of carbohydrate-degrading enzymes such as cellulases, xylanases and pectinases are linked to a central protein scaffold, a scaffoldin, through cohesin domains in the scaffoldin and dockerin domains on the enzymes. Most cellulosomal cellulases lack a CBM but a family-3 CBM is present in the scaffoldin. These cellulosomes are attached to the surface of the microorganism, helping the microorganism to bind cellulose in order to retain the hydrolyzed products efficiently. Cellulosomal cellulases also show synergy [14]. It was not known whether independent cellulases such as *T. fusca* Cel6A and Cel6B could be incorporated into active designer cellulosomes.

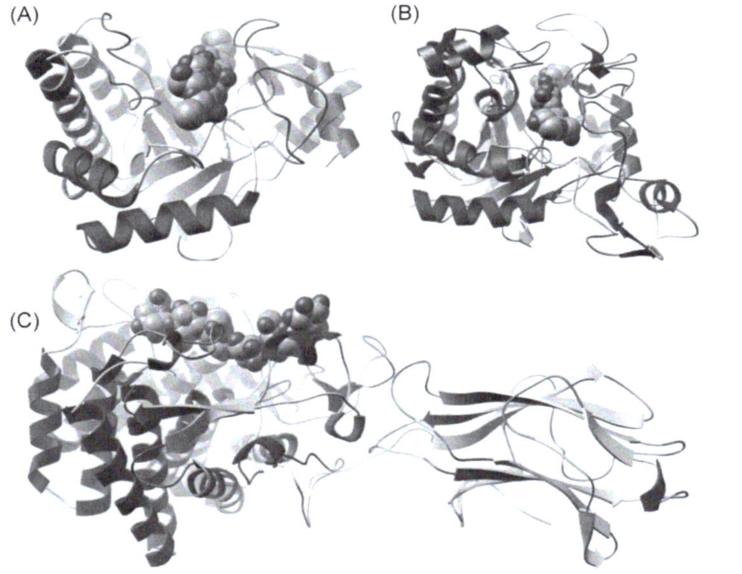

Figure 1. Structures of three T. fusca cellulases. (A) The catalytic domain of the inactive mutant endocellulase Cel6A Asp117Ala with cellopentaose bound in the open active site cleft (A. Larsson, personal communication), (B) The catalytic domain of exocellulase Cel6B modeled to Humicola insolens Cel6Acd (1OCB) with an oligosaccharide ligand in the active-site tunnel, and (C) Processive endocellulase Cel9A-68 (4TF4), showing the CBM3c module and the catalytic domain with six glucose residues in the active site cleft.

This chapter presents the results from engineering three *T. fusca* cellulases Cel6A, Cel6B and Cel9A, using mainly site-directed mutagenesis, for understanding their catalytic mechanisms, the roles of their CBM modules, as well as to try to improve activity on crystalline cellulose and enhance processivity. Additionally, this chapter will discuss the effectiveness of rational engineering of cellulases and other strategies for increasing cellulase activity.

ELUCIDATION OF CATALYTIC MECHANISMS

Understanding of the catalytic mechanisms of cellulases provides important information to help re-design them for better catalysis or different functions. TAll three *T. fusca* cellulases Cel6A, Cel6B and Cel9A were shown to use an inverting mechanism [10, 16, 17], which includes a single catalytic acid residue and a single catalytic base residue (Figure 2) [15, 18]. Site-directed mutagenesis followed by assays on a substrate with an excellent leaving group, 2,4-dinitrophenyl β-D-cellobioside, confirmed the conservation of the catalytic acid residue in each of these cellulases (Table 1). These catalytic residues are all acidic amino acids: Asp residues for the family-6 cellulases and a Glu residue for Cel9A. The catalytic acid residue Asp 117 of Cel6A is not located near the cleavage site in the crystal structure as it is in exocellulase Cel6B and processive endocellulase Cel9A, but more than 5.5Å from the glycosidic oxygen atom of the scissile bond. Molecular dynamics simulations suggest a proton transferring network from Asp117 through a water molecule to the glycosidic bond, and this water molecule is precisely positioned by both Tyr73 and Arg78 (Figure 3) [19]. Mutation of Tyr73 and Arg78 [16, 19, 20] support the crucial role of these residues in catalysis. Tyr73 also was found to participate in substrate distortion [21].

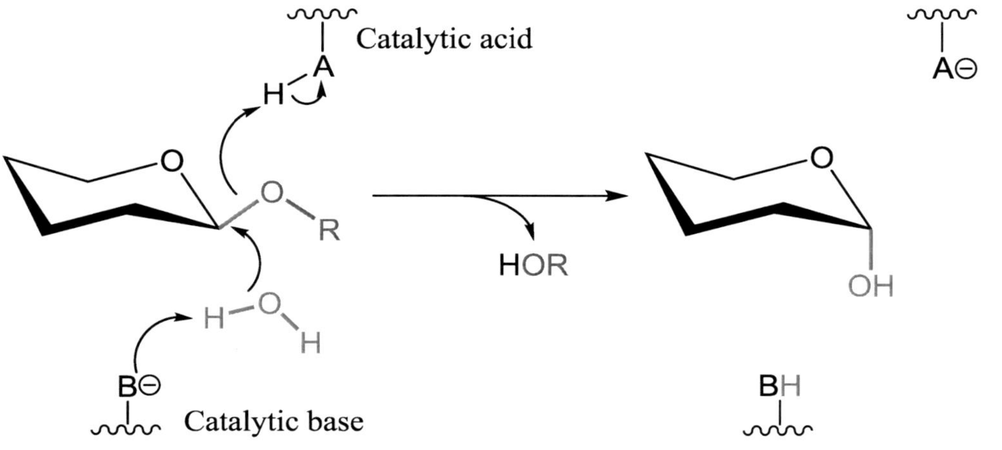

Figure 2. Proposed inverting mechanism [15]. A catalytic base residue (B^-) removes a proton from a water molecule, resulting in the attack of the nucleophilic water on the anomeric center to break the bond and invert the configuration while a catalytic acid residue (AH) donates a proton to the leaving group; R- carbohydrate derivative.

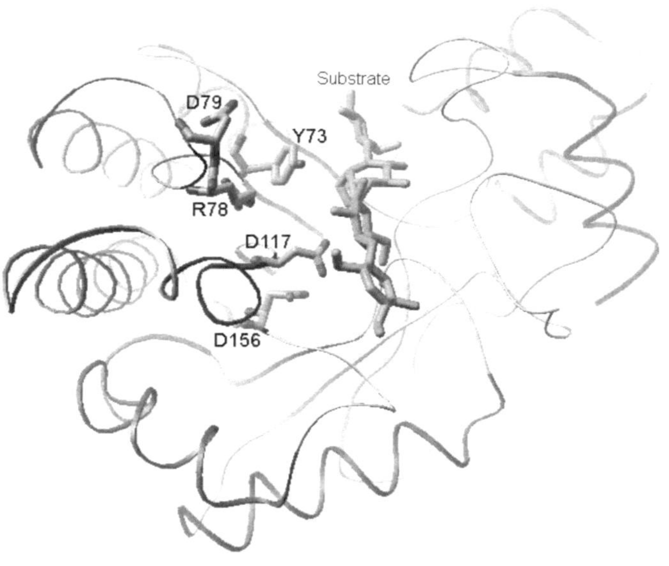

Figure 3. Location of *T. fusca* Cel6A key residues for catalysis. D79 and D156 control the pK_a of the catalytic acid residue D117 [16]. Y73 is involved in substrate distortion [21] and probably in a proton transferring network with R78 [19].

Table 1. Catalytic residues of T. fusca cellulases.

Enzyme	Catalytic acid residue	Catalytic base residue	Reference
Cel6A	D117	Unknown, a group of residues might function as the base	[2, 16]
Cel6B	D274	A proton-transferring network between D226 and S232	[2]
Cel9A	E424	D58, with the support of D55	[3, 22]

It is more complicated to identify a catalytic base residue in GHs and not all cellulases have a single catalytic base residue [23, 24]. In *T. fusca* Cel9A, two carboxylic residues (Asp55 and Asp58) were previously shown to be essential for catalysis, as both bind the catalytic water and mutating either drastically reduced enzymatic activity [22]. By chemical rescue assays, Asp58 was found to be the catalytic base and Asp55, an essential supporting residue [3]. No single base residue was found in either *T. fusca* Cel6A or Cel6B; but based on biochemical studies, a proton transferring network between Asp226 and Ser232 was proposed to function as a base in Cel6B [2]. This novel network supports a corresponding interaction, proposed by structural analysis in *Trichoderma reesei* Cel6A [25], a homolog of *T. fusca* Cel6B. These two residues are also present in *T. fusca* Cel6A; however, the corresponding Ser residue (Ser85) is located on a loop that is too flexible to be observed in Cel6A x-ray structures even in the presence of a ligand or substrate. Site-directed mutagenesis studies suggest the involvement of additional residues in catalysis. Two residues Asp156 and Asp79 in Cel6A (Figure 3) are involved in catalysis, at least partially by controlling the pK_a of the catalytic acid [16]. Both His125 and Tyr206 in Cel9A also are suggested to be involved in catalysis, due to their importance in hydrolysis [3,10]. These findings indicate that a simple model of a single catalytic base residue is not applicable for these *T. fusca* cellulases. The evolutional significance of using a network instead of a single residue as a base is not known, but there is no obvious advantage in catalysis between these two catalytic models.

The sugar bound in the -1 subsite of many glycosyl hydrolases is distorted in x-ray structures containing bound substrates [26, 27]. In the case of *T. fusca* Cel6A, substrate distortion was shown to be caused by Tyr73, as because it did not occur in a Tyr73Ser mutant enzyme [21]. This distortion was essential for activity, as the Ser mutant enzyme had less than 0.1% of wild-type activity while the Tyr73Phe mutant enzyme, which distorted the -1 subsite bound sugar, retained approximately 8% [21]. This mutant enzyme may not have completely recovered activity since Tyr73 also is involved in a proton transferring network as previously discussed.

The absence of a single base residue recently has been reported in several GH families. Some GH families use a proton-transferring network between several amino acid residues as the catalytic base [23, 24], others use the carbonyl oxygen of the 2-acetamide group in the substrate [28, 29] and some use an exogenous phosphate [30, 31]. Until now, it appears that these catalytic mechanisms are shared by all members of each family. The diversity of catalytic mechanisms probably reflects the evolution of microorganisms to deal with the wide range of glycosidic substrates present in different plant cell walls. Knowledge of the catalytic residues of glycoside hydrolases has been applied to engineer these enzymes for new functions: glycosynthases [24, 32], thioglycoligases [33] and thioglycosynthases [34].

ROLES OF CELLULOSE-BINDING MODULES

Beside the family-2 CBM, which is present in most *T. fusca* cellulases, Cel9A also contains two other non-catalytic modules: a family-3c CBM and a fibronectin III-like domain (FnIII) (Figure 4) [10]. Structures have been determined for CBM2s from *Pyrococcus furiosus* Chi18B [35] and two *Cellulomonas fimi* xylanases [36], revealing a common β-sandwich motif while *T. fusca* CBM3c was shown to have a 10β-stranded sandwich motif [10]; however, it lacks a strip of aromatic residues that provide tight binding in the other CBM3 sub-families [3, 37].

Table 2 shows that the absence of the family-2 CBM has little effect on hydrolysis of soluble cellulose, i.e. carboxymethyl cellulose (CMC), or amorphous cellulose except for Cel9A, but reduced enzymatic activity about 2 fold on crystalline substrates including bacterial microcrystalline cellulose (BMCC) and filter paper (FP). Synergism assays between *T. fusca* exocellulase and endocellulase constructs (with and without a CBM2) show that this CBM2 is more important for exocellulases than for endocellulases [4], supporting the importance of the CBM2 in crystalline cellulose degradation. When comparing the BMCC activity of the catalytic domains, Cel9Acd shows the lowest activity; however, the addition of the two CBMs boosts its activity tremendously. The addition of the CBM2 doubled BMCC activity for all enzymes while the presence of the CBM3c in Cel9A increases activity even more. However, only 15% of Cel9A-68, which contains the CBM3c but not the CBM2 (Figure 4), bound to BMCC [3]. In contrast, up to 80% of the native Cel9A-90 and Cel9A-74 (CBM3c deleted) bound to this substrate [38]. This finding indicates that the CBM3c has other functions besides binding.

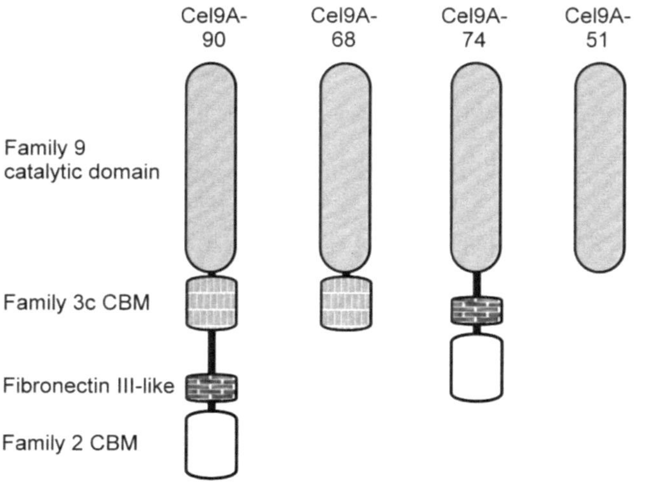

Figure 4. Different constructs of *T. fusca* Cel9A. Cel9A-68 does not have the CBM2 and an FnIII-like domain as seen in the native form, Cel9A-90.

Table 2. Properties of three T. fusca cellulases.

Mode of action	Enzyme	MW	Specific activity				Processivity	Ref.
			CMC	SC	BMCC	FP		
Endo	Cel6A	43.0	355	631	6.5	0.85	2.1	[39]
	Cel6Acd	30.4	330	560	3.2	0.50*	1.27	[40]
Exo	Cel6B	59.6	0.3*	1.5	2.0	0.13*	7.2-12.1	[12, 39]
	Cel6Bcd	45.7	0.5	1.1	1.0	0.05*	0.5	[4]
Processive endo	Cel9A	90.4	475	202	19.1	1.03	6.9	[38]
	Cel9Acd	51.4	108	6.3*	0.15*	0.13*	0.6	[3, 38]
	Cel9AΔFnIIIΔCBM2	68.0	488	54	6.1	0.24*	3.1-3.5	[3, 38]
	Cel9AΔCBM3c	74.0	121	23.2	0.29*	0.27*	1.3	[38]

* Target % digestion could not be achieved; thus, specific activity (μmole of cellobiose/ minute/ μmole of enzyme) was calculated at 1.5μM of enzyme. Processivity was calculated by the ratio of soluble to insoluble reducing sugars [39]. Endo- endocellulase, exo- exocellulase, cd- catalytic domain, CMC- carboxymethyl cellulose, SC- swollen cellulose, BMCC- bacterial microcrystalline cellulose, and FP- filter paper.

The CBM2 probably functions as an anchor, since up to 88% of Cel6B CBM2 bound to FP [41]. Replacing the CBM2 of Cel6A with a plant family-49 cellulose-binding module conferred equivalent binding [42]. The presence of a CBM2 helps keep the enzymes bound to the crystalline cellulose, increasing its hydrolytic activity; exocellulase Cel6B possessing a CBM2 decreased the crystalline index of native alfalfa cellulose by 34% [43].

The role of the CBM3c in Cel9A is not to lengthen the distance between the CBM2 and the catalytic domain, as Cel9A-74, which has the CBM2 located further from the catalytic domain than the CBM3c in Cel9A-68 (Figure 4) showed much lower activity on crystalline substrates (Table 2). To further investigate the role of the CBM3, a number of its residues, which are "in line" with the catalytic cleft, were mutated [3]. This was the first time a *T. fusca* binding module was engineered by site-directed mutagenesis. Some mutations significantly lowered BMCC activity while having little effect on CMC or swollen cellulose (SC) activity. However, these mutant enzymes did not affect the ratio of soluble/insoluble reducing sugars or processivity [3], which will be further discussed later in this chapter. The presence of the CBM3c in Cel9A is crucial for processivity, but this module does not play a key role in binding. The CBM3c might feed a single cellulose chain into the catalytic cleft after the CBM2 bound to a crystalline surface. This explains the need for both modules in Cel9A and the high activity of this *T. fusca* enzyme on crystalline cellulose, as accessibility of the catalytic domain to substrate is the rate-limiting step in the hydrolysis of crystalline cellulose [1].

Although the CBMs play a critical role in crystalline substrate binding, residues in the active sites are also involved in this process, as a number of mutations in the active site of Cel9A drastically affect substrate binding [3]. The catalytic domains of all three cellulases, particularly exocellulase Cel6B and processive endocellulase Cel9A, bind α-chitin (N-acetylated polymer of β-1,4-D-glucosamine) significantly; both enzymes showed up to 95% binding to α-chitin [44]. The catalytic domains of the *T. fusca* family-6 cellulases also bind to other polysaccharides including xylan (polymer of β-1,4-D-xylose), lichenan (polymer of β-1,3:1,4-D glucose) and pachyman (polymer of β-1,3-D glucose). The non-productive binding of these different catalytic domains is not similar; the catalytic domain of the endocellulase Cel6A bound much less to α-chitin than that of exocellulase Cel6B although both had similar weak binding to cellulose [44]. The balance of productive and non-productive binding probably controls the binding affinity of the active sites, as inactive *T. fusca* mutant cellulases bound more tightly to cellulose than their corresponding wild-type enzymes [2, 20]. A domain in a *C. thermocellum* carbohydrate esterase has been recently found to have a dual function for both acetyl esterase activity and cellulose binding, which reside within the same region in the protein [45]. Plant cell walls consist of different polysaccharides integrated with each other; therefore, it is possible that cellulases bind to other components of plant cell walls to disrupt the polysaccharide matrix before they locate, bind and hydrolyze cellulose. However, non-specific adsorption could limit the availability of cellulases for hydrolysis so that the relationship between productive binding and non-productive adsorption needs to be considered when engineering cellulases for biomass conversion.

IMPROVEMENT OF CRYSTALLINE CELLULOSE HYDROLYSIS

Besides determining the catalytic mechanism of cellulases, site-directed mutagenesis can help identify residues that directly participate in crystalline cellulose hydrolysis. The activities of certain mutant enzymes from all three cellulases show that the rate limiting step in crystalline cellulose hydrolysis differs from that on the other substrates [3, 12]. Mutation of a number of residues, including Trp209, Trp256, Arg317 and Asp261 in the Cel9A Glc(-2) to Glc(-4) subsites showed near wild-type activity on SC, but several-fold higher activity on CMC and only approximately 15% of wild-type activity on BMCC [3]. While many mutant enzymes from all three cellulases showed dramatic improvement in CMC activity, only a few showed higher activities on crystalline substrates (Table 3). Site-directed mutagenesis unfortunately did not improve activity on crystalline cellulose effectively, as no engineered cellulases showed greater than a two-fold increase in activity on crystalline cellulose, and higher improvement has not been reported by other groups. Furthermore, the increase in catalytic domain activity on crystalline substrates is not always seen in enzyme mixtures [12] or even in the intact mutant enzyme [46]. Trade-offs between catalytic activity and thermostability also happened sometimes [47]. Therefore, for industrial purpose, engineered enzymes must be assayed at different conditions on the desired substrate with the best combination of synergistic proteins to detect useful improvements.

Although activity enhancement for crystalline cellulose hydrolysis has not been very successful so far, site-directed mutagenesis has suggested some strategies for improvement. As enzymatic activity is substrate-specific, it is important to design a specific mixture of enzymes for a particular feedstock or biomass source. It seems likely that modifying residues that participate in the limiting step in crystalline cellulose hydrolysis might allow significant increases in activity on crystalline substrates once they are identified [4]. In Cel6B, the dramatic loss of BMCC activity seen in two Met514 mutant enzymes during storage suggests movement of a region needed for crystalline substrate hydrolysis. Circular dichroism spectra showed a global conformational change in these two mutant enzymes; however, their structures need to be determined to locate the specific residues involved in the loss of activity.

A study of *T. fusca* Cel6A [48] showed that substrate heterogeneity causes the non-linearity seen in the hydrolysis of cellulose. Therefore, disruption of crystalline cellulose is an important step to increase substrate accessibility, thus enhancing hydrolysis. Two small, non-catalytic cellulose-binding proteins of *T. fusca*, E7 and E8 were suggested to assist in this disrupting process [41]. Although these proteins displayed weaker binding than Cel6B CBM2, they increased initial FP activity in mixtures with other *T. fusca* cellulases at significantly lower concentrations than Cel6B CBM2. The co-regulation of *T. fusca* E7 and E8 was coupled with that of cellulases [41]; therefore, looking for other factors produced by *T. fusca* when grown on various cellulose substrates might provide valuable information on how microorganisms in nature hydrolyze crystalline cellulose effectively.

Table 3. T. fusca mutant enzymes resulting in higher crystalline substrate activity.

Enzyme	Mutation	Location	% corresponding wild-type activity on		% wild-type processivity*	Ref.
			BMCC	FP		
Cel6B	G234S	Glc(-2) subsite	108	141	104	[47]
	G284P	Beyond Glc(+3) subsite	125	151	100	
	G234S-G284P		111	195	127	
	N282D	Glc(+2) subsite	116	145	182	[12]
Cel9A-68	D513A	CBM3c	121	107	119	[3]
	I514H		112	104	110	

* Processivity was calculated by the ratio of soluble to insoluble reducing sugars [39]. Activity with BMCC and FP (μmole of cellobiose/ minute/ μmole of enzyme) was measured by the DNS method.

Non-catalytic factors helping in cellulose hydrolysis were found in other cellulolytic microorganisms. The anaerobic bacterium, *Fibrobacter succinogenes,* grows well on cellulose although it does not produce cellulosomes and only few cellulase-encoding genes with CBMs were found [49], suggesting involvement of other factors for cellulose degradation. In fact, a number of cellulose-binding proteins from this bacterium were identified [50]. Screening a genomic library of *C. stercorarium* identified only two cellulases [51]. This thermophilic bacterium grows very well on hemicellulose but grows significantly slower on cellulose than *C. thermocellum*, which is one of the most efficient cellulose degraders and produces a large number of cellulosomal cellulases [51]. In addition to cellulases, *T. reesei* also secretes expansin [52] and swollenin [53] to help degrade crystalline cellulose by disrupting cellulose microfibers.

IMPROVEMENT IN PROCESSIVITY

Exocellulases and processive endocellulases including Cel6B and Cel9A display the ability, known as processivity, to continue hydrolyzing a cellulose chain without dissociation. In this way, the enzyme remains close to the detached chain and prevents the chain from re-associating with adjacent cellulose chains. The ratio of soluble to insoluble reducing sugars is one method to evaluate the processivity of cellulases, as processive enzymes cleave many times along a cellulose chain producing only soluble reducing sugars [39]. Using this criterion along with the fact that Cel9A shows synergism with endocellulases and both types of exocellulases, Cel9A was identified as a new type of cellulase, a processive endocellulase [38, 39]. Engineering of Cel9A by removing the CBM3c revealed that this binding module is critical for processivity (Table 1). Site-directed mutagenesis of Cel9A Tyr206, Trp313 and Tyr318 indicated that binding affinity and hydrophobic interaction in the Glc(-3) to Glc(-1) subsites are essential for processivity since these mutations dramatically reduced both substrate binding and processivity [3, 10].

The deletion of a loop structure was performed to investigate the potential blocking effect of residues from 245 to 255 at the non-reducing end (beyond the Glc(-4) subsite) of the Cel9A active site cleft on processivity and product distribution [10]. This fragment is missing in many GH-9 members and has a high temperature factor (B-factor), which indicates high flexibility [10]. The removal of this loop did not change processivity and showed that the production of cellotetraose during processive hydrolysis is not due to the structural barrier [10].

We have recently used the ratio of oligosaccharide products to evaluate processivity, in addition to the ratio of soluble/insoluble reducing sugars [12]. Depending on the initial binding, the first bond cleavage can produce either cellobiose or cellotriose but the subsequent cleavages only produce cellobiose in exocellulases, or cellotetraose in processive endocellulases. Exocellulase Cel6B slowly hydrolyzes cellotriose to release cellobiose and glucose; therefore, the (G2-G1)/(G3+G1) ratio was used as an estimate of processivity of Cel6B. These approaches showed a good correlation with each other [12]; however, these approaches cannot provide an absolute measurement for processivity, as there is no standard for insoluble reducing sugars and products with an odd number of glucose units can also be produced at the end of a cellulose chain if the enzyme cleaves a cellulose molecule completely.

Improvement of processivity is a difficult task, as processivity in Cel9A requires coordination between the sliding of the substrate into the cleavage site and the release of the products. An Arg378Lys mutation produced the highest improvement in processivity among mutations in the catalytic domain; however, a Cel9A double mutant enzyme containing Arg378, which has two hydrogen bonds to Glc(+1) O2, and D261, which is located near Glc(-4) dramatically decreased processivity [3]. The combination of Arg378Lys and either of two other mutations in the CBM3c that increased processivity also lowered processivity [46], showing that an interaction between the catalytic domain and the CBM3c is important. Therefore, higher processivity requires a precise balance in binding between the catalytic domain and the CBM3c.

Binding in the catalytic cleft plays an important role in Cel9A processivity as many mutations in the catalytic domain dramatically lowered processivity [3]. In contrast, several residues in the active site of exocellulase Cel6B including Asn282 and Arg180, which are located at the Glc(+2) and Glc(-4) subsites, respectively, increased processivity; and this increase was confirmed by both assays [12]. It should be noted that higher processivity involves effectively keeping a cellulose chain detached from the crystalline surface and the optimized movement of an enzyme along a cellulose chain, rather than higher activity. Therefore, highly processive exocellulases may display lower activity on easily accessible substrates or be unable to provide higher synergism in mixtures with endocellulases [12].

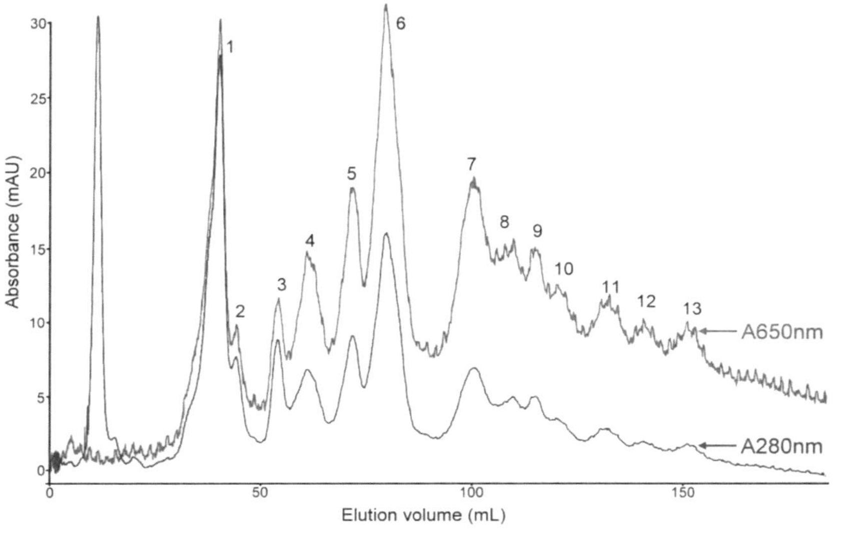

Figure 5. FPLC chromatogram showing thirteen fractions of an AF647-labeled Cel6B mutant enzyme (L230A) with different degrees of labeling, ranging from 1.0 to 3.3. Unlabeled protein (absorbance at 280nm) elutes earlier than labeled products (absorbance at 650nm).

It is still unclear what causes dissociation of a processive enzyme. Fluorescence labeling of enzymes to track their movement may answer this question while offering an optical approach for measuring binding and processivity. In an ongoing project, Cel6B wild-type and mutant enzymes with higher processivity have been labeled with the amine-reactive Alexa Fluor 647® (AF647) succinimidyl ester dye, which reacts with lysine side chains. Each enzyme gave multiple labeled products (Figure 5), which have been separated, purified and assayed for activity on crystalline cellulose. The labeled enzymes will be tracked at the single molecule level with total internal reflection fluorescence microscopy. Tracking a quantum dot-labeled CBM2 from *Acidothermus cellulolyticus* using single-molecule spectroscopy indicated a linear motion of this CBM along the cellulose fiber [54].

Converting Free Cellulases into Cellulosome Components

Designer cellulosomes were first proposed by Bayer and colleagues [55]. As both family-6 cellulases have been studied extensively and natural cellulosomes do not contain any cellulases from family 6 [56], it was interesting to investigate the addition of dockerins to *T.*

fusca family-6 cellulases. Recently, *T. fusca* free cellulases were modified and incorporated into artificial cellulosomes. The cellulose binding module of each cellulase was replaced with a dockerin domain, which was then docked to the corresponding cohesion on an artificial scaffoldin.

The CBM2s of endocellulase Cel6A and exocellulase Cel6B were replaced with dockerins from *C. cellulolyticum* and *C. thermocellum*, producing chimeras 6A-c and t-6B, respectively. Activity of the t-6B chimeric cellulases was reduced on most substrates even when the chimeras were docked to their matching dockerins on artificial scaffoldins [56, 57]. The GH-6 cellulases appear to function better as free enzymes. Surprisingly, t-6B showed about 14-fold higher activity on amorphous cellulose than the native enzyme. The mechanism for this increase is unknown, but it is not due to a change in mode of action of t-6B to that of an endocellulase as t-6B showed lower CMC activity than intact Cel6B [56].

Other Cellulase Properties for Biomass Conversion

Besides being engineered for high catalytic efficiencies, cellulases for biomass conversion can benefit from lower end-product inhibition and higher thermal stability. Mutating Tyr 245 of *A. cellulolyticus* Cel5A significantly decreased inhibition by cellobiose [58], demonstrating the feasibility of activity enhancement by lowering end-product inhibition.

Because cellulase activity on all substrates increases with temperature, increasing their thermostability can increase activity. Efforts to increase the thermostability of Cel6A and Cel6B by introducing disulfide bonds were not successful [47, 59]. Recently, a SCHEMA structure-guided protein recombination approach was used to generate a large pool of diverse, highly thermostable cellulases from three fungal family-6 cellulases including *H. insolens* Cel6A, *T. reesei* Cel6A and *Chaetomium thermophilum* Cel6A, by swapping blocks of sequences from these parent enzymes and minimizing the number of broken contacts upon recombination [60]. Some thermostable chimeras showed an increase in activity on swollen cellulose, particularly at high temperatures [60]. The stable chimeras are diverse and differ from the parents by an average of 50 changes [61]; however, subsequent analyses showed that a single mutation in the thermostabilizing sequence block is responsible for the entire thermostability of the chimeras as well as the parent enzymes [61]. This study showed that site-directed mutagenesis can enhance thermostability. Surprisingly, the thermostabilizing mutation was the substitution of a Cys for a Ser. The corresponding residues in the thermophilic *T. fusca* Cel6A and Cel6B are Ala264 and Ser496, respectively. It would be interesting to check whether mutating Ala264 to Ser or Cys can increase the thermostability of *T. fusca* Cel6A. SCHEMA is also a potential approach to improve other properties of cellulases [60].

Plant cell walls consist of various polysaccharides including cellulose, hemicellulose, and pectin. *T. fusca* cellulases can hydrolyze β-1,4-glycosidic bonds in cellulose effectively, but not in xylan or other polysaccharides. A number of bifunctional and multifunctional cellulases are produced by fungi and bacteria; for instance, one bifunctional enzyme from a bacterium *Cellulomonas flavigena* showed both cellulase and xylanase activity [62]. Synergism between cellulases and xylanases acting on biomass has been reported [63, 64]. It is too early to conclude that multifunctional cellulases is a better approach than mixing free enzymes to enhance activirty, as there are only a few studies on this direction. However, multifunctional enzymes could simplify the enzyme mixture required for hydrolysis. Artificial multifunctional cellulases can be generated by gene fusion, similar to the production of chimeras for designer cellulosomes.

Conclusion

Site-directed mutagenesis has shown that a subtle change in a single residue can significantly change the property of cellulases, and the precise conformation and structure of domains and side chains are critical for crystalline substrate hydrolysis and processivity. Site-directed mutagenesis can produce enzymes with higher activity and processivity; however, no general rules for improvement have been found and the effects so far are not large. The combination of this approach with directed evolution (error-prone PCR, DNA shuffling, family shuffling or their combinations) might provide a more powerful tool to engineer cellulases, as site-directed mutagenesis can produce an initial pool of mutant enzymes for directed evolution. The aid of computational approaches in designing and identifying optimized models may be useful helpful to create improved enzymes for cellulose degradation. The increasing availability of genomic sequences certainly expands candidates for engineering; therefore, mining genomes and metagenomes for novel enzymes with higher activity also may be useful, although extensive searches have not revealed promising candidates yet. As the correlation between activity on soluble substrates and insoluble substrates is low, higher activity on a soluble substrate such as CMC is rarely a good indicator of higher activity on crystalline substrates. Therefore, an effective high-throughput screening/selection method using insoluble substrates and testing candidate enzymes thoroughly in different conditions including in mixtures are critical factors in the strategy for producing more active cellulases.

Acknowledgments

We would like to thank Diana Irwin and Maxim Kostylev for valuable comments. This work was supported by the DOE Office of Biological and Environmental Research-Genomes to Life Program through the BioEnergy Science Center (BESC).

Reviewed by

Emma Master, Department of Chemical Engineering and Applied Chemistry, University of Toronto.

REFERENCES

[1] Wilson, DB. *Cellulases and biofuels.* Curr Opin Biotechnol, 2009, 20(3), 295-99.

[2] Vuong, TV; Wilson, DB. *The absence of an identifiable single catalytic base residue in Thermobifida fusca exocellulase Cel6B.* FEBS J, 2009, 276(14), 3837-45.

[3] Li, Y; Irwin, DC; Wilson, DB. *Processivity, substrate binding, and mechanism of cellulose hydrolysis by Thermobifida fusca Cel9A.* Appl Environ Microbiol, 2007, 73(10), 3165-72.

[4] Wilson, BD. *Studies of Thermobifida fusca plant cell wall degrading enzymes.* Chem Rec., 2004, 4, 72-82.

[5] Bayer, EA; Lamed, R; White, BA; Flint, HJ. *From cellulosomes to cellulosomics.* Chem Rec., 2008, 8(6), 364-77.

[6] Xie, G; Bruce, DC; Challacombe, JF; Chertkov, O; Detter, JC; Gilna, P; et al. *Genome sequence of the cellulolytic gliding bacterium Cytophaga hutchinsonii.* Appl Environ Microbiol, 2007, 73(11), 3536-46.

[7] Qi, M; Jun, HS; Forsberg, CW. *Characterization and synergistic interactions of Fibrobacter succinogenes glycoside hydrolases.* Appl Environ Microbiol, 2007, 73(19), 6098-105.

[8] Cantarel, BL; Coutinho, PM; Rancurel, C; Bernard, T; Lombard, V; Henrissat, B. *The Carbohydrate-Active EnZymes database (CAZy): an expert resource for Glycogenomics.* Nucleic Acids Res., 2009, 37(Database issue), D233-8.

[9] Sasvari, Z; Posta, K; Hornok, L. *Expression patterns of Cel5A-Cel5B, two endoglucanase encoding genes of Thermobifida fusca.* Acta Microbiol Immunol Hung, 2008, 55(4), 437-46.

[10] Sakon, J; Irwin, D; Wilson, DB; Karplus, PA. *Structure and mechanism of endo/exocellulase E4 from Thermomonospora fusca.* Nat Struct Biol, 1997, 4(10), 810-18.

[11] Spezio, M; Wilson, DB; Karplus, PA. *Crystal structure of the catalytic domain of a thermophilic endocellulase.* Biochemistry, 1993, 32(38), 9906-16.

[12] Vuong, TV; Wilson, DB. *Processivity, synergism and substrate specificity of Thermobifida fusca Cel6B.* Appl Environ Microbiol, 2009, 75 (21), p. 6655-6661(in press).

[13] Arai, T; Matsuoka, S; Cho, HY; Yukawa, H; Inui, M; Wong, SL; et al. *Synthesis of Clostridium cellulovorans minicellulosomes by intercellular complementation.* PNAS, 2007, 104(5), 1456-60.

[14] Doi, RH; Kosugi, A. *Cellulosomes: plant-cell-wall-degrading enzyme complexes.* Nat Rev Micro, 2004, 2(7), 541-51.

[15] Koshland, DEJ. *Stereochemistry and the mechanism of enzymatic reactions.* Biological Reviews, 1953, 28(4), 416-36.

[16] Wolfgang, DE; Wilson, DB. *Mechanistic studies of active site mutants of Thermomonospora fusca endocellulase E2.* Biochemistry, 1999, 38(30), 9746-51.

[17] Wilson, DB; Irwin, DC. *Genetics and properties of cellulases.* In: Scheper T, editor. Advances in Biochemical Engineering/Biotechnology. Heidelberg: Springer-Verlag, 1999, 1-21.

[18] Gebler, J; Gilkes, NR; Claeyssens, M; Wilson, DB; Beguin, P; Wakarchuk, WW; et al. *Stereoselective hydrolysis catalyzed by related beta-1,4-glucanases and beta-1,4-xylanases.* J Biol Chem., 1992, 267(18), 12559-61.

[19] Andre, G; Kanchanawong, P; Palma, R; Cho, H; Deng, X; Irwin, D; et al. *Computational and experimental studies of the catalytic mechanism of Thermobifida fusca cellulase Cel6A (E2).* Protein Eng., 2003, 16(2), 125-34.

[20] Barr, BK; Wolfgang, DE; Piens, K; Claeyssens, M; Wilson, DB. *Active-site binding of glycosides by Thermomonospora fusca endocellulase E2.* Biochemistry, 1998, 37(26), 9220-29.

[21] Larsson, AM; Bergfors, T; Dultz, E; Irwin, DC; Roos, A; Driguez, H; et al. *Crystal structure of Thermobifida fusca endoglucanase Cel6A in complex with substrate and inhibitor: The role of tyrosine Y73 in substrate ring distortion.* Biochemistry, 2005, 44(39), 12915-22.

[22] Zhou, W; Irwin, DC; Escovar-Kousen, J; Wilson, DB. *Kinetic studies of Thermobifida fusca Cel9A active site mutant enzymes.* Biochemistry, 2004, 43(30), 9655-63.

[23] Kitamura, M; Okuyama, M; Tanzawa, F; Mori, H; Kitago, Y; Watanabe, N; et al. *Structural and functional analysis of a glycoside hydrolase family 97 enzyme from Bacteroides thetaiotaomicron.* J Biol Chem., 2008, 283(52), 36328-37.

[24] Honda, Y; Fushinobu, S; Hidaka, M; Wakagi, T; Shoun, H; Taniguchi, H; et al. *Alternative strategy for converting an inverting glycoside hydrolase into a glycosynthase.* Glycobiology, 2008, 18(4), 325-30.

[25] Koivula, A; Ruohonen, L; Wohlfahrt, G; Reinikainen, T; Teeri, TT; Piens, K; et al. *The active site of cellobiohydrolase Cel6A from Trichoderma reesei: the roles of aspartic acids D221 and D175.* J Am Chem Soc, 2002, 124(34), 10015-24.

[26] Guerin, DM; Lascombe, MB; Costabel, M; Souchon, H; Lamzin, V; Beguin, P, et al. *Atomic (0.94Å) resolution structure of an inverting glycosidase in complex with substrate.* J Mol Biol, 2002, 316(5), 1061-9.

[27] Kitago, Y; Karita, S; Watanabe, N; Kamiya, M; Aizawa, T; Sakka, K; et al. *Crystal structure of Cel44A, a glycoside hydrolase family 44 endoglucanase from Clostridium thermocellum.* J Biol Chem., 2007, 282(49), 35703-11.

[28] Langley, DB; Harty, DWS; Jacques, NA; Hunter, N; Guss, JM; Collyer, CA. *Structure of N-acetyl-β-D-glucosaminidase (GcnA) from the endocarditis pathogen Streptococcus gordonii and its complex with the mechanism-based inhibitor NAG-thiazoline.* J Mol Biol., 2008, 377(1), 104-16.

[29] Ling, Z; Suits, MDL; Bingham, RJ; Bruce, NC; Davies, GJ; Fairbanks, AJ; et al. *The xX-ray crystal structure of an Arthrobacter protophormiae endo-β-N-acetylglucosaminidase reveals a (β/α)8 catalytic domain, two ancillary domains and active site residues key for transglycosylation activity.* J Mol Biol, 2009, 389(1), 1-9.

[30] Egloff, MP; Uppenberg, J; Haalck, L; van Tilbeurgh, H. *Crystal structure of maltose phosphorylase from Lactobacillus brevis: Unexpected evolutionary relationship with glucoamylases.* Structure, 2001, 9(8), 689-97.

[31] Hidaka, M; Kitaoka, M; Hayashi, K; Wakagi, T; Shoun, H; Fushinobu, S. *Structural dissection of the reaction mechanism of cellobiose phosphorylase.* Biochem J, 2006, 398(1), 37-43.

[32] Wada, J; Honda, Y; Nagae, M; Kato, R; Wakatsuki, S; Katayama, T; et al. *1,2-α-L-fucosynthase: A glycosynthase derived from an inverting α-glycosidase with an unusual reaction mechanism.* FEBS Lett, 2008, 582(27), 3739-43.

[33] Kim, YW; Chen, H; Kim, JH; Withers, SG. *Catalytic properties of a mutant βb-galactosidase from Xanthomonas manihotis engineered to synthesize galactosyl-thio-βb-1,3 and -βb-1,4-glycosides.* FEBS Lett, 2006, 580(18), 4377-81.

[34] Jahn, M; Chen, HM; Mullegger, J; Marles, J; Warren, RAJ; Withers, SG. *Thioglycosynthases: Double mutant glycosidases that serve as scaffolds for thioglycoside synthesis.* Chem Commun, 2004(3), 274-75.

[35] Nakamura, T; Mine, S; Hagihara, Y; Ishikawa, K; Ikegami, T; Uegaki, K. *Tertiary structure and carbohydrate recognition by the chitin-binding domain of a hyperthermophilic chitinase from Pyrococcus furiosus.* J Mol Biol, 2008, 381(3), 670-80.

[36] Simpson, PJ; Xie, H; Bolam, DN; Gilbert, HJ; Williamson, MP. *The structural basis for the ligand specificity of family 2 carbohydrate-binding modules.* J Biol Chem., 2000, 275(52), 41137-42.

[37] Tormo, J; Lamed, R; Chirino, AJ; Morag, E; Bayer, EA; Shoham, Y; et al. *Crystal structure of a bacterial family-III cellulose-binding domain: a general mechanism for attachment to cellulose.* EMBO J, 1996, 15(21), 5739-51.

[38] Irwin, D; Shin, DH; Zhang, S; Barr, BK; Sakon, J; Karplus, PA; et al. *Roles of the catalytic domain and two cellulose binding domains of Thermomonospora fusca E4 in cellulose hydrolysis.* J Bacteriol, 1998, 180(7), 1709-14.

[39] Irwin, DC; Spezio, M; Walker, LP; Wilson, DB. *Activity studies of eight purified cellulases: Specificity, synergism, and binding domain effects.* Biotech Bioeng, 1993, 42(8), 1002-13.

[40] Zhang, S; Barr, BK; Wilson, DB. *Effects of noncatalytic residue mutations on substrate specificity and ligand binding of Thermobifida fusca endocellulase Cel6A.* Eur J Biochem, 2000, 267(1), 244-52.

[41] Moser, F; Irwin, D; Chen, S; Wilson, BD. *Regulation and characterization of Thermobifida fusca carbohydrate-binding module proteins E7 and E8.* Biotech Bioeng, 2008, 100(6), 1066-77.

[42] Urbanowicz, BR; Catalá, C; Irwin, D; Wilson, DB; Ripoll, DR; Rose, JKC. *A tomato endo-β-1,4-glucanase, SlCel9C1, represents a distinct subclass with a new family of carbohydrate binding modules (CBM49).* J Biol Chem., 2007, 282(16), 12066-74.

[43] Bae, HJ; Turcotte, G; Soo Kim, Y; Vezina, LP; Laberge, S. *Cel6B of Thermobifidus fusca and a Cel5-CBM6 of Ruminococcus albus containing a cellulose binding site show synergistic effect on hydrolysis of native plant cellulose.* FEMS Microbiol Lett, 2004, 233(2), 325-31.

[44] Li, Y; Wilson, DB. *Chitin binding by Thermobifida fusca cellulase catalytic domains.* Biotech Bioeng, 2008, 100(4), 644-52.

[45] Montanier, C; Money, VA; Pires, VMR; Flint, JE; Pinheiro, BA; Goyal, A; et al. *The active site of a carbohydrate esterase displays divergent catalytic and noncatalytic binding functions.* PLoS Biol, 2009, 7(3), 0687-97.

[46] Li, Y; Irwin, DC; Wilson, DB. *Combinations of mutations in the family 9 catalytic domain and family 3c cellulose binding module of Thermobifida fusca Cel9A can increase crystalline cellulose activity.* (in progress).
[47] Zhang, S; Irwin, DC; Wilson, DB. *Site-directed mutation of noncatalytic residues of Thermobifida fusca exocellulase Cel6B.* Eur J Biochem, 2000, 267(11), 3101-15.
[48] Zhang, S; Wolfgang, DE; Wilson, DB. *Substrate heterogeneity causes the nonlinear kinetics of insoluble cellulose hydrolysis.* Biotechnol and Bioeng, 1999, 66(1), 35-41.
[49] Seon Park, J; Russell, JB; Wilson, DB. *Characterization of a family 45 glycosyl hydrolase from Fibrobacter succinogenes S85.* Anaerobe, 2007, 13(2), 83-8.
[50] Jun, HS; Qi, M; Gong, J; Egbosimba, EE; Forsberg, CW. *Outer membrane proteins of Fibrobacter succinogenes with potential roles in adhesion to cellulose and in cellulose digestion.* J Bacteriol, 2007, 189(19), 6806-15.
[51] Vladimir VZ; Wolfgang HS. *Bacterial cellulose hydrolysis in anaerobic environmental subsystems Clostridium thermocellum and Clostridium stercorarium, thermophilic plant-fiber degraders.* In: Wiegel J, Maier RJ, Adams MWW, editors. Incredible anaerobes from physiology to genomics to fuels: Ann NY Acad Sci, 2008,298-307.
[52] Verbeke J; Coutinho, P; Mathis, H; Quenot, A; Record, E; Asther, M; et al. *Transcriptional profiling of cellulase and expansin-related genes in a hypercellulolytic Trichoderma reesei.* Biotechnol Lett, 2009, 31(9), 1399-405.
[53] Brotman, Y; Briff, E; Viterbo, A; Chet, I. *Role of swollenin, an expansin-like protein from Trichoderma, in plant root colonization.* Plant Physiol, 2008, 147(2), 779-89.
[54] Liu, YS; Zeng, Y; Luo, Y; Xu, Q; Himmel, M; Smith, S; et al. *Does the cellulose-binding module move on the cellulose surface?* Cellulose, 2009, 16(4), 587-97.
[55] Bayer, EA; Morag, E; Lamed, R. *The cellulosome- A treasure-trove for biotechnology.* Trends Biotechnol 1994,12(9),379-86.
[56] Caspi, J; Irwin, D; Lamed, R; Shoham, Y; Fierobe, HP; Wilson, DB; et al. *Thermobifida fusca family-6 cellulases as potential designer cellulosome components.* Biocatalysis Biotransform, 2006, 24(1), 3-12.
[57] Caspi, J; Irwin, D; Lamed, R; Li, Y; Fierobe, HP; Wilson, DB; et al. *Conversion of Thermobifida fusca free exoglucanases into cellulosomal components: Comparative impact on cellulose-degrading activity.* J Biotech, 2008, 135(4), 351-57.
[58] Baker, JO; McCarley, JR; Lovettt, R; Yu, CH; Adney, WS; Rignall, TR; et al. *Catalytically enhanced endocellulase Cel5A from Acidothermus cellulolyticus.* Appl Biochem Biotech, 2005, 121, 129-48.
[59] Ai, YC; Zhang, S; Wilson, DB. *Positional expression effects of cysteine mutations in the Thermobifida fusca cellulase Cel6A and Cel6B catalytic domains.* Enzyme Microb Technol, 2003, 32(2), 331-36.
[60] Heinzelman, P; Snow, CD; Wu, I; Nguyen, C; Villalobos, A; Govindarajan, S; et al. *A family of thermostable fungal cellulases created by structure-guided recombination.* PNAS, 2009, 106(14), 5610-15.
[61] Heinzelman, P; Snow, CD; Smith, MA; Yu, X; Kannan, A; Boulware, K; et al. *SCHEMA recombination of a fungal cellulase uncovers a single mutation that contributes markedly to stability.* J Biol Chem., 2009, 284(39), 26229-33.
[62] Pérez-Avalos, O; Sánchez-Herrera, L; Salgado, L; Ponce-Noyola, T. *A bifunctional endoglucanase/endoxylanase from Cellulomonas flavigena with potential use in industrial processes at different pH.* Curr Microbiol, 2008, 57(1), 39-44.

[63] Selig, MJ; Knoshaug, EP; Adney, WS; Himmel, ME; Decker, SR. *Synergistic enhancement of cellobiohydrolase performance on pretreated corn stover by addition of xylanase and esterase activities*. Bioresour Technol, 2008, 99(11), 4997-5005.

[64] Murashima, K; Kosugi, A; Doi, RH. *Synergistic effects of cellulosomal xylanase and cellulases from Clostridium cellulovorans on plant cell wall degradation*. J Bacteriol, 2003, 185(5), 1518-24.

INDEX

C

F

G

H

N

O

P

Q

R

S

T